AF292515

HARDY SPACES

The theory of Hardy spaces is a cornerstone of modern analysis. It combines techniques from functional analysis, the theory of analytic functions, and Lesbesgue integration to create a powerful tool for many applications, pure and applied, from signal processing and Fourier analysis to maximum modulus principles and the Riemann zeta function.

This book, aimed at beginning graduate students, introduces and develops the classical results on Hardy spaces and applies them to fundamental concrete problems in analysis. The results are illustrated with numerous solved exercises which also introduce subsidiary topics and recent developments. The reader's understanding of the current state of the field, as well as its history, are further aided by engaging accounts of the key players and by the surveys of recent advances (with commented reference lists) that end each chapter. Such broad coverage makes this book the ideal source on Hardy spaces.

Nikolaï Nikolski is Professor Emeritus at the Université de Bordeaux working primarily in analysis and operator theory. He has been co-editor of four international journals and published numerous articles and research monographs. He has also supervised some 30 PhD students, including three Salem Prize winners. Professor Nikolski was elected Fellow of the AMS in 2013 and received the Prix Ampère of the French Academy of Sciences in 2010.

Hardy Spaces

NIKOLAÏ NIKOLSKI
Université de Bordeaux

CAMBRIDGE
UNIVERSITY PRESS

CAMBRIDGE
UNIVERSITY PRESS

University Printing House, Cambridge CB2 8BS, United Kingdom

One Liberty Plaza, 20th Floor, New York, NY 10006, USA

477 Williamstown Road, Port Melbourne, VIC 3207, Australia

314–321, 3rd Floor, Plot 3, Splendor Forum, Jasola District Centre,
New Delhi – 110025, India

79 Anson Road, #06–04/06, Singapore 079906

Cambridge University Press is part of the University of Cambridge.

It furthers the University's mission by disseminating knowledge in the pursuit of
education, learning, and research at the highest international levels of excellence.

www.cambridge.org
Information on this title: www.cambridge.org/9781107184541
DOI: 10.1017/9781316882108

Originally published in French as *Éléments d'analyse avancée:
1. Espaces de Hardy* by Belin, 2012. © Éditions Belin, 2012

This publication is in copyright. Subject to statutory exception
and to the provisions of relevant collective licensing agreements,
no reproduction of any part may take place without the written
permission of Cambridge University Press.

First published in English by Cambridge University Press 2019
English translation © Cambridge University Press 2019

Printed in the United Kingdom by TJ International Ltd. Padstow Cornwall

A catalogue record for this publication is available from the British Library.

Library of Congress Cataloging-in-Publication Data
Names: Nikolski, N. K. (Nikolai Kapitonovich), author.
Title: Hardy spaces : elements of advanced analysis /
Nikolai Nikolski (Universite de Bordeaux).
Other titles: Elements d'analyse avancee. 1, Espaces de Hardy. English | Espaces de Hardy
Description: Cambridge ; New York, NY : Cambridge University Press, 2019. |
Series: Cambridge studies in advanced mathematics ; 179 | Originally published in French:
Elements d'analyse avancee : 1, Espaces de Hardy (Paris : Editions Belin, 2012). |
First English translation. | Includes bibliographical references and index.
Identifiers: LCCN 2018049103 | ISBN 9781107184541 (hardback : alk. paper)
Subjects: LCSH: Hardy spaces. | Functions of complex variables. | Holomorphic functions.
Classification: LCC QA331.7 .N5513 2019 | DDC 515/.98–dc23
LC record available at https://lccn.loc.gov/2018049103

ISBN 978-1-107-18454-1 Hardback

Cambridge University Press has no responsibility for the persistence or accuracy
of URLs for external or third-party internet websites referred to in this publication
and does not guarantee that any content on such websites is, or will remain,
accurate or appropriate.

Every effort has been made to secure necessary permissions to reproduce copyright material
in this work, though in some cases it has proved impossible to trace copyright holders.
If any omissions are brought to our notice, we will be happy to include appropriate
acknowledgements on reprinting.

Less is more

Robert Browning,
"Andrea del Sarto," 1855

Contents

Preface

The *introduction to Hardy spaces* proposed in this book covers the basic techniques of modern analysis, conceived and developed at the beginning of the twentieth century over a very short period (a kind of "Silver Age" for mathematical analysis; Exercise 1: which was the "Golden Age"?), by a talented group of mathematical geniuses including Henri Lebesgue, Frigyes Riesz, G. H. Hardy, Andrey Kolmogorov, and Norbert Wiener. Over time, this cluster of ideas became the source of extremely powerful techniques for a variety of applications: from Fourier series to the Wiener theory of stationary filtering, not to mention the Euler ζ function and the Riemann hypothesis.

The contents of this text correspond to a course at the "Master 2" level given several times during the years 1990–2010 at the University of Bordeaux 1, and represent an introduction and invitation to the entire domain of modern analysis. The book is devoted to a multi-faceted subject: it involves harmonic analysis (since it concerns a unitary representation of the group $\mathbb{Z}$), but also complex analysis (as we restrict ourselves most often to the semigroup $\mathbb{Z}_+$), the theory of operators (by the nature of the representation, but also by a hidden universality that we will explore in future volumes), as well as the theme of signals and filtering, with a bit of number theory thrown in. It is for this superposition of major disciplines of mathematics (more a "roundabout" than a "crossroads") that the subject can be described as "classical" ("classical" $\neq$ "old-fashioned"!). The conjunction between the different facets of the subject is most fruitful and successful in the *Hilbert* framework of the spaces $L^2(\mathbb{T}, \mu)$; this is why we have developed the theory, and its applications, principally in the space H^2 (which is also closely linked with H^1 and H^∞), whereas the other H^p spaces appear only occasionally.

The prerequisites are a standard course in integration and functional analysis (or in Hilbert/Banach spaces) along with a few notions of complex analysis. A summary/reminder of all the necessary information (as well as certain notations) are gathered in the appendices at the end of the book. Within the text, we include a large number of historical details – on the subjects developed, their founders, and the diverse circumstances of their creation. We hope that this will help the reader to better understand Hardy spaces, along with the dramaturgy of mathematics (and mathematical life).[1]

Each chapter contains exercises and their solutions (75 in total) at different levels: to use a musical analogy from Glazman and Lyubich (1969), from exercises on open strings up to virtuoso pieces using double harmonics ("double flageolet tones").

Each chapter concludes with a section entitled "Notes and Remarks" which discusses the history of the main subjects of the chapter, certain recent results, and (at times) the open questions; this discussion is sometimes addressed to non-novice readers.

The reader will rapidly become aware that this text contains only a few elementary aspects of the techniques of harmonic analysis, linked particularly with an approach to Hardy spaces via complex analysis. Even if at times we delve into quite refined questions of analysis (such as the geometry of finite bases, in Chapter 4), our text is not meant to be a research monograph, but more a source of basic knowledge. This is why "less is more." Nonetheless, in principle, students reaching the end of the book should be capable of tackling independent research projects (the author can affirm this from experience). For such an endeavor they will need the aid of experts, but this can be found in the dozens of existing monographs devoted to Hardy spaces and the "hard analysis" that was developed around them. Several are mentioned at the end of Chapter 1, in the section Notes and Remarks 1.9. Good luck!

---

[1] The biographical details – which, given the technical and financial constraints, are sketched here at best – are drawn from various sources, notably the *MacTutor* website of the University of St Andrews (Scotland), www-history.mcs.st-and.ac.uk/, and the free encyclopedia *Wikipedia*, https://en.wikipedia.org/wiki/.

Acknowledgments

Acknowledgements for the French Edition

This book could not have seen the light of day without the generous and wide-ranging aid of my colleague at Bordeaux, Éric Charpentier, whose enthusiasm and availability supported me at several difficult points in the editing. I am also grateful to my colleagues from Saint Petersburg: Anton Baranov (who read a preliminary version and detected several "bugs"), and Andreï Khrabrov (who mastered all the "TEXnical" problems).

I thank Éditions Belin (and especially the editor responsible for this work, C. Counillon) for accepting my project and seeing it through to fruition.

And of course, I owe an eternal debt towards my young family for their infinite patience (and many sacrifices).

Acknowledgements for the English Edition

The author warmly thanks the translators Danièle Gibbons and Greg Gibbons for their high-quality job, for attention to all shades of meaning of the French text, and for a friendly collaboration at all stages of the work.

The author is also sincerely grateful to CUP for including the book in this prestigious series, and to the whole CUP editorial team for their highly professional preparation of the manuscript and for their patience during his numerous slowdowns due to many other projects.

Biographies

Figures

The Origins of the Subject

Prehistory. Cauchy – Fourier – Poisson – Weierstrass – Stieltjes – Fatou – Lebesgue – Hilbert – Parseval – Jensen.

History. Lebesgue – Hardy – Luzin – Privalov – Schur – the Riesz brothers – Szegő – Nevanlinna – Smirnov – Littlewood – Kolmogorov – Paley – Wiener – Zygmund.

Legacy/Continuation. Stein – Fefferman – de Branges – Helson – Kahane – Garnett – Gamelin – Carleson – Sarason – Havin – Douglas – Arveson – Sz.-Nagy – Foias – Fuhrmann – Lax – Phillips – Lacey, etc.

The birth of Hardy spaces dates back to the year 1915, at Cambridge University. At the time, it went virtually unnoticed. Admittedly, the year 1915 can be considered as "unremarkable" only for their creator, the British mathematician G. H. Hardy (1877–1947). Sure enough, as usual, he had published a dozen (!) articles and research notes, but apparently no salient result emerged from his efforts that year, with one exception – if we equate a definition with a result.

Godfrey Harold (G. H.) Hardy (1877–1947) was one of the founding fathers of modern "hard" analysis, and the author of several fundamental ideas that transformed such disciplines as Diophantine analysis, Tauberian theory, the summation of divergent series, Fourier series, the distribution of prime numbers, and the theory of the Euler ζ function. David Hilbert called him "the best mathematician in England." Several theorems and mathematical creations are named after Hardy. His book *A Mathematician's Apology* (1940) is a masterpiece on the philosophy and psychology of a mathematician. His remarkable essay "Orders of infinity: The 'Infinitärcalcül' of Paul Du Bois-Reymond" (1910) inspired a chapter in Bourbaki's treatise. He was a friend of the novelist and scientist C. P. Snow and a co-author with Littlewood, Ramanujan, Titchmarsh, Ingham, Landau, and Marcel Riesz.

Trinity College, Cambridge.

Specifically, in part of a short nine-page article published in the 1915 *Proceedings of the London Mathematical Society*, Hardy defined a family of spaces ("function classes") of holomorphic functions. At the time, the event was barely noticed: either by the general public (preoccupied by the

THE MEAN VALUE OF THE MODULUS OF AN ANALYTIC FUNCTION. 269

THE MEAN VALUE OF THE MODULUS OF AN ANALYTIC FUNCTION

By G. H. HARDY.

[Read November 12th 1914.—Received December 10th, 1914.—
Received, in revised form, February 10th, 1915.]

1. Suppose that $f(x)$ is an analytic function of the complex variable x, regular for $|x| < \rho$, and that $M(r)$ denotes, as usual, the maximum of $|f(x)|$ on the circle $|x| = r < \rho$. Then it is known that $M(r)$ possesses the following properties :—

(i) $M(r)$ is a steadily increasing function of r;

(ii) $\log M(r)$ is a convex function of $\log r$, so that

$$\log M(r) \leqslant \frac{\log (r_2/r)}{\log (r_2/r_1)} \log M(r_1) + \frac{\log (r/r_1)}{\log (r_2/r_1)} \log M(r_2),$$

if
$$0 < r_1 \leqslant r \leqslant r_2 < \rho.$$

Further, when $f(x)$ is an integral function, so that $\rho = \infty$, it is known that

(iii) $M(r)$ tends to infinity with (r), and, unless $f(x)$ is a polynomial, more rapidly than any power of r.*

It was suggested to me by Dr. H. Bohr and Prof. E. Landau, rather more than a year ago, that the property (i) is possessed also by the *mean* value of $|f(x)|$ on the circle $|x| = r$, *i.e.*, by the function

$$\mu(r) = \frac{1}{2\pi} \int_0^\pi |f(re^{i\theta})| \, d\theta.$$

---

* The theorems (i) and (iii) are classical. Theorem (ii) was discovered independently by Blumenthal (*Jahresbericht der Deutschen Math.-Vereinigung*, Vol. 16, p. 97), Faber (*Math. Annalen*, Vol. 63, p. 549), and Hadamard (*Bulletin de la Soc. Math. de France*, Vol. 24, p. 186). The first statement of the theorem was due to Hadamard and the first proof to Blumenthal. The theorem is a corollary of one concerning the associated radii of convergence of a power series in two variables, due to Fabry (*Comptes Rendus*, Vol. 134, p. 1190), and Hartogs (*Math. Annalen*, Vol. 62, p. 1).

The first page of Hardy's nine-page paper of 1915 defining "Hardy classes." Who could have prophesied that this acorn would grow into such a mighty oak?

First World War), or by the scientific world (1915 was above all the year of Einstein's General Relativity, as well as Wegener's theory of *Pangaea*), or even by mathematicians. Nevertheless, it was a turning point for a number of disciplines linked to mathematical analysis: complex analysis (then flourishing), harmonic analysis, signal processing, and in particular several theories non-existent at the time, but crucial today – the theory of operators, optimal control, diffusion theory, random processes.

Later on in his career, Hardy himself returned several times to the theory of the spaces he had defined in 1915, which, at first glance, seemed to be merely an auxiliary tool. However, for its transformation into an indispensable, extremely powerful technique of analysis and for the majority of its applications, we are highly indebted to the efforts of the "Golden Team" of analysts of that time (such as Schur, Marcel Riesz, Frigyes Riesz, Szegő, Nevanlinna, Luzin, Privalov, Smirnov, Kolmogorov, Paley, Wiener, Zygmund), and to their equally brilliant successors (such as Beurling, Stein, Fefferman, de Branges, Helson, Carleson, Kahane, Garnett, Gamelin, Sarason, Havin, Douglas, Sz.-Nagy, Foias, Fuhrmann, Lax, Phillips).

The explanation for its success can perhaps be summed up in just a few points: (1) the dynamics of the Hardy space $e^{inx}H^2$, $n \in \mathbb{Z}$, generates an orthonormal basis $e^{inx} \in e^{inx}H^2 \ominus e^{i(n+1)x}H^2$ in the Lebesgue space $L^2(-\pi, \pi)$; (2) the space H^2 is the "analytic half" of $L^2(-\pi, \pi)$; (3) in H^2, there is a property of factorization into elementary factors, similar to that of polynomials (in a sense, H^2 is a "factorial ring"). First of course come the definition and the basic properties.

A remark for the experts: the current dominant approach to Hardy spaces is via real harmonic analysis (maximal functions, Hilbert transforms, etc.); thus it is unnecessary to differentiate between H^2 and H^p, $p \neq 2$, or between the groups where the space is defined ($\mathbb{T}$, $\mathbb{T}^n$, $\mathbb{R}$, $\mathbb{R}^n$, etc., and even without any group structure). In this book, I follow a combination of the "genetic" approach based on analysis of a single complex variable, and the spectral analysis of a unitary representation of $\mathbb{Z}$. Why this choice? It is indeed the most elementary and direct route to obtain all the results of the theory needed for applications. Let us add that, so far, the true value of the powerful methods of real variables remains purely theoretical. As soon as we are faced with practical applications of Hardy spaces, we use the complex presentation and its techniques – beginning with signal processing and operator theory, and then H^∞ optimal control and diffusion theory, or even stochastic processes or the Euler ζ function. Our work is especially concerned with the spaces H^2, H^1, and H^∞.

The memorable events of 1915

- Einstein's theory of General Relativity.
- Wegener's theory of *Pangaea*.
- The use of chemical weapons by Germany on a massive scale (Second Battle of Ypres).
- The Mexican Revolution.
- The birth of Paul Tibbets (future pilot in the US Air Force, to be assigned the task of dropping the first atomic bomb on Hiroshima on August 6, 1945).
- The thesis of Nikolai Luzin (future founder of the Moscow school of analysis), written in Paris and defended in Moscow.
- G. H. Hardy's definition of H^p spaces.

1

The space $H^2(\mathbb{T})$: An Archetypal Invariant Subspace

Topics. Lebesgue spaces $L^p(\mathbb{T},\mu)$, Hardy spaces $H^p(\mathbb{T})$, lattice of invariant subspaces, the shift operator (reducing subspaces – Wiener's theorem – and invariant subspaces – Helson's theorem), uniqueness theorem, and inner and outer functions.

In this chapter we mainly work in the context of the Hilbert spaces $L^2(\mathbb{T},\mu)$, $L^2(\mathbb{T})$, $H^2(\mathbb{T})$; the other H^p appear occasionally.

1.1 Notation and Terminology of Operators

Let H be a Hilbert space (always over the field of complex numbers $\mathbb{C}$) and let $T: H \to H$ be a bounded linear operator on H. The space (the algebra) of operators on H is denoted $L(H)$. Let $E \subset H$ be a subspace of H (= closed linear subspace). E is said to be *invariant* for $T \in L(H)$ if

$$x \in E \Rightarrow Tx \in E$$

(in short, $TE \subset E$). The set

$$\mathrm{Lat}(T)$$

of invariant subspaces is a *lattice* with respect to the operations $\cap$ and span (= closed linear hull). If $\mathcal{T}$ is a family of operators on H, we set $\mathrm{Lat}(\mathcal{T}) = \bigcap_{T \in \mathcal{T}} \mathrm{Lat}(T)$.

In the particular case of $\mathcal{T} = \{T, T^*\}$, where $T \in L(H)$ and T^* is the adjoint operator of T (see Appendix E), a subspace $E \in \mathrm{Lat}(T, T^*)$ is said to be *reducing*.

The goal of this section is to describe the lattice $\mathrm{Lat}(M_z)$ where M_z is the operator of multiplication by an "independent variable" in the space $L^2(\mathbb{T},\mu)$,

5

with μ a finite Borel measure on the circle $\mathbb{T} = \{\zeta \in \mathbb{C}: |\zeta| = 1\}$,

$$M_z f = z f(z), z \in \mathbb{T}.$$

The operator M_z is called the bilateral shift operator.

1.2 Reducing Subspaces of the Bilateral Shift M_z

In the years 1920–1930, Norbert Wiener developed the mathematical theory of stationary filters. Since the tools he needed could not be found in the Analysis of the time, he created them himself, thus profoundly enriching harmonic analysis and spectral theory.

Norbert Wiener (1894–1964) was an American mathematician (MIT: Massachusetts Institute of Technology), creator of cybernetics (1948) and communication theory (co-founded with Kotelnikov and Shannon). He also created the theories of stochastic processes and generalized harmonic analysis (1930, the Wiener measure and Brownian motion), Tauberian theory, and also, independently of Stefan Banach, invented *Banach spaces* (1923). He authored innovative works in mathematical physics, in potential theory and the optimal prediction of random processes (with applications to the automatic correction of the firing of anti-aircraft guns, shared with Kolmogorov). An admirer of Leibniz, Lebesgue, and Hadamard, Norbert Wiener was one of the geniuses of the twentieth century, who revolutionized mathematics and science. The reader can find a remarkable overview of Wiener's scientific impact (as well as a biographical article by Norman Levinson) in vol. 72, issue 1-II (1966) of the *Bulletin of the American Mathematical Society*. Having received his Bachelor's degree at the age of 14, Wiener followed a Master's program in zoology at Harvard, in philosophy at Cornell, and then in mathematics at Harvard. After submitting his thesis in 1912 (at the age of 17), he came to Europe for post-doctoral studies. Upon his return to the USA, Wiener was denied a position at Harvard because of the anti-Semitic atmosphere of the establishment (George Birkhoff is often cited as one of his principal opponents, behind the scenes). Unlike other top-level scientists, Wiener was not invited to participate in the Manhattan Project. A

Massachusetts Institute of Technology.

confirmed pacifist, he systematically refused all government financing of his research after the Second World War and never participated in military projects.

In particular, for filtering theory, Wiener needed to solve the problem of the recognition (identification) of filters (see the details below in Chapter 5). As a first step, he proved the following theorem (in the case where $\mu = m$, the normalized Lebesgue measure on the circle $\mathbb{T}$; 80 years later, we prove it in a somewhat more general form).

Theorem 1.2.1 (Wiener, 1932) *Let μ be a positive Borel measure in $\mathbb{C}$ with compact support and E a (closed) subspace of $L^2(\mu)$. The following assertions are equivalent.*

(1) *$E \in \mathrm{Lat}(M_z, M_z^*)$.*
(2) *There exists a Borel set $A \subset \mathbb{C}$ such that*

$$E = \chi_A L^2(\mu) = \{f \in L^2(\mu): f = 0 \ \mu\text{-a.e. on the complement } A' = \mathbb{C} \setminus A\}.$$

The set A in (2) is unique modulo μ: $\chi_A L^2(\mu) = \chi_B L^2(\mu)$ if and only if $\chi_A = \chi_B$ μ-a.e., i.e. if and only if $\mu(A \triangle B) = 0$, where $A \triangle B = (A \setminus B) \cup (B \setminus A)$ is the symmetric difference.

Proof First observe that $M_z^* = M_{\bar{z}}$ and $\frac{1}{2}(z + \bar{z}) = X$, $\frac{1}{2i}(z - \bar{z}) = Y$ imply that a subspace E is reducing for M_z if and only if, for every polynomial $p = p(X, Y)$, we have $p \cdot E \subset E$. Let $\mathcal{P}$ denote the set of polynomials in X and Y.

Let us show (1) $\Rightarrow$ (2). Let $f \in E$ and $g \in E^\perp = \{g \in L^2(\mu): (h, g) = 0, \ \forall h \in E\}$ (orthogonal complement of E). Then

$$0 = (pf, g) = \int p f \bar{g} \, d\mu, \quad \forall p \in \mathcal{P}.$$

Since $\mathcal{P}$ is dense in the space $C(\text{supp}(\mu))$ of continuous functions on a compact set $\text{supp}(\mu)$ (Weierstrass's theorem), we obtain $f\bar{g}\,d\mu = 0$ (the null measure), hence $f\bar{g} = 0$ μ-a.e. Then, as $L^2(\mu)$ is separable, so is $E^\perp$. By taking a sequence (g_n) dense in $E^\perp$, we set

$$A = \bigcap_n Z(g_n), \quad Z(g_n) = \{z : g_n(z) = 0\}.$$

(More rigorously, we define $Z(g_n)$ by choosing a measurable representative in the equivalence class g_n of $L^2(\mu)$; another choice of representative would lead to a set A' differing from A only by a negligible set, hence $\chi_A = \chi_{A'}$ in the space $L^2(\mu)$.) We obtain, for any $f \in E$ and every n, $f\bar{g}_n = 0$ μ-a.e., and thus $f = 0$ a.e. on the set $\bigcup_n Z(g_n)^c = A^c$. This means that $f \in \chi_A L^2(\mu)$, and hence $E \subset \chi_A L^2(\mu)$.

Conversely, if $f \in \chi_A L^2(\mu)$, then (clearly) $f = 0$ μ-a.e. on A^c. Since $g_n = 0$ on A, we have $f\bar{g}_n = 0$ μ-a.e., thus $(f, g_n) = 0$, $\forall n$. By the density of (g_n) in $E^\perp$, we obtain $f \perp E^\perp$, hence $f \in E$. The two inclusions give $E = \chi_A L^2(\mu)$.

The implication $(2) \Rightarrow (1)$ is evident.

For the uniqueness, the equality $\chi_A L^2(\mu) = \chi_B L^2(\mu)$ implies $\chi_A \in \chi_B L^2(\mu)$, thus $\chi_A = 0$ a.e. on B^c, meaning that $A \subset B$ up to a μ-negligible set (i.e., $\mu(A \setminus B) = 0$). Similarly, $\mu(B \setminus A) = 0$, which completes the proof. $\blacksquare$

1.3 Non-reducing Subspaces of the Bilateral Shift M_z

In order to catalog the non-reducing subspaces of M_z, we use two related (but not coincident) orthogonal decompositions. The first is given by Lemma 1.3.1 below and concerns an invariant subspace of an arbitrary operator. The second is the *Radon–Nikodym decomposition* (see Appendix A)

$$L^2(\mu) = L^2(\mu_a) \oplus L^2(\mu_s),$$

where μ is a Borel measure on the circle $\mathbb{T}$, and μ_a, μ_s denote, respectively, the absolutely continuous and singular components of μ with respect to the normalized Lebesgue measure m, $m\{e^{it} : \theta_1 \leq t \leq \theta_2\} = (\theta_2 - \theta_1)/2\pi \leq 1$.

Lemma 1.3.1 *Let $T : H \to H$ be a bounded linear operator on a Hilbert space H and let $E \subset H$ be a closed subspace.*

(1) $E \in \text{Lat}(T) \Leftrightarrow E^\perp \in \text{Lat}(T^*)$.
(2) $E \in \text{Lat}(T, T^*) \Leftrightarrow E \in \text{Lat}(T), E^\perp \in \text{Lat}(T)$.
(3) *For every $E \in \text{Lat}(T)$,*

$$E = E_R \oplus E_N,$$

where $E_R \in \mathrm{Lat}(T, T^*)$ (a reducing subspace of T) and $E_N \in \mathrm{Lat}(T)$ is a completely non-reducing subspace, i.e. such that $E' \subset E_N$, $E' \in \mathrm{Lat}(T, T^*) \Rightarrow E' = \{0\}$. This representation is unique.

Proof (1) We first show the implication "$\Rightarrow$". Let $y \in E^{\perp}$. Then, $(T^*y, x) = (y, Tx) = 0$ for every $x \in E$, and hence $T^*y \in E^{\perp}$. It ensues that $T^*E^{\perp} \subset E^{\perp}$.

The implication "$\Leftarrow$" is immediate since $T = (T^*)^*$.

(2) It is immediate by (1) since $T = (T^*)^*$.

(3) Clearly the "span" (closed linear hull) of a family of reducing subspaces is still in $\mathrm{Lat}(T, T^*)$. Set

$$E_R = \mathrm{span}\,(E' : E' \subset E, E' \in \mathrm{Lat}(T, T^*)), \quad E_N = E \ominus E_R.$$

Then $E = E_R \oplus E_N$ and $E_R \in \mathrm{Lat}(T, T^*)$. Moreover, $E_N = E \cap (E_R)^{\perp}$ and hence, by (1), $E_N \in \mathrm{Lat}(T)$. If $E' \subset E_N$ and $E' \in \mathrm{Lat}(T, T^*)$, then $E' \subset E_R$ by the definition of the latter. Thus $E' = \{0\}$. The uniqueness is also immediate. ∎

Lemma 1.3.2 *Let μ be a finite Borel measure on $\mathbb{T}$, with $\mu = \mu_a + \mu_s = w \cdot m + \mu_s$ its Radon–Nikodym decomposition (see Appendix A), and let $E \subset L^2(\mu)$ be a NON-reducing invariant subspace of $M_z : L^2(\mu) \to L^2(\mu)$. Then:*

(1) *There exists a function $q \in E$ such that $|q|^2 w = 1$ m-a.e.*
(2) *$E_R \subset L^2(\mu_s)$, where E_R is the reducing part of E according to Lemma 1.3.1.*

Proof (1) Our subspace E satisfies the properties $M_z E \subset E$, $M_z E \neq E$; indeed, if we had $M_z E = E$, then $M_z^* E = M_{\bar{z}} M_z E = E$, hence $E \in \mathrm{Lat}(M_z, M_z^*)$ which is not the case. Moreover, M_z is an isometric (and even unitary) operator, and thus the image $M_z E$ is closed. Let

$$q \in E \ominus M_z E = E \cap (M_z E)^{\perp}, \quad \|q\| = 1.$$

Since $q \in E$ and $M_z^n q \in M_z E$ for all $n \geq 1$, we obtain

$$0 = (z^n q, q) = \int_{\mathbb{T}} z^n q \bar{q}\, d\mu = \int_{\mathbb{T}} z^n |q|^2\, d\mu, \quad n \geq 1.$$

We conclude, by complex conjugation, that all the Fourier coefficients of the measure $|q|^2\, d\mu$, except for one, are zero, and hence there exists a constant c such that $(\widehat{|q|^2\, d\mu})(n) = c\hat{m}(n)$ for all n, $n \in \mathbb{Z}$. By the theorem of uniqueness (see Appendix A), $|q|^2\, d\mu = m$ ($c = 1$ by the normalization

$\|q\| = 1$). Thus, $|q|^2 \, d\mu_a + |q|^2 \, d\mu_s = m$, and by the uniqueness of the Radon–Nikodym decomposition $m = |q|^2 \, d\mu_a = |q|^2 wm$, which is equivalent to $|q|^2 w = 1$ m-a.e.

(2) Let $f \in E_R$. Given that E_R is reducing and $M_z^* = M_{\bar{z}} = M_z^{-1}$, we have $z^n f \in E_R \subset E$ for all $n \in \mathbb{Z}$. Then $z^n f = z(z^{n-1} f) \in M_z E$, and by the definition of q we obtain

$$0 = (z^n f, q) = \int_{\mathbb{T}} z^n f \bar{q} \, d\mu, \quad \forall n \in \mathbb{Z}.$$

Therefore, $f\bar{q} = 0$ μ-a.e., hence μ_a-a.e., and thus (given that $m = |q|^2 \, d\mu_a$) $f\bar{q} = 0$ m-a.e. However $q \neq 0$ m-a.e., hence $f = 0$ m-a.e., which means $f \in L^2(\mu_s)$. We thus obtain $E_R \subset L^2(\mu_s)$. ∎

Corollary 1.3.3 *Every invariant subspace of $L^2(\mu)$ contained in $L^2(\mu_s)$ is reducing and can be written $E = \chi_A L^2(\mu_s)$ with A a Borel set.*

Indeed, if E were not reducing, it would contain a function q satisfying $|q|^2 \neq 0$ m-a.e., which is impossible. ∎

Definition 1.3.4 (the space $H^2(\mathbb{T})$, the generic non-reducing subspace) *Let $L^2(\mathbb{T}) = L^2(\mathbb{T}, m)$ (normalized Lebesgue measure). The Hardy space $H^2(\mathbb{T})$ is defined as the following subspace of $L^2(\mathbb{T})$:*

$$H^2(\mathbb{T}) = \{f \in L^2(\mathbb{T}) : \hat{f}(n) = 0 \text{ for all integers } n < 0\}.$$

Reminder The exponentials $(z^n)_{n\in\mathbb{Z}} = (e^{int})_{n\in\mathbb{Z}}$ form an orthonormal basis of the space $L^2(\mathbb{T})$, and hence every function $f \in L^2(\mathbb{T})$ is the sum of its Fourier series

$$f = \sum_{n\in\mathbb{Z}} \hat{f}(n) z^n,$$

norm-$L^2(\mathbb{T})$ convergent for the symmetric partial sums $\sum_{n=-N}^{N} \hat{f}(n) z^n$ (for $N \to \infty$), or even for "disordered" sums $\sum_{n\in\sigma(N)} \hat{f}(n) z^n$ where $\sigma(N) \subset \mathbb{Z}$, $\sigma(N) \nearrow \mathbb{Z}$ for $N \to \infty$:

$$\lim_N \left\| f - \sum_{n\in\sigma(N)} \hat{f}(n) z^n \right\|_{L^2(\mathbb{T})} = 0.$$

With this reminder, we can say

$$H^2(\mathbb{T}) = \left\{ f \in L^2(\mathbb{T}) : f = \sum_{n\geq 0} \hat{f}(n) z^n \right\} = \left\{ \sum_{n\geq 0} a_n z^n : \sum_{n\geq 0} |a_n|^2 < \infty \right\}.$$

Moreover, the use of properties of orthogonal bases leads to

$$H^2(\mathbb{T}) = \text{span}_{L^2(\mathbb{T})} \left(z^n : n = 0, 1, \dots \right).$$

By the above, clearly

$$M_z H^2(\mathbb{T}) \subset H^2(\mathbb{T}) \text{ and } M_z H^2(\mathbb{T}) \neq H^2(\mathbb{T})$$

(hence, $H^2(\mathbb{T})$ is a non-reducing invariant subspace). Moreover, $H^2(\mathbb{T})$ is completely non-reducing since for every $f \in H^2(\mathbb{T})$, $f \neq 0$, there exists a positive integer n such that $M_z^{*n} f = \bar{z}^n f \notin H^2(\mathbb{T})$. The following theorem shows that this is a generic example: any other completely non-reducing subspace coincides with $H^2(\mathbb{T})$ up to a factor of correction. This was proved in the 1960s by Henry Helson, professor at the University of California (Berkeley).

Henry Helson (1927–2010), one of the primary experts of his generation in harmonic analysis, was a professor at the University of California (Berkeley), 1955–2010. His work on Hardy spaces and "abstract Hardy spaces" (1960–1965, in collaboration with David Lowdenslager, a mathematician from Yale), as well as his perfectly written research monographs (such as *Lectures on Invariant Subspaces* (Helson, 1964)) profoundly influenced the development of the subject. A rich personality with an extraordinary range of talent (in particular, he was a violinist and cellist at a professional level), he had a truly singular career: as a dedicated Quaker, he turned down a position in California in 1948, because he refused to take a "loyalty oath" (mandatory in California in the era of McCarthyism), and left for Europe where he continued his studies in Poland (with Szpilrajn), then in Sweden (with Beurling) and in France, at Nancy (with Schwartz, Dieudonné, Godement, and Grothendieck).

Theorem 1.3.5 (Helson, 1964) *Let μ be a finite Borel measure on $\mathbb{T}$, $\mu = \mu_a + \mu_s = wm + \mu_s$, and let $E \subset L^2(\mu)$ be an invariant subspace of M_z. Then:*

(1) *either $M_z E = E$, and then $E = \chi_A L^2(\mu)$ with A a Borel set,*

(2) *or $M_z E \neq E$, and then $E = \chi_A L^2(\mu_s) \oplus q H^2(\mathbb{T})$, where $|q|^2 w = 1$ m-a.e. and A is a Borel set; such a function q is unique, and so is A (meaning $q = \lambda q'$ with $\lambda \in \mathbb{T}$ and $\chi_A = \chi_{A'}$ μ_s-a.e.).*

Conversely, each A and q satisfying this equation generate a reducing subspace by the formula $E = \chi_A L^2(\mu)$, and a non-reducing subspace by $E = \chi_A L^2(\mu_s) \oplus q H^2(\mathbb{T})$. The latter subspace is completely non-reducing if and only if $\chi_A L^2(\mu_s) = \{0\}$ ($\Leftrightarrow \chi_A = 0$ μ_s-a.e.).

Proof (1) This is Wiener's Theorem 1.2.1.

(2) Let $E \in \mathrm{Lat}(M_z)$, $E \neq M_z E$. By Lemma 1.3.2, there exists a function $q \in E \cap L^2(\mu_a)$, $q \perp M_z E$ such that $|q|^2 w = 1$ m-a.e. The sequence $(z^n q)_{n \in \mathbb{Z}}$ is orthonormal:

$$(z^n q, z^m q) = \int_{\mathbb{T}} z^{n-m} |q|^2 \, d\mu = \int_{\mathbb{T}} z^{n-m} |q|^2 w \, dm = \int_{\mathbb{T}} z^{n-m} \, dm = \delta_{m,n},$$

where $\delta_{m,n}$ is the *Kronecker delta* (= 0 if $m \neq n$, and = 1 if $m = n$). Consequently,

$$\mathrm{span}_{L^2(\mu)} \left(z^n q : n \geq 0 \right) = \left\{ \sum_{n \geq 0} a_n z^n q : \sum_{n \geq 0} |a_n|^2 < \infty \right\}.$$

Moreover, clearly the mapping $U: f \longmapsto qf$ is a linear isometry of $L^2(\mathbb{T}) = L^2(\mathbb{T}, m) \to L^2(\mathbb{T}, \mu_a)$:

$$\|Uf\|^2_{L^2(\mu_a)} = \int_{\mathbb{T}} |f|^2 |q|^2 w \, dm = \int_{\mathbb{T}} |f|^2 \, dm = \|f\|^2_{L^2(\mathbb{T})}.$$

Hence $\mathrm{span}_{L^2(\mu)} \left(z^n q : n \geq 0 \right) = U(H^2(\mathbb{T})) = qH^2(\mathbb{T}) \subset E$. Let $E = E' \oplus qH^2(\mathbb{T})$ where $E' = E \cap (qH^2(\mathbb{T}))^\perp$ (orthogonal complement in $L^2(\mu)$). For an arbitrary function $f \in E' \subset E$, we have

$$\int_{\mathbb{T}} z^n f \bar{q} \, d\mu = (q, z^n f) = 0 \text{ for } n \geq 1, \text{ and } \int_{\mathbb{T}} f \bar{z}^n \bar{q} \, d\mu = (f, qz^n) = 0 \text{ for } n \geq 0,$$

so $f \bar{q} \, d\mu = 0$, and hence $f \bar{q} = 0$ μ-a.e. However $q \neq 0$ μ_a-a.e., thus $f = 0$ μ_a-a.e., and then $f \in L^2(\mu_s)$. We have shown that $E' \subset L^2(\mu_s)$, and – because the converse $E \cap L^2(\mu_s) \subset E'$ is clear – $E' = E \cap L^2(\mu_s)$. As both E and $L^2(\mu_s)$ are M_z-invariant, then Corollary 1.3.3 leads to $E' = \chi_A L^2(\mu_s)$.

For the uniqueness, let $\chi_A L^2(\mu_s) \oplus qH^2(\mathbb{T}) = \chi_{A'} L^2(\mu_s) \oplus q' H^2(\mathbb{T})$ where $|q'|^2 w = 1$ m-a.e. Then clearly $\chi_A L^2(\mu_s) = \chi_{A'} L^2(\mu_s)$ and $qH^2(\mathbb{T}) = q' H^2(\mathbb{T})$, hence $\chi_A = \chi_{A'}$ μ_s-a.e. The second equation implies $q/q' \in H^2(\mathbb{T})$ and $q'/q \in H^2(\mathbb{T})$, and since $|q| = |q'|$ m-a.e., all the Fourier coefficients of q/q', with the exception of $(q/q')\hat{}(0)$, are zero. Hence $q/q' = constant = \lambda$; clearly $|\lambda| = 1$.

The rest of the statement is also evident. ∎

Corollary 1.3.6 *The space $L^2(\mathbb{T}, \mu)$ contains a non-reducing invariant subspace E (i.e. $M_z E \subset E$, $M_z E \neq E$) if and only if $m \ll \mu$ (i.e. $w > 0$ m-a.e. on $\mathbb{T}$).*

Indeed, according to Theorem 1.3.5(2), it is a question of the existence of a function q such that $|q|^2 w = 1$ m-a.e., which is equivalent to the condition of the corollary. ∎

1.3.1 $H^p(\mathbb{T})$ Spaces

Let $1 \leq p \leq \infty$. The *Hardy space $H^p(\mathbb{T})$* is defined similarly to the space $H^2(\mathbb{T})$

$$H^p(\mathbb{T}) = \{f \in L^p(\mathbb{T}) : \hat{f}(n) = 0 \text{ for any integer } n < 0\}.$$

The $H^p(\mathbb{T})$ spaces share many of the properties of the space $H^2(\mathbb{T})$, but of course not all of them, and always after some modifications. For example, the exponentials $(z^n)_{n \in \mathbb{Z}}$ no longer form an unconditional basis in $L^p(\mathbb{T})$, $p \neq 2$, but form a Schauder basis for $1 < p < \infty$, as will be seen in Chapter 2, Exercise 2.8.4(f)).

Here, we limit ourselves to a short list of initial properties. For more information, see Chapter 2, in particular Exercise 2.8.1 (concerning an analog in $L^p(\mathbb{T})$ of the theorems of Beurling and Helson).

(1) *$H^p(\mathbb{T})$ is a closed vector subspace of $L^p(\mathbb{T})$.*
(2) *If $f, \overline{f} \in H^p(\mathbb{T})$, then f = constant.*

Indeed, all the Fourier coefficients of f are zero, except perhaps $\hat{f}(0)$.   ∎

(3) *If $f \in H^p(\mathbb{T})$ and $f = 0$ on a set $A \subset \mathbb{T}$, with $m(A) > 0$, then $f = 0$.*

For a proof, see Corollary 2.3.3 below.

(4) *Let*

$$H^p_-(\mathbb{T}) = \{f \in L^p(\mathbb{T}) : \hat{f}(n) = 0 \text{ for every integer } n \geq 0\}.$$

Then, $H^p(\mathbb{T}) \cap H^p_-(\mathbb{T}) = \{0\}$ and, if $p < \infty$, $\mathrm{clos}_{L^p}(H^p(\mathbb{T}) + H^p_-(\mathbb{T})) = L^p(\mathbb{T})$.

(For $p = \infty$ see Exercise 2.8.4(i). In fact, for $1 < p < \infty$, the sum is already closed, by Marcel Riesz's Theorem 2.8.4(e)).

Indeed, the first equation holds for the reason used for (2), the second because $H^p(\mathbb{T}) + H^p_-(\mathbb{T}) \supset \mathcal{P}$.   ∎

(5) *The invariant subspaces of $L^p(\mathbb{T})$ were described by Srinivasan (1963): let E be a subspace invariant under the shift operator $M_z : L^p(\mathbb{T}) \to L^p(\mathbb{T})$. Then:*
   (a) *either $zE = E$, and then $E = \chi_A L^p(\mathbb{T})$ for some Borel set A,*
   (b) *or $zE \neq E$, and then there exists q, measurable on $\mathbb{T}$, with $|q| = 1$ m-a.e., such that $E = qH^p(\mathbb{T})$.*

The parameters A and q are uniquely defined by E in the same sense as in Theorem 1.3.5.

For the proof, see Exercise 2.8.1 where this analog of Theorem 1.3.5 is a corollary of a more general proposition.

1.4 Beurling "Inner Functions"

The special case $\mu = m$ is particularly important.

Corollary 1.4.1 (invariant subspaces of $L^2(\mathbb{T})$) *Let E be a subspace invariant under the shift operator $M_z \colon L^2(\mathbb{T}) \to L^2(\mathbb{T})$. Then:*

(1) *either $zE = E$, and then $E = \chi_A L^2(\mathbb{T})$ for some Borel set A,*
(2) *or $zE \neq E$, and then there exists a function $q \in L^\infty(\mathbb{T})$, with $|q| = 1$ m-a.e., such that $E = qH^2(\mathbb{T})$. The parameters A and q are uniquely defined by E in the same sense as in Theorem 1.3.5.*

Indeed, this is Theorem 1.3.5 with $\mu_s = 0$ and $w = 1$.   ∎

The following even more specialized case, called "Beurling's theorem," is important for not only its consequences, but also for its role in the development of the theory of Hardy spaces. Even though the proof given below is totally different from the original proof, we still need the following definition introduced by Beurling: that of a special class of "inner functions" in $H^2(\mathbb{T})$ which, today, plays a fundamental role in the entire theory. See also the historical remarks in the biographical sketch below, and in §§ 1.9, 2.9, 3.6.

Arne Beurling (1905–1986), was a Swedish mathematician, the author of numerous remarkable works in mathematical analysis and cryptography, and a professor (1937–1954) at the University of Uppsala (a university founded in 1477, where Carl Linnaeus and Anders Celsius worked), and later at the Institute for Advanced Study at Princeton, USA. Simultaneously with Gelfand, he discovered the fundamental principles of Banach algebras and introduced an important class of weighted algebras ("Beurling algebras"), described the invariant subspaces of the isometric shift operator, and (with Malliavin) resolved the problem of the completeness radius of families of exponentials by proving the "multiplier theorem," important for the uncertainty principle in harmonic analysis. His doctoral students include Carleson, Domar, Esseen, Hall, and Nyman.

Beurling is also famous for having single-handedly (in 1940) deciphered the German Nazi secret code known as *Geheimfernschreiber* ("secret teleprinter"), based on a machine that could create 10^{18} different combinations (many more than the Enigma machine, famous for its role in Operation Overlord!). This feat allowed the Swedish secret service to systematically decipher coded messages that were passing through Sweden via a cable linking Norway with Nazi Germany. The invasion plan Barbarossa and the date at which it was to start (June 22, 1941) were intercepted and communicated to the Soviets, but they did not believe the information as its source was not revealed.

Definition 1.4.2 (Beurling inner functions) *A function on the circle* $\mathbb{T}$ *is said to be inner (in the sense of Beurling) if*

$$\varphi \in H^2(\mathbb{T}) \ and \ |\varphi| = 1 \ m\text{-}a.e.$$

Corollary 1.4.3 (Beurling's Theorem, 1949) *Let* $E \subset H^2(\mathbb{T}) \subset L^2(\mathbb{T})$ *be a subspace of* $H^2(\mathbb{T})$, $E \neq \{0\}$. *Then,* E *is* M_z-*invariant if and only if there exists an inner function* q *such that*

$$E = qH^2(\mathbb{T}).$$

There is a bijective *correspondence between* $\mathrm{Lat}(M_z|H^2(\mathbb{T}))$ *and the set of inner functions* q *whose first non-zero Fourier coefficient is positive.*

Indeed, by applying Corollary 1.4.1, we see that case (1) is impossible: if $zE = E$, then we would have $\bar{z}^n f \in E \subset H^2(\mathbb{T})$ for all $n \geq 0$ and every function $f \in E$; however if $f \neq 0$, $f = \sum_{k \geq n} a_k z^k$ with $a_n \neq 0$, we obtain $\bar{z}^{n+1} f = a_n \bar{z} + \sum_{k>n} a_k z^{k-n-1}$ and hence $\bar{z}^{n+1} f \notin H^2(\mathbb{T})$. Consequently, $zE \neq E$. In case (2) of Corollary 1.4.1, we have $E = qH^2(\mathbb{T}) \subset H^2(\mathbb{T})$, thus $q \in H^2(\mathbb{T})$, and the result follows. ∎

Corollary 1.4.4 (boundary uniqueness theorem) *If* $f \in H^2(\mathbb{T})$ *and* $f = 0$ *on a set* $A \subset \mathbb{T}$ *such that* $m(A) > 0$, *then* $f = 0$.

Indeed, let

$$E_f := \mathrm{span}_{L^2(\mathbb{T})}(z^n f : n \geq 0) = \mathrm{clos}_{L^2(\mathbb{T})}(f\mathcal{P}_a),$$

the smallest M_z-invariant subspace containing f, where $\mathcal{P}_a$ is the space of analytic polynomials,

$$\mathcal{P}_a := \mathcal{P} \cap H^2(\mathbb{T}).$$

If we suppose $f \neq 0$, then by Corollary 1.4.3, $E_f = qH^2(\mathbb{T})$ with an inner function q (hence, $|q| = 1$ m-a.e.). In particular, $q \in E_f$, which is impossible, since for any polynomial p,

$$\|q - pf\|_2^2 \geq \int_A |q - pf|^2 \, dm = \int_A |q|^2 \, dm = m(A) > 0,$$

thus a contradiction. $\blacksquare$

Furthermore, with regard to the uniqueness theorem (proved by Frigyes and Marcel Riesz in 1916: see the biographical sketch in § 1.5), we can add that in Chapter 3 a more complete (even definitive) description of the subject will be presented. Finally, note that numerous examples of inner functions are known (see Exercises § 1.8.3), and better still, that there exists an intelligible description of all inner functions. This was well known long before Beurling's theorem (see §§ 1.9, 2.9 for details).

1.5 $H^2(\mu)$ Spaces and the Riesz Brothers' Theorem

We begin with the definition of the space H^2 associated with an arbitrary Borel measure μ on the circle $\mathbb{T}$ (in place of the Lebesgue measure m) and the Radon–Nikodym decomposition lemma of invariant subspaces.

Definition 1.5.1 *Let μ be a finite measure on $\mathbb{T}$. The Hardy space associated with μ is defined by*

$$H^2(\mu) := \mathrm{span}_{L^2(\mu)} (z^n : n \geq 0) = \mathrm{clos}_{L^2(\mu)} \mathcal{P}_a,$$

where $\mathcal{P}_a = \mathcal{L}in(z^n : n \geq 0)$ is again the space of analytic polynomials. Clearly, $H^2(m) = H^2(\mathbb{T})$.

Lemma 1.5.2 *Let μ be a (finite) Borel measure on $\mathbb{T}$. Then:*

(1) *For every $E \in \mathrm{Lat}(M_z)$, with $M_z : L^2(\mu) \to L^2(\mu) = L^2(\mu_a) \oplus L^2(\mu_s)$, we have*

$$E = E_a \oplus E_s \text{ where } E_a = E \cap L^2(\mu_a), \quad E_s = E \cap L^2(\mu_s).$$

(2) $H^2(\mu) = H^2(\mu_a) \oplus L^2(\mu_s).$

Proof (1) By Helson's Theorem 1.3.5, either $E = \chi_A L^2(\mu) = \chi_A L^2(\mu_a) \oplus \chi_A L^2(\mu_s)$, or $E = \chi_A L^2(\mu_s) \oplus qH^2(\mathbb{T})$ and $qH^2(\mathbb{T}) \subset L^2(\mu_a)$, and the result follows.

(2) Since $H^2(\mu) \in \mathrm{Lat}(M_z)$, by (1) we have $E := H^2(\mu) = E_a \oplus E_s$. Then by
Corollary 1.3.3, $E_s = \chi_A L^2(\mu_s)$ with a Borel set A. As $1 \in H^2(\mu)$, we have
$\chi_A = 1$ μ_s-a.e., and hence $E_s = L^2(\mu_s)$.

Moreover, by writing $1 = 1_a \oplus 1_s \in E_a \oplus E_s$, we obtain $1_a \in E_a$ ($1_a = 1$ μ_a-a.e.), and hence $H^2(\mu_a) \subset E_a$. However the reverse inclusion is evident, since
for every $f \in E_a$ and every sequence of polynomials $p_n \in \mathcal{P}_a$ converging to f,
we have

$$\|f - p_n\|^2_{L^2(\mu_a)} \leq \|f - p_n\|^2_{L^2(\mu_a)} + \|p_n\|^2_{L^2(\mu_s)} = \|f - p_n\|^2_{L^2(\mu)} \to 0,$$

hence $f \in H^2(\mu_a)$. $\blacksquare$

Remark 1.5.3 What equality (2) in the lemma means is that there is a
simultaneous polynomial approximation: $\forall f \in H^2(w \cdot m)$, $\forall g \in L^2(\mu_s)$, there
exists a sequence of polynomials $(p_n) \subset \mathcal{P}_a$ such that, simultaneously, $p_n \to f$
(in $L^2(\mu_a)$) and $p_n \to g$ (in $L^2(\mu_s)$).

The following theorem (usually called the Riesz Brothers Theorem) is a
cornerstone in the construction of Hardy spaces and of harmonic analysis on
the circle $\mathbb{T}$ (moreover, there exist analogs of this statement for other groups
such as $\mathbb{R}$, $\mathbb{T}^n$, $\mathbb{R}^n$; see § 1.9). *A priori*, this is somewhat unexpected: certain
restrictions on the *Fourier spectrum* $\sigma_{\mathcal{F}}(\mu)$ of a complex measure μ, where

$$\sigma_{\mathcal{F}}(\mu) = \mathrm{supp}(\hat{\mu}) = \{n \in \mathbb{Z}: \hat{\mu}(n) \neq 0\}, \quad \hat{\mu}(n) = \int_{\mathbb{T}} \bar{z}^n \, d\mu,$$

imply consequences on the size of the (Borel) support of μ.

Theorem 1.5.4 (Riesz and Riesz, 1916) *Let μ be a complex Borel measure on
$\mathbb{T}$, assumed "analytic," i.e. its Fourier coefficients of negative index are zero:*

$$\hat{\mu}(-n) := \int_{\mathbb{T}} z^n \, d\mu = 0, \quad n \geq 0.$$

*Then, $\mu \ll m$ (μ is absolutely continuous with respect to m) and $\mu = hm$ with
$h \in H_0^1$, where*

$$H_0^1 := \{f \in L^1(\mathbb{T}): \hat{f}(k) = 0 \text{ for } k \leq 0\}.$$

Proof Let $|\mu|$ be the variation of the measure μ (see Appendix A). Clearly
$\mu \ll |\mu|$; let ϵ be the corresponding Radon–Nikodym derivative: $\mu = \epsilon|\mu|$. It is
well known that $|\epsilon| = 1$ $|\mu|$-a.e. (see Appendix A). Since $\int_{\mathbb{T}} z^n \, d\mu = \int_{\mathbb{T}} z^n \epsilon \, d|\mu|$,
the hypothesis on μ means that $\bar{\epsilon} \perp H^2(|\mu|)$ in the space $L^2(|\mu|)$. However,
by Lemma 1.5.2, $H^2(|\mu|) = H^2(|\mu|_a) \oplus L^2(|\mu|_s)$, which implies $\bar{\epsilon} \perp L^2(|\mu|_s)$,
and hence $\epsilon = 0$ $|\mu|_s$-a.e. Since, at the same time, $|\epsilon| = 1$ $|\mu|_s$-a.e., we obtain

$|\mu|_s = 0$. Then, the measure $\mu = \epsilon|\mu| = \epsilon|\mu|_a$ is absolutely continuous with respect to m, and hence by Radon–Nikodym there exists $h \in L^1(m)$ such that $\mu = hm$. Clearly the hypothesis on μ can be translated to $h \in H_0^1$.   ∎

The brothers **Frigyes (Frédéric)** and **Marcell (Marcel) Riesz** were two pillars of analysis in the twentieth century. They founded various domains of analysis, thus offering a rare example of familial scientific endeavor at such a high level. The elder, Frigyes Riesz (1880–1956), laid the foundations of functional analysis and operator theory as separate disciplines (1910); he was strongly influenced by the ideas of Fréchet, Lebesgue, and Hilbert. The representation theorem of linear functionals, as well as the *Riesz–Fischer theorem*, bear his name. He also founded the János Bolyai Mathematical Institute and the journal *Acta Scientiarum Mathematicarum* (Szeged), and with his student Béla Sz.-Nagy co-authored an influential text, *Leçons d'analyse fonctionelle*.

Marcel Riesz (1886–1969) spent (almost) all of his career at the University of Lund (Sweden). His contribution to analysis was enormous: his discoveries include *Riesz transformation*, the *Riesz potential*, the *Riesz (–Bochner) mean*, and the *Riesz–Thorin theorem*. Curiously, in his search for a permanent position, he was classed in second place twice in a row (for different positions), each time behind Torsten Carleman. His doctoral students included Thorin, Cramér, Hille, Frostman, and Hörmander.

Frigyes and Marcel Riesz wrote only one article together (Riesz and Riesz, 1916): it contains the *Riesz brothers' theorem*, which subsequently became so important for harmonic analysis and its applications.

Furthermore, with regard to Theorem 1.5.4, it can be mentioned that the original proof was much more complicated than that above; however, our proof depends on invariant subspaces and on some of the already developed theory, hence it is indirect. An alternative proof is presented below, which is completely elementary and depends only on the definition of absolutely continuous measures.

1.5.1 Elementary Proof of Theorem 1.5.4 (Øksendal, 1971)

First note that the hypothesis on μ implies $\int_{\mathbb{T}} p\,d\mu = 0$ for every $p \in \mathcal{P}_a$, and then $\int_{\mathbb{T}} f\,d\mu = 0$ for every function f defined and holomorphic in a disk $(1 + \epsilon)\mathbb{D} = D(0, 1 + \epsilon)$, $\epsilon > 0$: indeed such a function f is the sum of a power series normally convergent on $\mathbb{T}$, $f(z) = \sum_{k\geq 0} \hat{f}(k)z^k$ (with radius of convergence $\geq 1 + \epsilon$), hence the series can be integrated term by term.

In particular, for every rational function $f = p/q$ ($p, q \in \mathcal{P}_a$) having poles (zeros of the denominator q) in $\mathbb{C} \setminus \overline{\mathbb{D}}$, we have

$$\int_{\mathbb{T}} f \, d\mu = 0.$$

By the definition of $\mu \ll m$, we must show that, for every Borel set $A \subset \mathbb{T}$, $m(A) = 0 \Rightarrow \mu(A) = 0$. In fact, it is sufficient to do this only for a closed $A = F$, because of the regularity of the variation $|\mu|$ (see Appendix A).

So, let $F \subset \mathbb{T}$, with $F = \overline{F}$, $m(F) = 0$. We are going to construct a sequence of rational functions (h_n) such that

(1) $|h_n(z)| \leq 2$ on $\mathbb{T}$,
(2) $\lim_n h_n(z) = \chi_F(z)$ for all $z \in \mathbb{T}$.

Then, by the dominated convergence theorem,

$$0 = \int_{\mathbb{T}} h_n \, d\mu \to \int_{\mathbb{T}} \chi_F \, d\mu = \mu(F) \quad \text{(when } n \to \infty\text{),}$$

thus $\mu(F) = 0$, which will complete the proof.

Construction of a sequence (h_n): since $m(F) = 0$, for every $n \geq 1$, there exist disks $D(z_i, r_i)$, $i = 1, \ldots, N$, such that

$$z_i \in F \subset \mathbb{T}, \quad F \subset \bigcup_i D(z_i, r_i), \quad \text{and} \quad \sum_{i=1}^{N} r_i < \frac{1}{n^2}.$$

Set

$$f_n = \prod_{i=1}^{N} \frac{z - z_i}{z - z_i - nr_iz_i}.$$

The f_n satisfy the following properties.

(1) The functions f_n are rational, with poles $z = (1 + nr_i)z_i$ in $\mathbb{C} \setminus \overline{\mathbb{D}}$, hence $\int_{\mathbb{T}} f_n \, d\mu = 0$.
(2) For $z \in \mathbb{T}$, by elementary geometry we have $|z - z_i| < |z - z_i(1 + nr_i)|$, thus $|f_n(z)| < 1$.

(3) For $z \in \mathbb{T} \cap D(z_i, r_i)$, we obtain $|z - z_i| < r_i$ and
$|z - z_i(1 + nr_i)| > |z_i|nr_i - |z - z_i| > nr_i - r_i = (n-1)r_i$, hence

$$\left| \frac{z - z_i}{z - z_i - z_i nr_i} \right| < \frac{r_i}{(n-1)r_i} = \frac{1}{n-1}.$$

However, the other factors of f_n are bounded above by 1 (see (2)): hence $|f_n(z)| < 1/(n-1)$ for every point of F.

(4) For $z \in \mathbb{T} \setminus F$, let $d = \operatorname{dist}(z, F)$; we have $|z - z_i| \geq d > 0$ for every i. Writing

$$\frac{1}{f(z)} = \prod_{i=1}^{N} \left(1 - \frac{nr_i z_i}{z - z_i}\right) = \exp\left\{ \sum_{i=1}^{N} \log\left(1 - \frac{nr_i z_i}{z - z_i}\right) \right\},$$

we observe that for $n > 2/d$,

$$\left| \frac{nr_i z_i}{z - z_i} \right| \leq \frac{nr_i}{d} \leq \frac{1}{nd} < \frac{1}{2},$$

and by using $|\log(1 - w)| \leq 2|w|$ for $|w| \leq 1/2$, we obtain

$$\left| \sum_{i=1}^{N} \log\left(1 - \frac{nr_i z_i}{z - z_i}\right) \right| \leq 2 \sum_{i=1}^{N} \frac{nr_i}{|z - z_i|} \leq \frac{2n}{d} \sum_{i=1}^{N} r_i < \frac{2}{dn},$$

and hence $\lim_n (1/f_n(z)) = 1$.

In conclusion, the functions $h_n = 1 - f_n$ satisfy all the required properties, which completes the proof. ∎

1.6 The Past and the Future: The Prediction Problem

The problems of prediction, prognosis, and extrapolation of stochastic (random) processes have played an extraordinary role in the history of Hardy spaces.

Definition 1.6.1 *A discrete time stationary process (also known as a stationary sequence) is a sequence $(x_n)_{n \in \mathbb{Z}}$ in a Hilbert space H such that the elements of its correlation matrix $\{(x_n, x_k)_H\}$ depend only on the difference $n - k$, i.e.*

$$(x_n, x_k) = (x_{n+j}, x_{k+j}), \quad \forall n, k, j \in \mathbb{Z},$$

and $H = \operatorname{span}_H(x_n : n \in \mathbb{Z})$.

A subspace $E_- = \operatorname{span}_H(x_n : n < 0)$ is said to be the past of the process, and $E_+ = \operatorname{span}_H(x_n : n \geq 0)$ the future of the process. The process is said to *be singular (or deterministic) if $E_- = H$, and regular (or non-deterministic) if $E_- \neq H$.*

The problem of optimal (quadratic, one step ahead) prediction is to calculate

$$\mathrm{dist}_H(x_n, H_n) = \inf_{x \in H_n} \|x_n - x\|,$$

where $H_n = \mathrm{span}_H(x_k : k < n)$ is the past of x_n.

The main problem concerning random processes is to study the "dependence of the future of a process on its past," and in particular, to measure the best prediction of its state one or several step(s) ahead. The following theorem introduces the central concept of the theory.

Theorem 1.6.2 (Kolmogorov, 1939) *Let $(x_n)_{n \in \mathbb{Z}}$ be a stationary random process, $H = \mathrm{span}_H(x_n : n \in \mathbb{Z})$. Then, there exist a unique Borel measure μ on $\mathbb{T}$ and a unitary operator $U : H \to L^2(\mu)$ such that*

$$U x_n = z^n, \quad n \in \mathbb{Z}.$$

Conversely, for any μ and every unitary operator $U : H \to L^2(\mu)$, the sequence $(U^{-1} z^n)_{n \in \mathbb{Z}}$ is a stationary process.

The measure μ is called the *spectral measure* of the process.

Proof First, observe that for every linear combination of x_n, we have

$$\left\| \sum a_n x_n \right\|^2 = \sum_{n,k} a_n \bar{a}_k (x_n, x_k) = \sum_{n,k} a_n \bar{a}_k (x_{n+1}, x_{k+1}) = \left\| \sum a_n x_{n+1} \right\|^2.$$

This means that the mapping defined by

$$V x_n = x_{n+1}, \quad n \in \mathbb{Z},$$

can be extended by linearity $V(\sum a_n x_n) = \sum a_n x_{n+1}$ to an isometric mapping $H \to H$ such that VH is dense in H; hence $VH = H$ and V is unitary. Moreover, $x_n = V^n x_0$ for every $n \in \mathbb{Z}$. By the spectral theorem (Appendix E), there exists a unique Borel measure μ on $\mathbb{T}$ and a unitary mapping $U : H \to L^2(\mu)$ such that $U x_0 = 1$ and $V = U^{-1} M_z U$, where M_z is the shift operator on $L^2(\mu)$. The rest of the statement is immediate. ∎

Corollary 1.6.3 *A stationary process $(x_n)_{n \in \mathbb{Z}}$ is singular if and only if $H^2_-(\mu) = L^2(\mu)$, with $H^2_-(\mu) = \mathrm{span}_{L^2(\mu)}(z^n : n < 0)$ and μ the spectral measure of $(x_n)_{n \in \mathbb{Z}}$.*

The corollary is immediate by the definitions and the theorem. ∎

Andrey Nikolaevich Kolmogorov (1903–1987) was a Russian mathematician, one of the greatest geniuses in mathematics of the twentieth century, creator of the modern mathematical theory of probability, the KAM (Kolmogorov–Arnold–Moser) theory, the Kolmogorov complexity theory, turbulence theory, etc. Dozens of concepts of mathematics and their applications bear Kolmogorov's name: the *Kolmogorov A-integral*, *Kolmogorov's inequality*, the *Kolmogorov–Smirnov test* in statistics, the *Kolmogorov 0–1 law*, the *Chapman–Kolmogorov equation*, the entropy of a dynamic system, etc. Originally a member of Nikolai Luzin's famous group of students (at the University of Moscow), throughout his career he founded a number of scientific schools in different domains, eventually training a total of 69 doctoral students, of which 18 became members of the Academies of Science of various countries.

Among other achievements, Kolmogorov is famous for his solution (with Vladimir Arnold) of Hilbert's 13th problem (1957). He published more than 300 articles, as well as several books that became classics. He was awarded the Chebyshev Prize (1950), the Balzan Prize (1962), the Wolf Prize (1980), and the Lobachevsky Prize (1986), and was a member of dozens of scientific academies and societies.

Luzin wrote to him: Вам дан высокий дух, и я хочу, чтобы Вы его силы берегли для вещей, которые под силу очень немногим … ("You were given a great spirit, and I want you to save its strength to achieve exploits accessible by only a very few"). A caveat from Kolmogorov: "Beware of those said to be 'good mathematicians' by engineers and 'good engineers' by mathematicians."

Lomonosov Moscow State University.

The study of the problem of prediction starts with a few lemmas, somewhat technical but very useful. This study will be continued in Chapters 2 and 3.

In the lemmas, μ always stands for the spectral measure of a stationary process $(x_n)_{n\in\mathbb{Z}}$.

Lemma 1.6.4 *For every $n \in \mathbb{Z}$, we have*

$$\mathrm{dist}_H(x_n, H_n) = \mathrm{dist}_H(x_0, H_0) = \mathrm{dist}_{L^2(\mu)}(1, H^2_-(\mu))$$
$$= \mathrm{dist}_{L^2(\mu)}(1, H^2_0(\mu)) := d,$$

where $H^2_0(\mu) =:\ \mathrm{span}_{L^2(\mu)}(z^n : n > 0) = \mathrm{clos}_{L^2(\mu)}(z\mathcal{P}_a)$.

Proof　We first use the isometric nature of the operators U and V in the proof of Theorem 1.6.2; then, because $p \in z\mathcal{P}_a \Leftrightarrow \overline{p} \in \mathcal{L}in(z^n : n < 0)$, for every polynomial $p \in \mathcal{P}$, we have

$$\|1 - p\|_{L^2(\mu)} = \|1 - \overline{p}\|_{L^2(\mu)}. \qquad\blacksquare$$

Lemma 1.6.5 *Let μ be a finite Borel measure on $\mathbb{T}$. The following assertions are equivalent.*

(1) $d = 0$ *(d is defined in Lemma 1.6.4).*
(2) $1 \in H_0^2(\mu) := \operatorname{span}_{L^2(\mu)}(z^n : n > 0)$.
(3) $1 \in H_-^2(\mu) := \operatorname{span}_{L^2(\mu)}(z^n : n < 0)$.
(4) $\bar{z} \in H^2(\mu) := \operatorname{span}_{L^2(\mu)}(z^n : n \geq 0)$.
(5) $H^2(\mu) = L^2(\mu)$.
(6) $H_-^2(\mu) = L^2(\mu)$.
(7) $zH_0^2(\mu) = H_0^2(\mu)$ *and/or* $zH^2(\mu) = H^2(\mu)$.

Proof (1) $\Leftrightarrow$ (2) since in a metric space X, $x_0 \in \operatorname{clos}(A) \Leftrightarrow \operatorname{dist}_X(x_0, A) = 0$.
(2) $\Leftrightarrow$ (3) by Lemma 1.6.4.
(2) $\Leftrightarrow$ (4) since the mapping $f \longmapsto \bar{z}f$ is unitary on $L^2(\mu)$.

    (4) implies $\lim_n \|\bar{z} - p_n\|_{L^2(\mu)} = 0$ for a sequence $p_n \in \mathcal{P}_a$, and hence $\lim_n \|\bar{z}q - p_n q\|_{L^2(\mu)} = 0$ for every $q \in \mathcal{P}_a$, thus $\bar{z}\mathcal{P}_a \subset H^2(\mu)$. Since $\lim_n \|\bar{z}^k - \bar{z}^{k-1} p_n\|_{L^2(\mu)} = 0$ for every $k \geq 1$, then by induction $\bar{z}^n \mathcal{P}_a \subset H^2(\mu)$, for every $n \geq 0$. Therefore, $\mathcal{P} \subset H^2(\mu)$, and we obtain (5): $H^2(\mu) = L^2(\mu)$.

(5) $\Rightarrow$ (4) is evident.
(3) $\Leftrightarrow$ (6) for the same reason that (2) $\Leftrightarrow$ (5).

Finally, clearly, (5) $\Rightarrow$ (7); and (7) $\Rightarrow$ (2), since $z \in zH_0^2(\mu)$ implies $1 \in H_0^2(\mu)$ (and the same manipulation with $H^2(\mu)$). $\blacksquare$

Lemma 1.6.6 *Let $\mu = wm + \mu_s$ be the Radon–Nikodym decomposition of a finite Borel measure on $\mathbb{T}$. Then,*

$$d^2 := \operatorname{dist}^2_{L^2(\mu)}(1, H_0^2(\mu)) = \operatorname{dist}^2_{L^2(wm)}(1, H_0^2(wm)) = \inf_{p \in z\mathcal{P}_a} \int_{\mathbb{T}} |1 - p|^2 w\, dm.$$

Proof By Lemma 1.5.2, $H^2(\mu) = H^2(\mu_a) \oplus L^2(\mu_s)$; hence

$$H_0^2(\mu) = zH^2(\mu) = H_0^2(\mu_a) \oplus zL^2(\mu_s) = H_0^2(\mu_a) \oplus L^2(\mu_s).$$

Writing $1 = 1_a \oplus 1_s$ (according to the decomposition $L^2(\mu) = L^2(\mu_a) \oplus L^2(\mu_s)$), we have

$$d^2 = \mathrm{dist}^2_{L^2(\mu)}(1_a \oplus 1_s, H_0^2(\mu_a) \oplus L^2(\mu_s))$$

$$= \mathrm{dist}^2_{L^2(\mu_a)}(1_a, H_0^2(\mu_a)) + \mathrm{dist}^2_{L^2(\mu_s)}(1_s, L^2(\mu_s)) = \mathrm{dist}^2_{L^2(\mu_a)}(1_a, H_0^2(\mu_a)).$$

∎

The spaces $H^2(\mu)$ are the principal tools used in § 1.7, but the final conclusion on the subject will be made in Chapter 2, § 2.7.2.

1.7 Inner–Outer Factorization and Szegő's Infimum

Recall that in Definition 1.4.2 we defined the inner functions, in the sense of Beurling. We now complete this terminology as follows.

Definition 1.7.1 *Let $f \in H^2(\mathbb{T})$. It is said to be outer if $E_f = H^2(\mathbb{T})$, where*

$$E_f := \mathrm{span}_{H^2}(z^n f : n \geq 0),$$

i.e. E_f is the smallest (closed) invariant subspace of M_z containing f.

Theorem 1.7.2 (Smirnov, 1928a,b) *Every function $f \in H^2(\mathbb{T})$, $f \neq 0$, can be factorized as*

$$f = f_{in} f_{out},$$

where f_{in} is an inner function and f_{out} is outer. This factorization is unique up to a constant factor: if $f = f'_{in} f'_{out}$ is another inner–outer factorization, then $f_{in} = \lambda f'_{in}$, $f_{out} = \overline{\lambda} f'_{out}$ with some $\lambda \in \mathbb{T}$.

Proof By Corollary 1.4.3, there exists an inner function q such that $E_f = qH^2(\mathbb{T})$. In particular, $f = qg$ where $g \in H^2(\mathbb{T})$. Let us show that g is outer. Indeed, for every function $h \in H^2(\mathbb{T})$ there exist polynomials $p_n \in \mathcal{P}_a$ such that $\lim_n \|p_n f - qh\| = 0$. However, $\|p_n f - qh\|^2 = \|q p_n g - qh\|^2 = \int_{\mathbb{T}} |q(p_n g - h)|^2 \, dm = \|p_n g - h\|^2$, which shows that $h \in E_g$, and hence $E_g = H^2(\mathbb{T})$ (i.e. g is outer). By setting $f_{in} = q$, $f_{out} = g$, we obtain the desired factorization.

For the uniqueness, suppose there is another factorization: then $f_{in} f_{out} = f'_{in} f'_{out}$, hence $f_{in} \overline{f'_{in}} f_{out} = f'_{out}$. Let (p_n) be a sequence of polynomials such that $\lim_n \|p_n f_{out} - 1\| = 0$. Since $f_{in} \overline{f'_{in}}$ is a unimodular function, we obtain $\|p_n f_{out} - 1\| = \|p_n f_{out} f_{in} \overline{f'_{in}} - f_{in} \overline{f'_{in}}\| \to 0$ as $n \to \infty$. However, $p_n f_{out} f_{in} \overline{f'_{in}} = p_n f'_{out} \in H^2(\mathbb{T})$, and consequently $f_{in} \overline{f'_{in}} \in H^2(\mathbb{T})$. Similarly, $f'_{in} \overline{f_{in}} \in H^2(\mathbb{T})$, which gives $f_{in} \overline{f'_{in}} = \mathrm{constant}$ (compare with the proof of Theorem 1.3.5), and the result follows. ∎

Vladimir Ivanovich Smirnov (1887–1974) was a Russian mathematician, a representative of the Saint Petersburg school (founded by Chebyshev), and a founder of modern complex analysis at Saint Petersburg. He obtained numerous important results on Hardy spaces (canonical factorization, the Smirnov "class D," Hardy spaces on *Smirnov domains*, etc.), as well as in ordinary differential equations and mathematical physics. He is also known for his five-volume *Course of Higher Mathematics*, which for years dominated the teaching of mathematics at university level in Russia/USSR. He co-authored works with Friedman, Tamarkin, Lebedev, and others.

His notable students include Goluzin, Havin, Kantorovich (Nobel Prize in Economics, 1975), Lozinsky, Sobolev, and Yakubovich.

Moreover, Smirnov was renowned for his exceptional personality; he was irreproachable for his nobility, kindness, and generosity, even under the unforgiving circumstances of Russian/Soviet reality in the twentieth century.

The "Twelve Colleges" of the University of Saint Petersburg (the rightmost building).

Corollary 1.7.3 (Beurling, 1949) *Let* $f \in H^2(\mathbb{T})$, $f \neq 0$. *Then,* $E_f = f_{in}H^2(\mathbb{T})$.

Indeed:

$$
\begin{aligned}
E_f &= \mathrm{clos}_{H^2}(f_{in}f_{out}\mathcal{P}_a) \\
&= f_{in}\,\mathrm{clos}_{H^2}(f_{out}\mathcal{P}_a) && \text{(since } f_{in} \text{ is unimodular)} \\
&= f_{in}H^2(\mathbb{T}) && (f_{out} \text{ is outer}). \qquad\blacksquare
\end{aligned}
$$

Theorem 1.7.2 also leads to a crucial development in the problem of L^2 optimal prediction (see Theorem 1.7.6 below), i.e. in the expression of the quantity d in Lemma 1.6.6 as a function of the measure $\mu = wm + \mu_s$ (more precisely: of the Radon–Nikodym derivative $w = d\mu/dm$),

$$
d^2 = \mathrm{dist}^2_{L^2(\mu)}(1, H_0^2(\mu))^2 = \inf_{p \in z\mathcal{P}_a} \int_{\mathbb{T}} |1 - p|^2 \, d\mu.
$$

This last extremal problem appeared in the research of Gábor Szegő in the 1920s, and bears his name: the *Szegő infimum*.

However, we first need a property of outer functions of the type "maximum principle" (for details see Chapter 3 below).

Theorem 1.7.4 (Smirnov, 1932) *Let* $f \in H^2(\mathbb{T})$, $f \neq 0$. *Then the following assertions are equivalent.*

(1) *f is an outer function.*
(2) *$\forall g \in H^2(\mathbb{T})$, $g/f \in L^2(\mathbb{T}) \Rightarrow g/f \in H^2(\mathbb{T})$.*

Proof (1) $\Rightarrow$ (2) Let $p_n \in \mathcal{P}_a$ such that $\lim_n \|p_n f - 1\|_2 = 0$, and suppose that $g \in H^2(\mathbb{T})$ such that $g/f \in L^2$, i.e. $g = fh$ where $h \in L^2$. Then

$$
\int_{\mathbb{T}} |p_n g - h| \, dm = \int_{\mathbb{T}} |p_n fh - h| \, dm \leq \|h\|_2 \|p_n f - 1\|_2 \to 0 \quad (\text{for } n \to \infty).
$$

The convergence in $L^1(\mathbb{T})$ implies the convergence of the Fourier coefficients: $\forall k \in \mathbb{Z}$ we have $\hat{h}(k) = \lim_n (p_n g)\hat{}\,(k)$. However $p_n g \in H^2(\mathbb{T})$ (because $H^2(\mathbb{T})$ is M_z-invariant), and hence $\hat{h}(k) = 0$ for every $k < 0$. Thus $h \in H^2(\mathbb{T})$.

(2) $\Rightarrow$ (1) Let $f = f_{in}f_{out}$ be the inner–outer factorization of f. Then, $f_{out} \in H^2(\mathbb{T})$ and $f_{out}/f = \overline{f}_{in} \in L^2(\mathbb{T})$; hence, by (2), $\overline{f}_{in} \in H^2(\mathbb{T})$ and of course $f_{in} \in H^2(\mathbb{T})$. As seen several times earlier (for example, in Theorem 1.3.5), this implies $f_{in} = $ constant: hence f is an outer function. $\qquad\blacksquare$

Corollary 1.7.5

(1) *If $f \in H^2(\mathbb{T})$ is simultaneously inner and outer, then $f = $ constant.*

(2) *If $f, g \in H^2(\mathbb{T})$ are outer and $|f| = |g|$ a.e. on $\mathbb{T}$, then $f = \lambda g$ for some unimodular constant λ.*

Indeed, for (1), we apply Theorem 1.7.4 to 1 and f, and obtain $1/f = \overline{f} \in H^2(\mathbb{T})$, which, with $f \in H^2(\mathbb{T})$, again implies $f = $ constant.

For (2), by setting $h = f/g$ and applying the theorem, we obtain $h \in H^2(\mathbb{T})$, and, by switching the roles of f and g, $\overline{h} \in H^2(\mathbb{T})$. Hence, $h = $ constant (clearly unimodular).   ∎

Gábor Szegő (1895–1985), a Hungarian–German–American mathematician, is known for his work in classical analysis, such as orthogonal polynomials and Toeplitz operators. After obtaining his doctorate in Budapest under the supervision of Fejér, he went to Berlin and Königsberg, but then, pressured by the Nazis, he emigrated to the USA. His famous collection of solved problems, with Pólya, *Aufgaben und Lehrsätze aus der Analysis* (1925), served for years as an essential source of training for generations of analysts. He is the author of several other reference monographs. His experiences in Budapest included tutoring the young child prodigy Johannes von Neumann. According to witnesses, Szegő was moved to tears by his first meeting with the young Johannes, so rapid and profoundly complete were the responses of his new student.

Theorem 1.7.6 (Szegő, 1920; Verblunsky, 1936; Kolmogorov, 1941) *Let $\mu = wm + \mu_s$ be the Radon–Nikodým decomposition of a finite Borel measure on $\mathbb{T}$. Then:*

(1) *either there does not exist any $f \in H^2(\mathbb{T})$ such that $|f|^2 = w$, and then*

$$d = \mathrm{dist}_{L^2(\mu)}(1, H_0^2(\mu)) = 0,$$

(2) *or there exists a (unique) outer function $F \in H^2(\mathbb{T})$ such that $|F|^2 = w$, and then*

$$d = \operatorname{dist}_{L^2(\mu)}(1, H_0^2(\mu)) = |\hat{F}(0)| > 0.$$

Proof Suppose $d > 0$. Then, $\operatorname{dist}_{L^2(wm)}(1, H_0^2(wm)) > 0$ (Lemma 1.6.6), and hence $zH_0^2(wm) \neq H_0^2(wm)$ (Lemma 1.6.5), which implies that $H_0^2(wm)$ is an invariant non-reducing subspace of $L^2(wm)$. Helson's theorem (Theorem 1.3.5) provides a function q such that $H_0^2(\mu) = qH^2(\mathbb{T})$ and $|q|^2 w = 1$ a.e. on $\mathbb{T}$. In particular, $z = qf$ where $f \in H^2(\mathbb{T})$, which implies $|f|^2 = |z/q|^2 = w$. Setting $F = f_{out}$, we obtain $|F|^2 = w$ and

$$d^2 = \inf_{p \in z\mathscr{P}_a} \int_{\mathbb{T}} |1 - p|^2 |F|^2 \, dm = \inf_{p \in z\mathscr{P}_a} \int_{\mathbb{T}} |F - pF|^2 \, dm$$
$$= \operatorname{dist}_{H^2}^2(F, zH^2(\mathbb{T})) = \|P_{H^2 \ominus zH^2} F\|^2 = |\hat{F}(0)|^2.$$

Conversely, if $w = |F|^2$ with an outer function $F \in H^2(\mathbb{T})$, then the last formula shows again that $d = |\hat{F}(0)|$. It remains to remark that $\hat{F}(0) \neq 0$ for every outer function F. Indeed, if we suppose $\hat{F}(0) = 0$, we would have $F = \sum_{n \geq 1} \hat{F}(n)z^n = z(\sum_{k \geq 0} \hat{F}(k+1)z^k) \in zH^2(\mathbb{T})$, which implies $E_F \subset zH^2(\mathbb{T}) = H_0^2(\mathbb{T}) \neq H^2(\mathbb{T})$. Thus we obtain a contradiction. $\blacksquare$

In fact, the last theorem does not resolve the prediction problem: expressing an error $d(\mu)$ of the best quadratic prediction of a process as a function of the spectral measure μ. In order to obtain the famous Szegő–Verblunsky–Kolmogorov formula

$$d(\mu) = \exp\left(\int_{\mathbb{T}} \log\left|\frac{d\mu}{dm}\right| dm\right),$$

we need to develop a theory of "canonical factorization" of functions $H^2(\mathbb{T})$. This is the goal of Chapter 2.

1.8 Exercises

1.8.1 The Wold–Kolmogorov Decomposition

Let $T\colon H \to H$ be a linear isometry in a Hilbert space H, $E \in \operatorname{Lat}(T)$ and $W = E \ominus TE$. Prove the following.

(a) *$T^n W \perp T^m W$ for every $n \neq m$ $(n, m \geq 0)$ (W is said to be a "wandering subspace").*

SOLUTION: Let $x, y \in W$ and $n > m \geq 0$. Then, $(T^n x, T^m y) = (T^{n-m} x, y) = 0$ because $T^{n-m} x \in TE$ and $y \in E$. ∎

(b) *The subspace $E_\infty = \bigcap_{n \geq 0} T^n E$ reduces T, and the restriction $T|E_\infty$ is unitary.*

SOLUTION: Let $x \in E_\infty$. For every $n \geq 0$, there exists $x_n \in E$ such that $x = T^n x_n$, which implies $Tx = T^{n+1} x_n$, and hence, $Tx \in E_\infty$. Moreover, $T^n x_n = T^{n+k} x_{n+k} \Rightarrow x_n = T^k x_{n+k}$, which in turn implies $x_n \in E_\infty$, and in particular, $x \in TE_\infty$. Hence, $TE_\infty = E_\infty$ and $T|E_\infty$ is a unitary mapping of E_∞ onto itself. For the reduction property, we have $T^* x = T^* T^{n+1} x_{n+1} = T^n x_{n+1}$, hence $T^* x \in E_\infty$. ∎

(c) *The subspace $E_0 = \sum_{n \geq 0} \oplus T^n(W)$ is T-invariant and $T|E_0$ is completely non-unitary (i.e. if $E' \subset E_0$, $TE' \subset E'$ and $T|E'$ is unitary, then $E' = \{0\}$).*

SOLUTION: $E_0 = \{x: x = \sum_{n \geq 0} T^n w_n: w_n \in W, \sum_{n \geq 0} \|T^n w_n\|^2 = \sum_{n \geq 0} \|w_n\|^2 < \infty\}$ (convergence in norm, unique representation, see Appendix C). This implies $TE_0 \subset E_0$ and $\bigcap_{n \geq 0} T^n E_0 = \{0\}$. If $E' \subset E_0$, $TE' \subset E'$ and $T|E'$ is unitary, then $E' = \bigcap_{n \geq 0} T^n E' \subset \bigcap_{n \geq 0} T^n E_0 = \{0\}$. ∎

(d) The Wold–Kolmogorov decomposition (1939): $E = E_0 \oplus E_\infty$.

SOLUTION: Let $x \in E$; then, $x \in E \ominus E_0 \Leftrightarrow x \in E$, $x \perp T^n E \ominus T^{n+1} E$ (for every $n \geq 0$) $\Leftrightarrow$ (consecutively, with $n = 0, 1, \dots$) $x \in E$, $x \in TE$, $x \in T^2 E, \dots \Leftrightarrow x \in E_\infty$. ∎

1.8.2 The Shift Operator M_z on $L^2(\mathbb{T}, \mu)$

Let μ be a finite Borel measure on $\mathbb{T}$ and $M_z: L^2(\mathbb{T}, \mu) \to L^2(\mathbb{T}, \mu)$ the shift operator (translation), $M_z f = zf$.

(a) *Let $E \in \mathrm{Lat}(M_z)$. Describe its Wold–Kolmogorov decomposition (using Helson's Theorem 1.3.5).*

SOLUTION: If E is reducing, $M_z E = E$, then $E_\infty = E$, $W = \{0\}$. Otherwise, by Theorem 1.3.5, $E = \chi_A L^2(\mu_s) \oplus q H^2(\mathbb{T})$ where $|q|^2 w = 1$ m-a.e. ($\mu = \mu_s + wm$ is the Radon–Nikodym decomposition of μ). Then clearly $M_z(\chi_A L^2(\mu_s)) = \chi_A L^2(\mu_s)$ and $W = E \ominus M_z E = q H^2(\mathbb{T}) \ominus qz H^2(\mathbb{T}) = q\mathbb{C}$ (a subspace of dim $= 1$ containing q). Consequently, $E_\infty = \chi_A L^2(\mu_s)$ and the completely non-unitary portion of M_z is $M_z|q H^2(\mathbb{T})$. ∎

(b) *Let μ_i ($i = 1, 2$) be finite Borel measures on $\mathbb{T}$. Find a necessary and sufficient condition on μ_i so that the shift operators $S_i := M_z: L^2(\mu_i) \to L^2(\mu_i)$*

($i = 1, 2$) are unitarily equivalent (i.e. there exists a unitary $U: L^2(\mu_1) \to L^2(\mu_2)$ such that $US_1 = S_2 U$).

SOLUTION: Suppose S_1 and S_2 are equivalent and U is a unitary operator such that $US_1 = S_2 U$. Then, $US_1^k = S_2^k U$ for every $k \in \mathbb{Z}$, and hence, for any polynomial $p \in \mathcal{P}$, we have $Up = p \cdot U1$. By a passage to the limit in the last equation (for the norm L^2 on the left, and for the norm L^1 on the right) we obtain $Uf = fU1$ for any $f \in L^2(\mu_1)$. Then U is unitary, and therefore

$$\int_{\mathbb{T}} |f|^2 |U1|^2 \, d\mu_2 = \int_{\mathbb{T}} |f|^2 \, d\mu_1, \quad \forall f \in L^2(\mu_1),$$

which implies that $\mu_1 = |U1|^2 \mu_2$, hence $\mu_1 \ll \mu_2$. By swapping the roles of S_1 and S_2, we obtain $\mu_2 \ll \mu_1$ (thus, the measures are equivalent: $\mu_1 \sim \mu_2$). *Conversely, if* $\mu_1 \sim \mu_2$, then $\mu_1 = h\mu_2$ where $h \in L^1(\mu_2)$ and $1/h \in L^1(\mu_1)$ ($\Leftrightarrow h \neq 0$ μ_2-a.e.), and the mapping $Uf = f\sqrt{h}$ is unitary: $U: L^2(\mu_1) \to L^2(\mu_2)$ and satisfies $US_1 = S_2 U$. ∎

(c) *The same question as in (b) but for restrictions $S_i|H^2(\mu_i)$.*

SOLUTION: The operators $S_i|H^2(\mu_i)$ are isometric: they are simultaneously unitary or not, and this is the case if and only if $H^2(\mu_i) = L^2(\mu_i)$. If this last equality holds, the question is already answered in (b); if not, we extend the operator $U: H^2(\mu_1) \to H^2(\mu_2)$ such that $US_1 = S_2 U$ to a mapping $U: L^2(\mu_1) \to L^2(\mu_2)$ with the same relation of commutation by the equation $U(\bar{z}^n f) := \bar{z}^n Uf$, $f \in H^2(\mu_1)$. *The final answer is: $S_i|H^2(\mu_i)$, $i = 1, 2$, are unitarily equivalent if and only if $\mu_1 \sim \mu_2$ and $H^2(\mu_i)$ simultaneously coincide (or not) with $L^2(\mu_i)$ for $i = 1, 2$.* ∎

(d) *Describe the finite Borel measures μ on $\mathbb{T}$ for which all the invariant subspaces of the shift operator $M_z: L^2(\mu) \to L^2(\mu)$ are reducing, i.e. $\mathrm{Lat}(M_z) \subset \mathrm{Lat}(M_z^*)$ ($\Leftrightarrow \mathrm{Lat}(M_z) = \mathrm{Lat}(M_z^*)$).*

SOLUTION: By Theorem 1.3.5, there exists a non-reducing invariant subspace if and only if there exists a measurable function q such that $|q|^2 w = 1$ m-a.e. on $\mathbb{T}$, with $w = d\mu/dm$. Clearly the last property is equivalent to $w > 0$ m-a.e. on $\mathbb{T}$. *The answer to (d) is:* it is necessary and sufficient that $w = 0$ on a set $\sigma \subset \mathbb{T}$ having $m(\sigma) > 0$ (which is equivalent to $m \not\ll \mu$). ∎

1.8.3 Inner and Outer Functions

A few "bare-hands" examples, without using the theory of Chapter 2, but nonetheless using knowledge of the multipliers of $H^2(\mathbb{T})$ (part (a) below).

(a) Multipliers, algebra H^∞. *Here $L^2 = L^2(\mathbb{T})$ and $H^\infty(\mathbb{T}) := H^2(\mathbb{T}) \cap L^\infty(\mathbb{T})$. The multiplier spaces are defined by*

$$\mathrm{Mult}(L^2) = \{h: f \in L^2 \Rightarrow hf \in L^2\},$$
$$\mathrm{Mult}(H^2(\mathbb{T})) = \{h: f \in H^2(\mathbb{T}) \Rightarrow hf \in H^2(\mathbb{T})\}.$$

(i) *Show that* $\mathrm{Mult}(L^2) = L^\infty(\mathbb{T})$, $\mathrm{Mult}(H^2(\mathbb{T})) = H^\infty(\mathbb{T})$.

SOLUTION: The inclusion $L^\infty(\mathbb{T}) \subset \mathrm{Mult}(L^2)$ is evident. For the converse, let $h \in \mathrm{Mult}(L^2)$, then $\forall f \in L^2 \int_{\mathbb{T}} |f|^2 |h|^2 \, dm < \infty$. However, $g = |f|^2$ is an arbitrary positive function of $L^1(\mathbb{T})$, hence $h \in L^\infty(\mathbb{T})$ (Appendix A). For the case of $\mathrm{Mult}(H^2(\mathbb{T}))$, clearly $\mathrm{Mult}(H^2(\mathbb{T})) \subset H^2(\mathbb{T})$ and thus, for any $h \in \mathrm{Mult}(H^2(\mathbb{T}))$, the multiplication operator $M_h f = hf$ is continuous $H^2(\mathbb{T}) \to L^1(\mathbb{T})$, hence it is closed for $H^2(\mathbb{T}) \to H^2(\mathbb{T})$, thus bounded (by the closed graph theorem). Consequently, the formula $M_h(f) = \bar{z}^n M_h(z^n f)$ extends M_h on the subspace $\bar{z}^n H^2(\mathbb{T}) \subset L^2$ (with the same norm), and by approximation, on the whole space L^2. Thus $\mathrm{Mult}(H^2(\mathbb{T})) \subset L^\infty(\mathbb{T})$, which leads to $\mathrm{Mult}(H^2(\mathbb{T})) \subset H^\infty(\mathbb{T})$. For the converse, note that $h \in H^\infty(\mathbb{T})$, $p \in \mathcal{P}_a \Rightarrow hp \in H^2(\mathbb{T})$, and again by approximation (letting $\|p_n - f\|_2 \to 0$), we obtain $hf \in H^2(\mathbb{T})$ for any $f \in H^2(\mathbb{T})$, which shows that $H^\infty(\mathbb{T}) \subset \mathrm{Mult}(H^2(\mathbb{T}))$. ∎

(ii) $H^\infty(\mathbb{T})$ *is a Banach algebra for standard multiplication on $\mathbb{T}$. Moreover, for every function $f \in L^2$, we have $f \cdot H^\infty(\mathbb{T}) \subset E_f$.*

SOLUTION: The space of multipliers $\mathrm{Mult}(X) = \{h: f \in X \Rightarrow hf \in X\}$ of a function space X is clearly an algebra. Moreover, the inequality $\|fg\|_\infty \leq \|f\|_\infty \|g\|_\infty$ for $f, g \in H^\infty$ is also evident, and the result follows for the algebra $H^\infty(\mathbb{T})$. For the rest, clearly $f\mathcal{P}_a \subset E_f$. It only remains to show that $(f\mathcal{P}_a)^\perp \subset (fH^\infty)^\perp$ (orthogonal complement in L^2). Let $g \in (f\mathcal{P}_a)^\perp$, i.e. $\int_{\mathbb{T}} \bar{g}fp \, dm = 0$ for any polynomial $p \in \mathcal{P}_a$. Thus for any $h \in H^\infty$, $\int_{\mathbb{T}} \bar{g}fh \, dm = 0$ because $\bar{g}f \in L^1$ and h is a weak limit $\sigma(L^\infty, L^1)$ of its Fejér polynomials (see Appendix A). ∎

(b) Examples of inner functions. *Show that the following functions are inner.*

(i) $b_\lambda = (\lambda - z)/(1 - \bar{\lambda}z)$ *where* $\lambda \in \mathbb{D} = \{z \in \mathbb{C}: |z| < 1\}$.

SOLUTION: $b_\lambda = (\lambda - z) \sum_{n \geq 0} \bar{\lambda}^n z^n$ ($|z| = 1$), and clearly $\hat{b}_\lambda(k) = 0$ for $k < 0$, and $\sum_{k \geq 0} |\hat{b}_\lambda(k)|^2 < \infty$: hence $b \in H^2(\mathbb{T})$. Moreover, for $|z| = 1$, we have $|\lambda - z| = |\bar{\lambda} - \bar{z}| = |1 - \bar{\lambda}z|$, thus $|b_\lambda(z)| = 1$. ∎

(ii) $f = \prod_{k=1}^{N} b_{\lambda_k}$ *where* $\lambda_k \in \mathbb{D}$.

SOLUTION: As $H^\infty(\mathbb{T}) \cdot H^\infty(\mathbb{T}) \subset H^\infty(\mathbb{T})$ (by part (ii) of (a)), a product of inner functions is inner. ∎

(iii) $s_{\zeta,a} = \exp(-a((\zeta + z)/(\zeta - z)))$ *where* $a > 0$, $\zeta \in \mathbb{T}$.

SOLUTION: As

$$\mathrm{Re}\left(\frac{\zeta + z}{\zeta - z}\right) = \frac{1 - |z|^2}{|\zeta - z|^2} \geq 0$$

for any $\zeta \in \mathbb{T}$, $|z| \leq 1$, $z \neq \zeta$, we obtain $|s_{\zeta,a}| = 1$ on $\mathbb{T}$. Moreover, for every $n > 0$, we have $\hat{s}_{\zeta,a}(-n) = \int_{\mathbb{T}} z^n s_{\zeta,a}(z)\, dm = \lim_{r \to 1} \int_{\mathbb{T}} f_r(z)\, dm = 0$ where $f(z) = z^n s_{\zeta,a}(z)$ and $f_r(z) = f(rz)$, $0 \leq r < 1$ ($\hat{f}_r(0) = 0$ since f_r is analytic in $|z| < 1/r$ and $f_r(0) = 0$). ∎

(iv) $f = \prod_{k=1}^{N} s_{\zeta_k,a_k}$ *where* $a_k > 0$, $\zeta_k \in \mathbb{T}$.

SOLUTION: See the solution of (ii) above. ∎

(c) Examples of outer functions. *Show that the following functions are outer.*

(i) $f \in H^2(\mathbb{T})$ *such that* $1/f \in H^\infty(\mathbb{T})$.

SOLUTION: By (a,ii), clearly $1 = f \cdot 1/f \in E_f$, hence $E_f = H^2(\mathbb{T})$. ∎

(ii) $f \in H^\infty$ *such that* $\mathrm{Re}(f) \geq 0$.

SOLUTION: For any $\epsilon > 0$ there exist (a large) $r > 0$ and (a small) $\delta > 0$ such that $|f + \epsilon - r| \leq (1 - \delta)r$ a.e. on $\mathbb{T}$ (to verify this, sketch the region in $\mathbb{C}$ where the values of $f(z) + \epsilon$, $|z| = 1$ are found), or $|(f + \epsilon)/r - 1| \leq (1 - \delta)$; this implies the normal convergence of the series

$$\frac{r}{f + \epsilon} = \sum_{k \geq 0} \left(1 - \frac{f + \epsilon}{r}\right)^k.$$

However, part (ii) of (a) implies $(1 - (f + \epsilon)/r)^k \in H^\infty$, hence $r/(f + \epsilon) \in H^\infty$. By (ii) we have $f/(f + \epsilon) \in E_f$ and even

$$\lim_{\epsilon \to 0} \int_{\mathbb{T}} \left|\frac{f}{f + \epsilon} - 1\right|^2 dm = 0$$

(by the dominated convergence theorem). Thus $1 \in E_f$, hence f is outer. ∎

(iii) $f = 1 + g$, $g \in H^\infty$, $\|g\|_\infty \leq 1$.

SOLUTION: This is a special case of (ii). ∎

(iv) $f \in H^2(\mathbb{T})$ *such that* $\mathrm{Re}(f) \geq 0$.

SOLUTION: The solution of (ii) above shows that it suffices to prove the inclusion $1/(f + \epsilon) \in H^\infty$, or the inclusion $1/(f + \epsilon) \in H^2(\mathbb{T})$ (since $1/(f + \epsilon) \in L^\infty$ is evident). To this end, fix $0 < r < 1$ and consider $f_r(z) = \sum_{k \geq 0} \hat{f}(k) r^k z^k$, $z \in \mathbb{T}$. Then, $f_r \in C(\mathbb{T})$

(hence bounded) and $\mathrm{Re}(f_r) \geq 0$ (because $f_r = f * P_r$, which is a convolution with the positive function

$$P_r(z) = \frac{1 - r^2}{|z - r|^2} = \sum_{k \in \mathbb{Z}} r^{|k|} z^k,$$

$z \in \mathbb{T}$, see Appendix A). By the solution (ii) above, $1/(f_r + \epsilon) \in H^\infty \subset H^2(\mathbb{T})$ and by the dominated convergence theorem,

$$\lim_{r \to 1} \left\| \frac{1}{f_r + \epsilon} - \frac{1}{f + \epsilon} \right\|_2 = 0.$$

Thus $1/(f + \epsilon) \in H^2(\mathbb{T})$.   ∎

(d) An extremal problem. *First, we justify Cauchy's formula for Fourier coefficients:*

(i) *Let $f, g \in L^2(\mathbb{T})$ (thus $fg \in L^1(\mathbb{T})$). Show that, for every $n \in \mathbb{Z}$, $\widehat{fg}(n) = \sum_{k \in \mathbb{Z}} \hat{g}(k) \hat{f}(n - k)$: the series converges absolutely.*

SOLUTION: By Cauchy's inequality $\|f(g - g')\|_1 \leq \|f\|_2 \|g - g'\|_2$, the multiplication $M_g f = fg$ is continuous $L^2(\mathbb{T}) \to L^1(\mathbb{T})$. Moreover, the Fourier series $g = \sum_{k \in \mathbb{Z}} \hat{g}(k) z^k$ converges for the norm of $L^2(\mathbb{T})$. Hence, $fg = \sum_{k \in \mathbb{Z}} \hat{g}(k) z^k f$ converges in $L^1(\mathbb{T})$, which implies $\widehat{fg}(n) = \sum_{k \in \mathbb{Z}} \hat{g}(k) \widehat{(z^k f)}(n)$. The calculation $\widehat{(z^k f)}(n) = \hat{f}(n - k)$ is elementary.   ∎

(ii) *Let $f = f_{in} f_{out} \in H^2(\mathbb{T})$. Show that*

$$\sup\{|\hat{g}(0)| : g \in H^2(\mathbb{T}),\ |g| \leq |f| \text{ a.e. on } \mathbb{T}\} = |\hat{f}_{out}(0)|.$$

SOLUTION: By (i), clearly $\widehat{\varphi\psi}(0) = \hat{\varphi}(0)\hat{\psi}(0)$ for all functions $\varphi, \psi \in H^2(\mathbb{T})$. Moreover, for every inner function h, we have $|\hat{h}(0)| \leq \|h\|_1 = 1$. Given $g \in H^2(\mathbb{T})$, $|g| \leq |f|$, which implies $|\hat{g}(0)| = |\hat{g}_{in}(0)\hat{g}_{out}(0)| \leq |\hat{g}_{out}(0)|$. Then by Theorem 1.7.6,

$$|\hat{g}(0)|^2 \leq |\hat{g}_{out}(0)|^2 = \inf_{p \in z\mathcal{P}_a} \int_{\mathbb{T}} |1 - p|^2 |g|^2 \, dm \leq \inf_{p \in z\mathcal{P}_a} \int_{\mathbb{T}} |1 - p|^2 |f|^2 \, dm = |\hat{f}_{out}(0)|^2.$$ ∎

1.9 Notes and Remarks

As already mentioned, Hardy spaces H^p were defined in 1915 (Hardy, 1915), and by 1930 the essentials of the theory had been constructed. At the time, it was a novel mix of fundamental ideas: complex analysis, the Lebesgue integral, and functional vector spaces. Very rapidly, Hardy spaces became one of the mainsprings of the development of analysis in the twentieth century. However, the theory had to wait another 30 years, until the 1960s, for the true magnitude of its potential to be revealed, via the discovery of the main source of its force: the invariant subspaces of the group of translations $(M_z^n)_{n \in \mathbb{Z}}$ and its semigroup $(M_z^n)_{n \in \mathbb{Z}_+}$.

Arne Beurling (1949) formulated the correspondence between the invariant subspaces and the inner–outer factorization (in fact, the latter had been known by Smirnov for more than 20 years (Smirnov, 1928a,b)). Beurling's work led to the discovery of the hidden heart of the theory (Helson and Lowdenslager, 1961; Helson, 1964)): the fact that analyticity is a consequence of the causality of the semigroup under consideration, and that the main feature of the subject is that this semigroup is linearly ordered (it is not very important whether we take $\mathbb{Z}$ or $\mathbb{R}$, as made clear in the years 1920–1940).

The presentation of this book is based on the novel version of the theory proposed by Helson (1964) (see also Nikolski (1980, 1986, 2002)): in his work, the point of view described above is accepted from the start as the cornerstone of the whole construction. This is a spectacular difference from the classical and/or post-modern presentations, i.e. Privalov (1941), Duren (1970), Garnett (1981), Stein (1993), Koosis (1980), and Pavlović (2004).

Formal references: for Theorem 1.2.1 see Wiener (1933), for Theorem 1.3.5 see Helson (1964), and for Corollary 1.4.3 see Beurling (1949). Historically, the astounding success of the approach by invariant subspaces led to the creation of an "abstract complex analysis" where analyticity is defined and studied with the aid of invariant subspaces with respect to a "semigroup" satisfying certain conditions. This theory is well-developed and is highly efficient for the study of functions of several variables, of almost periodic functions, etc.: see Gamelin (1969) and Barbey and König (1977).

The uniqueness theorem Corollary 1.4.4, as well as Theorem 1.5.4, is due to Riesz and Riesz (1916) (for the proof of § 1.5.1 see Øksendal (1971)). Theorem 1.5.4 plays an important role in several applications, in particular for different forms of the uncertainty principle in harmonic analysis. Numerous generalizations and improvements of this theorem are known; for all these subjects, see Havin and Jöricke (1994).

The contents of § 1.6 are taken from Kolmogorov (1941). The inner–outer factorization of § 1.7 was discovered by Smirnov (1928a,b) and published in a minor Russian journal (but in French! See the Russian translation in Smirnov (1988)). There, Smirnov (following Szegő (1921)) speaks of "maximal functions" instead of "outer functions" (he does not introduce a name for the "inner functions"), which finds a strong justification in several forms of the "maximum principle" (Theorem 1.7.4, found in Smirnov (1932), is one of them; for others, see § 3.3–3.4 below). Because of the isolation of Russia after the Bolshevik revolution, followed by Stalin's Iron Curtain, these results remained almost unknown until the 1960s. The other principal result of Beurling (1949) met with the same destiny: Corollary 1.7.3 is an almost immediate consequence of another article by Smirnov (1932).

We can also mention that the inner–outer factorization was rediscovered (practically independently of Smirnov or Beurling) by Wiener and Masani in the framework of the theory of linear prediction (by the generalized Wold–Kolmogorov decomposition), under the name of *optimal-residual factorization*: see Masani (1966).

Theorem 1.7.6 (with the formula mentioned at the end of this section) was proved by Szegő (1920) in the case $\mu = \mu_a$, and by Verblunsky (1936) and Kolmogorov (1941) in the general case. The role and significance of Verblunsky's 1936 paper was overlooked by the community for many decades and was restored by a thorough historical analysis in Barry Simon's book *Orthogonal Polynomials on the Unit Circle*, Part 1: *Classical Theory* (Simon, 2005; see especially pp. 141, 221).

Generalizations for continuous-time processes are also due to Kolmogorov, and for vector-valued processes to Kolmogorov, Matveev, and Rozanov (see Rozanov, 1963), as well as Wiener and Masani (1957, 1958). The Wold–Kolmogorov decomposition (Wold, 1938; Kolmogorov, 1941) plays an important role in the analysis of time series (in the prediction of random processes).

2

The $H^p(\mathbb{D})$ Classes: Canonical Factorization and First Applications

Topics. Spaces $H^p(\mathbb{D})$, Poisson extension, Jensen's inequality, Fatou's theorem, the Smirnov canonical factorization, a return to Szegő's "inf", weighted approximation, the Hilbert and Hardy inequalities, the harmonic conjugate, the Littlewood subordination principle.

2.1 Fejér and Poisson Means

First, recall the notion of 2π-periodic *convolution*, convolution on $\mathbb{T} = \mathbb{R}/2\pi\mathbb{Z}$ (see Appendix A for more details): if μ, ν are two complex Borel measures on $\mathbb{T}$, then $\mu * \nu$ is the unique complex measure satisfying

$$\int_{\mathbb{T}} f d(\mu * \nu) = \int_{\mathbb{T}} \int_{\mathbb{T}} f(s\bar{t}) \, d\mu(s) d\nu(t) \quad \text{for every function } f \in C(\mathbb{T}).$$

For measures with density $\mu = fm$, $\nu = gm$, where $f, g \in L^1(\mathbb{T}, m)$, the definition reduces to the convolution of f and g: $\mu * \nu = (f * g)m$, where, for almost all s,

$$f * g(s) = \int_{\mathbb{T}} f(s\bar{t})g(t) \, dm(t), \quad s \in \mathbb{T}.$$

For the Fourier coefficients (see Appendix A), $\widehat{\mu * \nu}(n) = \hat{\mu}(n)\hat{\nu}(n)$ for any $n \in \mathbb{Z}$.

In this chapter, two important approximate identities from harmonic analysis are frequently used (see Appendix A): those of Fejér and Poisson. Specifically, for $k, n \in \mathbb{Z}_+$ and $0 < r < 1$, we set

$$D_k = \sum_{j=-k}^{k} e^{ijx} = \frac{\sin(k + 1/2)\,x}{\sin(x/2)} \quad \text{(Dirichlet kernel)},$$

37

$$\Phi_n = \frac{1}{n+1}\sum_{k=0}^{n} D_k(x) = \sum_{j=-n}^{n}\left(1 - \frac{|j|}{n+1}\right)e^{ijx}$$

$$= \frac{1}{n+1}\left(\frac{\sin\frac{n+1}{2}x}{\sin(x/2)}\right)^2 \quad \text{(Fejér kernel)},$$

$$P_r(x) = P(re^{ix}) := \frac{1-r^2}{|1-re^{ix}|^2} = \sum_{j\in\mathbb{Z}} r^{|j|}e^{ijx} \quad \text{(Poisson kernel)}.$$

Notation For $f \in L^1(\mathbb{T})$ and $0 < r < 1$, let $f_r = f * P_r$.

Lemma 2.1.1 *Let $f, g \in L^1(\mathbb{T})$. Then:*

(1) $f * g = g * f$, $\|f * g\|_1 \le \|f\|_1 \|g\|_1$.

(2) *If $f \in L^p(\mathbb{T})$, $1 \le p \le \infty$, then $f * g \in L^p(\mathbb{T})$ and $\|f * g\|_p \le \|f\|_p \|g\|_1$.*

(3) *If $(E_\alpha) \subset L^1(\mathbb{T})$ is a family satisfying: (i) $C := \sup_\alpha \|E_\alpha\|_1 < \infty$ and (ii) $\lim_\alpha \hat{E}_\alpha(n) = 1$ for every $n \in \mathbb{Z}$, then*

$$\lim_\alpha \|f - f * E_\alpha\|_p = 0 \text{ for every function } f \in L^p(\mathbb{T}),$$

$$1 \le p < \infty \text{ (approximate identity of } L^p).$$

(4) *If*

(i) $C := \sup_\alpha \|E_\alpha\|_1 < \infty$,

(ii) $\lim_\alpha \hat{E}_\alpha(0) = 1$, *and*

(iii) *for every $\delta > 0$, $\lim_\alpha(\sup_{\delta \le |x| \le \pi} |E_\alpha(x)|) = 0$,*

then (E_α) satisfies conditions (i) and (ii) of (3), and hence is an approximate identity.

(5) *For every $n \in \mathbb{Z}_+$ and $0 < r < 1$, we have*

$$f * D_n = \sum_{j=-n}^{n} \hat{f}(j)e^{ijx} = s_n(f,x) \text{ (a partial sum of } f),$$

$$f * \Phi_n = \sum_{j=-n}^{n} \hat{f}(j)\left(1 - \frac{|j|}{n+1}\right)e^{ijx} = \frac{1}{n+1}\sum_{k=0}^{n} s_k(f,x),$$

$$f * P_r = \sum_{j\in\mathbb{Z}} \hat{f}(j)r^{|j|}e^{ijx}.$$

(6) *(Φ_n) and (P_r) satisfy properties (i) to (iii) of (4) (when, respectively, $n \to \infty$ and $r \to 1$), and hence are approximate identities. Moreover, $(P_r)_{0<r<1}$ is a semigroup: $P_r * P_\rho = P_{r\rho}$.*

Proof (1) See Appendix A.

(2) We have

$$|f * g(s)| \leq \int_{\mathbb{T}} |f(t)g(s\bar{t})| \, dm(t)$$

$$\leq \left(\int_{\mathbb{T}} |f(t)|^p |g(s\bar{t})| \, dm(t) \right)^{1/p} \left(\int_{\mathbb{T}} |g(s\bar{t})| \, dm(t) \right)^{1/p'}$$

$$= \left(\int_{\mathbb{T}} |f(t)|^p |g(s\bar{t})| \, dm(t) \right)^{1/p} \|g\|_1^{1/p'},$$

where $1/p + 1/p' = 1$, and thus

$$\|f * g\|_p \leq \|f\|_p \|g\|_1^{1/p} \|g\|_1^{1/p'} = \|f\|_p \|g\|_1.$$

(3) The hypothesis leads to, $\lim_\alpha \|f - f * E_\alpha\|_p = 0$ for every polynomial $f \in \mathcal{P}$. By (2), the convolution operation $T_\alpha f = f * E_\alpha$ is uniformly bounded for the norm of L^p: $\|T_\alpha\| \leq \|E_\alpha\|_1 \leq C$. As the set of polynomials is dense in $L^p(\mathbb{T})$, for $1 \leq p < \infty$ (Appendix A), the result follows from the Banach–Steinhaus theorem, i.e. the uniform boundedness principle (Appendix E).

(4) Replacing if necessary E_α by $E_\alpha / \hat{E}_\alpha(0)$, we can assume that $\hat{E}_\alpha(0) = 1$ for every α. To verify part (ii) of (3), we write, for any $n \in \mathbb{Z}$ and $\delta > 0$,

$$|\hat{E}_\alpha(n) - 1| = \left| \int_{-\pi}^{\pi} (e^{-inx} - 1) E_\alpha(e^{ix}) \frac{dx}{2\pi} \right| = \left| \int_{|x| \leq \delta} + \int_{\delta < |x| \leq \pi} \right|$$

$$\leq n\delta \|E_\alpha\|_1 + 2 \sup_{\delta \leq |x| \leq \pi} |E_\alpha(x)|.$$

By the hypothesis, $\overline{\lim}_\alpha |\hat{E}_\alpha(n) - 1| \leq n\delta C$, and, since $\delta > 0$ is arbitrary, $\overline{\lim}_\alpha |\hat{E}_\alpha(n) - 1| = 0$.

(5) It clearly ensues from the formulas for D_n, Φ_n, P_r.

(6) The formulas for Φ_n and P_r imply $\Phi_n \geq 0$ and $P_r \geq 0$; hence $\|\Phi_n\|_1 = \hat{\Phi}_n(0) = 1, \|P_r\|_1 = \hat{P}_r(0) = 1$. Moreover, for any $x, \pi \geq |x| \geq \delta > 0$, we have $\Phi_n(x) \leq ((n+1)\sin^2(\delta/2))^{-1}$ and

$$P_r(x) = \frac{1 - r^2}{(1 - r)^2 + 4r \sin^2(x/2)} \leq \frac{1 - r^2}{2 \sin^2(\delta/2)} \quad \text{for } 1/2 \leq r < 1.$$

The result follows. ∎

The following properties are immediate by Lemma 2.1.1.

Corollary 2.1.2

(1) *For every $f \in L^p(\mathbb{T})$, $1 \le p < \infty$, $\lim_n \|f - f * \Phi_n\|_p = 0$ ($f * \Phi_n$ are called the Fejér polynomials of f) and $\lim_{r \to 1} \|f - f_r\|_p = 0$; moreover, $0 < r < \rho < 1 \Rightarrow \|f_r\|_p \le \|f_\rho\|_p \le \|f\|_p$.*
(2) *If $f \in L^1(\mathbb{T})$ and $\hat{f}(n) = 0$, $\forall n \in \mathbb{Z}$, then $f = 0$.*

Notation Let $\mathrm{Hol}(\mathbb{D}) = \{f : f$ is defined and holomorphic in $\mathbb{D}\}$ the space of holomorphic functions in $\mathbb{D}$; if $f \in \mathrm{Hol}(\mathbb{D})$ and $0 < r < 1$, we set

$$f_{(r)}(z) = f(rz) \text{ for } |z| < 1/r.$$

Corollary 2.1.3 *For every function $f \in \mathrm{Hol}(\mathbb{D})$, for $1 \le p \le \infty$ and $0 < r < \rho < 1$, we have $\|f_{(r)}\|_{L^p(\mathbb{T})} \le \|f_{(\rho)}\|_{L^p(\mathbb{T})}$, and hence the following limit exists (finite or not): $\lim_{r \to 1} \|f_{(r)}\|_{L^p(\mathbb{T})} = \sup_{0 < r < 1} \|f_{(r)}\|_{L^p(\mathbb{T})}$.*

2.2 Definition of $H^p(\mathbb{D})$: Identification of $H^p(\mathbb{D})$ and $H^p(\mathbb{T})$

From this section on, we can think of the functions of Hardy spaces as being defined not only on the circle $\mathbb{T}$, but also in the disk $\mathbb{D}$:

Definition 2.2.1 $H^p(\mathbb{D}) = \{f \in \mathrm{Hol}(\mathbb{D}) : \sup_{0 < r < 1} \|f_{(r)}\|_{L^p(\mathbb{T})} < \infty\}$, $0 < p \le \infty$.

Theorem 2.2.2 *Let $1 \le p \le \infty$. Then:*

(1) *$\forall f \in H^p(\mathbb{D})$ the limit*

$$\lim_{r \to 1} f_{(r)}|\mathbb{T} := bf \in H^p(\mathbb{T})$$

exists (limit for the norm of $L^p(\mathbb{T})$, if $1 \le p < \infty$, and weak-$*$ limit if $p = \infty$).
(2) *The mapping $f \mapsto bf$ is an isometric bijection between $H^p(\mathbb{D})$ and $H^p(\mathbb{T})$.*
(3) *$\forall f \in H^p(\mathbb{D})$, $f_{(r)} = (bf)_r = bf * P_r$ (Poisson formula representing f in terms of bf).*

Proof (1) Let $f = \sum_{n \ge 0} a_n z^n$, $z \in \mathbb{D}$, the Taylor series of f. To abbreviate, we write $f_{(r)}$ instead of $f_{(r)}|\mathbb{T}$. For $f \in H^p(\mathbb{D})$, the family $(f_{(r)})_{0 < r < 1}$ is bounded in $L^p(\mathbb{T})$, hence we can use arguments of weak compactness; see Appendix D.

Case $1 < p \leq \infty$. Since $L^p = (L^{p'})^*$, $1/p + 1/p' = 1$, the weak compactness implies that there exists a sequence $r_k \nearrow 1$ such that $(f_{(r_k)})_{k \geq 1}$ converges $\sigma(L^p, L^{p'})$-weakly to a limit $bf \in L^p$. In particular, for every $n \in \mathbb{Z}$, $\lim_k \widehat{f_{(r_k)}}(n) = \widehat{(bf)}(n)$. However, by the uniform convergence of the series $f_{(r)} = \sum_{n \geq 0} a_n r^n z^n$, $z \in \mathbb{T}$, we have $\widehat{f_{(r)}}(n) = a_n r^n$ for $n \geq 0$ and $\widehat{f_{(r)}}(n) = 0$ for $n < 0$. Thus $bf \in H^p(\mathbb{T})$ and $\widehat{(bf)}(n) = a_n$, $n \geq 0$.

This implies $f_{(r)}|\mathbb{T} = (bf) * P_r$ for $0 < r < 1$, and hence

$$\lim_{r \to 1} \|f_{(r)} - bf\|_p = 0 \quad (p < \infty).$$

Case $p = 1$. We consider $L^1(\mathbb{T})$ as a (closed) subspace of the space $M(\mathbb{T})$ of complex Borel measures on $\mathbb{T}$. This latter space is a dual space, $M(\mathbb{T}) = (C(\mathbb{T}))^*$ (Riesz representation theorem: see Appendices A and D). Hence, by the same arguments as above, there exists a measure $bf \in M(\mathbb{T})$ such that $\widehat{(bf)}(n) = a_n$ for $n \geq 0$, and $\widehat{(bf)}(n) = 0$ for $n < 0$. By applying Theorem 1.5.4 of the Riesz brothers, we obtain more: $bf \in H^1(\mathbb{T})$. The rest of the argument is unchanged.

(2) For $p < \infty$, the mapping in question is clearly isometric, since by (1),
$$\|bf\|_p = \lim_{r \to 1} \|f_{(r)}\|_p = \|f\|_{H^p(\mathbb{D})}.$$

For $p = \infty$, the mapping is isometric because $bf = (\sigma^*) \lim_{r \to 1} f_{(r)}$ and $f_{(r)} = (bf)_r$, hence $\|bf\|_\infty \leq \underline{\lim}_{r \to 1} \|f_{(r)}\|_\infty = \overline{\lim}_{r \to 1} \|(bf)_r\|_\infty \leq \|bf\|_\infty$.

The surjectivity is also clear since, for $F \in H^p(\mathbb{T})$ and $\zeta \in \mathbb{T}$, $0 < r < 1$, we have $(F_r)(\zeta) = \sum_{n \geq 0} \hat{F}(n)(r\zeta)^n := f(r\zeta)$, where, by Corollary 2.1.2, $f \in H^p(\mathbb{D})$ and $bf = F$.

(3) Already verified. ∎

Convention 2.2.3   In most cases, by convention we *identify a function $f \in H^p(\mathbb{D})$ with its boundary values $bf \in H^p(\mathbb{T})$*, and hence also the spaces $H^p(\mathbb{T})$ and $H^p(\mathbb{D})$. Thus, in particular, we can write

$$f_{(r)} = f_r = f * P_r, \quad f = \sum_{n \geq 0} \hat{f}(n) z^n$$

(this gives a double sense to $\hat{f}(n)$ as a Fourier coefficient of bf and a Taylor coefficient of f). We denote

$$H^p = H^p(\mathbb{D}) = H^p(\mathbb{T}).$$

Of course, the language of Fourier coefficients (used in the definition of $H^p(\mathbb{T})$) is only applicable for $p \geq 1$. But as we will see in § 2.6, every function f of

$H^p(\mathbb{D})$, $p > 0$, admits limits on the boundary a.e. on $\mathbb{T}$, and we can identify f with its limits.

Corollary 2.2.4 *For every $z \in \mathbb{D}$, the evaluation functional $\varphi_z \colon f \mapsto f(z)$ is linear and continuous on H^1 (and hence on all the H^p) and we have*

$$\|\varphi_z\| \leq \frac{1 + |z|}{1 - |z|}.$$

In particular, the convergence in H^p implies the uniform convergence on the compact subsets of $\mathbb{D}$.

Indeed, if $z = r\zeta$, $\zeta \in \mathbb{T}$, then, for every function $f \in H^1$,

$$|\varphi_z(f)| = |f(z)| = |f * P_r(\zeta)| \leq \|f\|_1 \|P_r\|_\infty = \|f\|_1 \frac{1 + r}{1 - r}$$

(see the beginning of § 2.1), and the result follows. ∎

Corollary 2.2.5 *Let A be a set of complex numbers, $A \subset \mathbb{C}$, and let $f \in H^1(\mathbb{T})$ such that $f(\zeta) \in A$ for a.e. $\zeta \in \mathbb{T}$. Then $f(z) \in \mathrm{conv}(A)$ for every $z \in \mathbb{D}$, where $\mathrm{conv}(A)$ stands for the closed convex hull of A.*

Indeed, from Theorem 2.2.2(3),

$$f(z) = f(rw) = (f * P_r)(w) = \int_{\mathbb{T}} f(\zeta) P_r(\bar{\zeta} w) \, dm(\zeta)$$

where $z = rw$, $|w| = 1$. However, $\mathrm{conv}(A) = \bigcap H$ where the intersection is taken over all half-planes $H = \{z \in \mathbb{C} \colon \mathrm{Re}(az + b) \geq 0\}$ containing A $(a, b \in \mathbb{C})$. Since $P_r \geq 0$ and $\int_{\mathbb{T}} P_r \, dm(\zeta) = 1$, we see that the condition $\mathrm{Re}(af(\zeta) + b) \geq 0$ for a.e. $\zeta \in \mathbb{T}$ entails $\mathrm{Re}(af(z) + b) \geq 0$, and hence $f(z) \in \mathrm{conv}(A)$. ∎

2.3 Jensen's Formula and Jensen's Inequality: $\log |f| \in L^1(\mathbb{T})$

The following inequality and formula play an important role in complex analysis because they make it possible to "count" the zeros of a holomorphic function.

Lemma 2.3.1 (Jensen, 1899) *Let $f \in H^1$, $f(0) = \hat{f}(0) \neq 0$. Denote $\{\lambda_n \colon n = 1, 2, \ldots\}$ the sequence of zeros of f in the disk $\mathbb{D}$, counted with their multiplicity. Then,*

$$\log |f(0)| + \sum_{n \geq 1} \log \frac{1}{|\lambda_n|} \leq \int_{\mathbb{T}} \log |f| \, dm$$

(Jensen's inequality), and in particular,

$$\log|f(0)| \le \int_{\mathbb{T}} \log|f|\, dm.$$

Moreover, if f is analytic for $|z| < 1 + \epsilon$, $\epsilon > 0$, and $f(0) = \hat{f}(0) \ne 0$, then

$$\log|f(0)| + \sum_{n \ge 1} \log \frac{1}{|\lambda_n|} = \int_{\mathbb{T}} \log|f|\, dm$$

(Jensen's formula).

Proof We first consider the case where $f \in \mathrm{Hol}((1 + \epsilon)\mathbb{D})$, $\epsilon > 0$, and $f(z) \ne 0$ for $|z| = 1$. In this case, f has a finite family $\{\lambda_k\}$ of zeros in $\mathbb{D}$ (counted with their multiplicity), $\lambda_k \ne 0$. Let B be a product of conformal mappings b_{λ_j} of the disk $\mathbb{D}$ onto itself,

$$B = \prod_j b_{\lambda_j}, \quad b_\lambda = \frac{\lambda - z}{1 - \bar{\lambda}z} \cdot \frac{|\lambda|}{\lambda}.$$

Clearly $|B| = 1$ on $\mathbb{T}$ (see Exercise 1.8.3(b)) and the quotient f/B is holomorphic and different from zero on $(1 + \delta)\mathbb{D}$ for some $\delta > 0$. Hence, the mean value theorem for the harmonic functions (see Appendix B) leads to Jensen's formula:

$$\log|f(0)| + \sum_n \log \frac{1}{|\lambda_n|} = \log\left|\frac{f}{B}(0)\right| = \int_{\mathbb{T}} \log\left|\frac{f}{B}\right| dm = \int_{\mathbb{T}} \log|f|\, dm.$$

We now consider the general case. Let $f \in \mathrm{Hol}(\mathbb{D})$, $f \ne 0$. As the zeros of f are isolated, there exist numerous sequences $r_k \nearrow 1$ such that $f(r_k \zeta) \ne 0$ for every $\zeta \in \mathbb{T}$. By applying the above formula to f_r, with $r = r_k$, we obtain

$$\log|f(0)| + \sum_{|\lambda_j| < r} \log \frac{r}{|\lambda_n|} = \int_{\mathbb{T}} \log|f(rz)|\, dm.$$

Suppose $f \in \mathrm{Hol}((1 + \epsilon)\mathbb{D})$, $\epsilon > 0$. Then, a passage to the limit for $k \to \infty$ in the preceding formula is possible. Indeed, the zeros of f on $\mathbb{T}$ are isolated points, finite in number, say $\zeta_i \in \mathbb{T}$, and hence $f = pg$ with $p = \prod(z - \zeta_i)$ and g a holomorphic function such that g and $1/g$ are bounded on $(1 + \delta)\mathbb{D}$ with some $\delta > 0$. However, for every r, $0 < r < 1$, and $z \in \mathbb{D}$, we have

$$\frac{1}{2}|\zeta_i - z| \le |\zeta_i - rz| \le 2.$$

Indeed, for the first inequality: $|\zeta_i - z| \le |\zeta_i - rz| + |z(1 - r)| \le |\zeta_i - rz| + |1 - r| \le 2|\zeta_i - rz|$. Hence $|\log|f(r_k z)|| \le C + |\log|p(z)||$, with some suitable constant

$C > 0$ and k large enough. Consequently, the family $\log |f(r_k z)|$ admits an integrable majorant, thus by letting $k \to \infty$, we obtain Jensen's formula under the hypothesis $f \in \mathrm{Hol}((1 + \epsilon)\mathbb{D})$, $\epsilon > 0$.

If we suppose $f \in H^1$ (and $f(0) \neq 0$) and $\delta > 0$, we have

$$\log |f(0)| + \sum_{|\lambda_j| < r_k} \log \frac{r_k}{|\lambda_n|} = \int_{\mathbb{T}} \log |f(r_k z)| \, dm \leq \int_{\mathbb{T}} \log(|f_{r_k}| + \delta) \, dm.$$

By using $|\log(x) - \log(y)| \leq C_\delta |x - y|$ for $x, y \geq \delta$, we obtain on $\mathbb{T}$:

$$|\log(|f_{r_k}| + \delta) - \log(|f| + \delta)| \leq C_\delta |f_{r_k} - f|.$$

Since $\lim_k \|f_{r_k} - f\|_1 = 0$, we have $\lim_k \int_{\mathbb{T}} \log(|f_{r_k}| + \delta) \, dm = \int_{\mathbb{T}} \log(|f| + \delta) \, dm$. As the left side $\log |f(0)| + \sum_{|\lambda_j| < r_k} \log(r_k/|\lambda_n|)$ is increasing as $r_k \nearrow 1$, we obtain

$$\log |f(0)| + \sum_{n \geq 1} \log \frac{1}{|\lambda_n|} \leq \int_{\mathbb{T}} \log(|f| + \delta) \, dm.$$

Letting $\delta \searrow 0$ gives Jensen's inequality.   ∎

Corollary 2.3.2 (Jensen's inequality with the harmonic measure) *Let $g \in H^1$ and $\lambda \in \mathbb{D}$. Then*

$$\log |g(\lambda)| \leq \int_{\mathbb{T}} \frac{1 - |\lambda|^2}{|\lambda - t|^2} \log |g(t)| \, dm(t).$$

Indeed, first let $f \in \mathrm{Hol}((1 + \epsilon)\mathbb{D})$, $\epsilon > 0$. We apply Lemma 2.3.1 to $f = g \circ b$, where $b(z) = (\lambda - z)/(1 - \bar{\lambda}z)$. Since $b \circ b = \mathrm{id}$ (to be verified!) and the derivative of b on $\mathbb{T}$ is

$$|b'(z)| = \frac{1 - |\lambda|^2}{|1 - \bar{\lambda}z|^2} = \frac{1 - |\lambda|^2}{|\lambda - z|^2},$$

we obtain the result. In the general case $g \in H^1$, using the case already proved, we pass to g_{r_k}, $r_k \nearrow 1$, then to $\log(|g_{r_k}| + \delta)$, etc., as in the proof of Lemma 2.3.1.   ∎

Corollary 2.3.3 (boundary uniqueness theorem for H^1) *If $f \in H^1$ and $f \neq 0$, then $\log |f| \in L^1(\mathbb{T})$. In particular, $m\{t \in \mathbb{T}: f(t) = 0\} = 0$.*

Indeed, let $f = \sum_{k \geq 0} \hat{f}(k) z^k = \sum_{k \geq n} \hat{f}(k) z^k$ with $\hat{f}(n) \neq 0$. Then, $g = f/z^n \in H^1$ and $g(0) \neq 0$; hence, by Lemma 2.3.1:

$$-\infty < \log |g(0)| \leq \int_{\mathbb{T}} \log |g| \, dm = \int_{\mathbb{T}} \log |f| \, dm \leq \int_{\mathbb{T}} |f| \, dm < \infty.$$ ∎

Remark 2.3.4 (confrontation of two Jensen inequalities)   Curiously, Jensen's inequality of Lemma 2.3.1 and Corollary 2.3.2 for the holomorphic functions is, in a way, the opposite of the fundamental inequality of convexity in real analysis, which also bears the name of Johan Jensen (!); see also Jensen's biography. In fact, the Jensen convexity inequality states that

$$\varphi\left(\int_{\mathbb{T}} g\,dm\right) \le \int_{\mathbb{T}} (\varphi \circ g)\,dm$$

for every real integrable function g and any convex function φ ($\varphi'' \ge 0$). Setting $g = \log|f|$ and $\varphi(x) = e^x$, we obtain

$$\int_{\mathbb{T}} \log|f|\,dm \le \log \int_{\mathbb{T}} |f|\,dm = \log(|\widehat{f}|(0)).$$

There is no contradiction with Lemma 2.3.1.

Johan Ludvig Jensen (1859–1925) was a Danish mathematician with an atypical career. Indeed, he is one of the rare amateur mathematicians to have left a serious track record in the discipline. As a telecommunications engineer, he had a lifelong, highly successful professional career at the Bell Telephone Company in Copenhagen. He never held a research position, and his entire work in mathematics was done during his spare time. Even though he was self-taught to reach his professional level in mathematics, his mathematical style is notable for its exemplary clarity and rigor. He contributed to the study of the Euler zeta function: it was with this aim that he established the formula and inequality of Lemma 2.3.1. He also discovered the fundamental inequality of convex analysis named after him, the *Jensen convexity inequality* (*Acta Mathematica*, 1906); see Remark 2.3.4 and Appendix A.5.

2.4 Blaschke Products

In this short section, we take a first step towards the canonical factorization of H^p functions.

Lemma 2.4.1 (Blaschke condition) *Let $f \in \mathrm{Hol}(\mathbb{D})$, $f \neq 0$, and let $\{\lambda_n\}_{n\geq1}$ be the zeros of f in $\mathbb{D}$ (counted with their multiplicity). Suppose*

$$\varlimsup_{r\to1} \int_{\mathbb{T}} \log|f_r|\, dm < \infty.$$

Then,

$$\sum_{n\geq1}(1 - |\lambda_n|) < \infty \ (\text{Blaschke condition}).$$

In particular, the zeros of a function $f \in H^p(\mathbb{D})$, $p > 0$, $f \neq 0$, satisfy the Blaschke condition.

Proof Replacing if necessary f by f/z^n, we can assume that $f(0) \neq 0$. By Jensen's formula Lemma 2.3.1,

$$\log|f(0)| + \sum_{|\lambda_j|<r} \log\frac{r}{|\lambda_n|} = \int_{\mathbb{T}} \log|f(rz)|\, dm,$$

and hence

$$\sum_{n\geq1} \log\frac{1}{|\lambda_n|} = \lim_{r\to1} \sum_{|\lambda_j|<r} \log\frac{r}{|\lambda_n|} < \infty.$$

However, since $|\lambda_n| \to 1$, we have $\log(1/|\lambda_n|) \sim 1 - |\lambda_n|$ when $n \to \infty$.   ∎

Theorem 2.4.2 (Blaschke, 1915) *Let $\{\lambda_k\}_{k\geq1}$ be a sequence in $\mathbb{D}$ satisfying the Blaschke condition $\sum_{k\geq1}(1 - |\lambda_k|) < \infty$, and set, as above,*

$$b_\lambda = \frac{\lambda - z}{1 - \bar{\lambda}z} \cdot \frac{\bar{\lambda}}{|\lambda|} \quad (\lambda \in \mathbb{D} \setminus \{0\}), \quad b_0 = z.$$

Then the product

$$B = B_{\{\lambda_k\}} = \prod_{k\geq1} b_{\lambda_k}$$

converges uniformly on the compact subsets of $\mathbb{C} \setminus \mathrm{clos}\{1/\bar{\lambda}_k : k \geq 1\}$ to a holomorphic function $B \neq 0$ satisfying: (1) $|B| < 1$ in $\mathbb{D}$, (2) $|B| = 1$ a.e. on $\mathbb{T}$, (3) the sequence of zeros of B coincides with $\{\lambda_k\}_{k\geq1}$ (including multiplicities).

Proof First, we show the convergence of the partial products $B_n = \prod_{k=1}^{n} b_{\lambda_k}$ in H^2. For $1 \le n \le N$ we have

$$\|B_N - B_n\|_2^2 = \|B_N\|_2^2 + \|B_n\|_2^2 - 2\operatorname{Re}(B_N, B_n) = 2 - 2\operatorname{Re}\int_{\mathbb{T}} B_N \overline{B}_n \, dm$$

$$= 2 - 2\operatorname{Re}\int_{\mathbb{T}} \frac{B_N}{B_n} \, dm = 2 - 2\frac{B_N}{B_n}(0) = 2 - 2\prod_{n<k\le N} |\lambda_k|.$$

By the hypothesis, $\prod_{k\ge K} |\lambda_k| > 0$, with K such that $\lambda_j \ne 0$ for $j \ge K$; thus the partial products and the remainder tend to 1: $\lim_n \prod_{k>n} |\lambda_k| = 1$. Consequently, there exists $B \in H^2$ such that $\lim_n \|B - B_n\|_2 = 0$.

Since for every n, $|B_n| = 1$ a.e. on $\mathbb{T}$, so is B: $|B| = 1$ a.e. on $\mathbb{T}$. By Corollary 2.2.4, $\{B_n\}_{n\ge 1}$ converges (to B) uniformly on the compact subsets of $\mathbb{D}$. The same holds for $\{B/B_n\}_{n\ge 1}$: since

$$\left\| \frac{B}{B_n} - 1 \right\|_2 = \|B - B_n\|_2 \to 0$$

when $n \to \infty$, $B/B_n(z) \to 1$ uniformly on the compact subsets of $\mathbb{D}$. Therefore, the zeros of B coincide with $\{\lambda_k\}_{k\ge 1}$.

The justification of the convergence outside $\mathbb{D}$ is different: for any $z \in \mathbb{C} \setminus \operatorname{clos}\{1/\overline{\lambda_k} : k \ge 1\}$ and $\lambda = \lambda_k$ $(k \ge K)$, we have

$$|b_\lambda(z) - 1| = \left| \frac{(1 - |\lambda|)(|\lambda| + \overline{\lambda}z)}{|\lambda|(1 - \overline{\lambda}z)} \right| \le (1 - |\lambda|)\frac{1 + |z|}{|\lambda| \cdot |z - 1/\overline{\lambda}|} \le (1 - |\lambda|)\frac{1 + |z|}{c \cdot \operatorname{dist}(z, \Lambda)},$$

where $\Lambda = \operatorname{clos}\{1/\overline{\lambda_k} : k \ge 1\}$. The last estimate implies the normal convergence of the series $\sum_{k\ge 1} |b_{\lambda_k}(z) - 1|$ on the compact subsets of $\mathbb{C} \setminus \operatorname{clos}\{1/\overline{\lambda_k} : k \ge 1\}$. ∎

Corollary 2.4.3 (Frigyes Riesz, 1923) *Let $f \in H^p$ $(p \ge 1)$, $f \ne 0$, and let $\{\lambda_k\}_{k\ge 1}$ be the zeros of f in $\mathbb{D}$ (counted with their multiplicities) and $B = B_{\{\lambda_k\}}$, the Blaschke product of f. Then $f = Bg$ with $g \in H^p$, $\|g\|_p = \|f\|_p$ and $g(z) \ne 0$ for every $z \in \mathbb{D}$.*

Indeed, by Theorem 2.4.2, $f = Bg$ with $g \in \operatorname{Hol}(\mathbb{D})$ and $g(z) \ne 0$ for every $z \in \mathbb{D}$. Moreover, since for the partial products $B_n = \prod_{k=1}^{n} b_{\lambda_k}$ we have $|B_n(rt)| \to 1$ $(r \to 1)$ uniformly for $t \in \mathbb{T}$, we obtain

$$\|f/B_n\|_p = \lim_{r\to 1}\left(\int_{\mathbb{T}} \left| \frac{f}{B_n}(rt) \right|^p dm(t) \right)^{1/p} = \|f\|_p,$$

so that, for any r, $0 < r < 1$,

$$\left(\int_{\mathbb{T}} \left| \frac{f}{B_n}(rt) \right|^p dm(t) \right)^{1/p} \le \|f\|_p.$$

Letting $n \to \infty$, we obtain

$$\left(\int_{\mathbb{T}} |f(rt)|^p \, dm(t) \right)^{1/p} \leq \left(\int_{\mathbb{T}} |g(rt)|^p \, dm(t) \right)^{1/p} \leq \|f\|_p,$$

hence the conclusion. The adaptation to the case $p = \infty$ requires only a modification of the notation. $\blacksquare$

Wilhelm Blaschke (1885–1962), an Austrian mathematician, was one of the key figures of differential geometry in the twentieth century. He prepared his thesis under the supervision of Wirtinger, and completed his training with Bianchi, Klein, Hilbert, and Runge. He is the author of roughly 20 monographs, texts, and essays on history and mathematics, including influential works: *Kreis und Kugel* (1916), *Vorlesungen über Differentialgeometrie* (three volumes, 1921–1929), and *Einführung in die Geometrie der Waben* (1955). A colleague of Hecke, Artin, and Hasse at the University of Hamburg, President of the German Mathematical Society in 1934–1935, he was forced out by the Nazi regime in 1935 after a heated argument with Ludwig Bieberbach, a highly influential scientific figure of the Third Reich. Nonetheless, Blaschke was known as a sympathizer of the Nazi regime, member of the NSPD and "Nazi to the heart" (as he himself wrote) or "Mussolinetto" (as he was called by his colleagues in Hamburg). As Director of the Department of Mathematics in Hamburg (1919–1945), he used political pressure in matters of recruitment. After "de-Nazification," he was reinstated in this post in 1946.

Remark 2.4.4 (unconditional convergence; divisor of zeros)   A computation similar to Theorem 2.4.2 shows that the convergence of a Blaschke product is unconditional: whatever the bijection $\sigma : \mathbb{N} \to \mathbb{N}$, we have $B_{\{\lambda_k\}} = B_{\{\lambda_{\sigma(k)}\}}$. This suggests a move to a divisorial notation: for any function $f \in \mathrm{Hol}(\mathbb{D})$ and every $\lambda \in \mathbb{D}$ we define the *zero multiplicity function $k_f(\lambda)$ at the point λ* by

$k_f(\lambda) = 0$ if and only if $f(\lambda) \neq 0$;

$k_f(\lambda) = m$ if and only if $f^{(m)}(\lambda) \neq 0$ and $f(\lambda) = \cdots = f^{(m-1)}(\lambda) = 0$.

The function k_f is called the *zero divisor* of f. The *Blaschke condition* for the sequence of zeros of a function $f \in \mathrm{Hol}(\mathbb{D})$ can be written as follows:

$$\sum_{\lambda \in \mathbb{D}} k_f(\lambda)(1 - |\lambda|) < \infty.$$

Let $Z(f) = \{\lambda \in \mathbb{D}: f(\lambda) = 0\}$ denote the set of zeros of a function $f \neq 0$, $f \in H^1$. Then the *Blaschke product of f* can be written as

$$B = B_{k_f} := \prod_{\lambda \in \mathbb{D}} b_\lambda^{k_f(\lambda)} = \prod_{\lambda \in Z(f)} b_\lambda^{k_f(\lambda)}.$$

2.5 Fatou's Theorem (Non-tangential Boundary Values)

So far, the boundary values of a function $f \in H^p(\mathbb{D})$, i.e. $\lim_{r \to 1} f(r\zeta) = f(\zeta)$ ($\zeta \in \mathbb{T}$), have been considered in the sense of convergence in norm: $\lim_{r \to 1} \|f - f_r\|_p = 0$ ($p < \infty$). At present, they need to be given a sense in terms of the almost everywhere convergence on $\mathbb{T}$ (with respect to the measure m). Of course, by elementary integration theory, from any sequence $r_k \nearrow 1$, a subsequence $r_{k_j} \nearrow 1$ can be extracted with $\lim_j f(r_{k_j}\zeta) = f(\zeta)$, for m-a.e. $\zeta \in \mathbb{T}$; however for the majority of applications, this is not sufficient. In this section, a much stronger result is proved: the theorem of Pierre Fatou below.

Pierre Fatou (1878–1929), a French mathematician, graduated from the École Normale Supérieure in Paris in 1901, and submitted his thesis in 1907. The results of his thesis include the famous *Fatou's lemma,* $\int \liminf F(n)\, d\mu \leq \liminf \int F(n)\, d\mu$, as well as Theorem 2.5.1 (in the case of measures with density $\mu = f \cdot m$).

He never held a post in mathematics (he was a candidate twice), but worked as an associate astronomer at the Paris Observatory. Fatou is one of the founders of the theory of rational complex iterations: he discovered the sets now known as *Julia sets*, a name given by Mandelbrot in 1980. To compete for the Grand Prix de l'Académie des Sciences on the subject of iterations, he published his results in 1917, but the Prix was awarded for comparable results in 1918 to Gaston Julia, another brilliant young mathematician, hero of the First World War. Fatou died of a stomach ulcer, just after finishing a book on Fuchsian groups and automorphic functions. Initially, he was only supposed to provide additional material for a new edition of the course by Appell and Goursat; however, as stated by Goursat, *au lieu des quelques chapitres que nous attendions de lui, c'est un véritable traité … qu'il nous a laissé* ("instead of the few chapters we expected, he left us a veritable treatise").

First, a reminder and a definition are required. Recall (see also Appendix A) that a complex Borel measure on $\mathbb{T}$, with its Radon–Nikodym decomposition $\mu = hm + \mu_s$, is m-a.e. differentiable with respect to m, i.e. for almost every $\zeta \in \mathbb{T}$, the limit

$$\lim_{\Delta \to \zeta} \frac{\mu(\Delta)}{m(\Delta)} := \frac{d\mu}{dm}(\zeta) = h(\zeta)$$

exists (such a point ζ is called a *Lebesgue point of μ*).

Definition A *Stolz angle* at the point $\zeta \in \mathbb{T}$ is the set

$$S_\zeta = \mathrm{conv}\{\zeta,\ \sin(\theta) \cdot \mathbb{D}\}, \quad 0 < \theta < \pi/2.$$

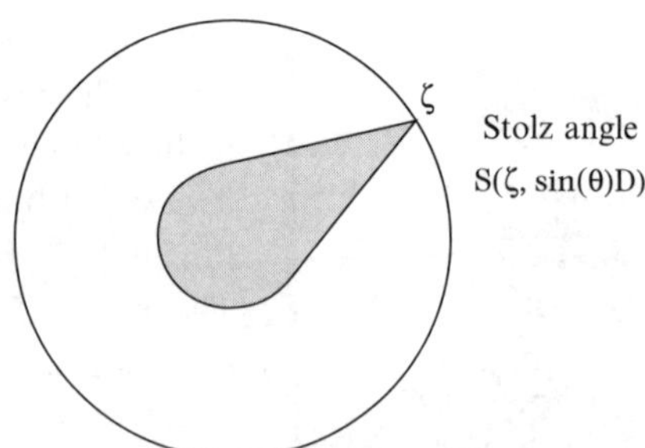

Stolz angle at the point ζ on the unit circle.

A limit along a Stolz angle, $\lim_{z \in S_\zeta, z \to \zeta} f(z)$, is called a *non-tangential limit at the point ζ.*

Theorem 2.5.1 (Fatou, 1906) *Let $\mu \in \mathcal{M}(\mathbb{T})$ and let $\zeta \in \mathbb{T}$ be a Lebesgue point of μ. Then, the Poisson integral of μ,*

$$P * \mu(z) = \int_{\mathbb{T}} \frac{1 - |z|^2}{|t - z|^2}\, d\mu(t), \quad z \in \mathbb{D},$$

has a non-tangential limit at the point ζ, which is equal to $\frac{d\mu}{dm}(\zeta)$. In particular,

$$\lim_{r\to 1} P * \mu(r\zeta) = \frac{d\mu}{dm}(\zeta) \text{ m-a.e. on } \mathbb{T}.$$

Proof Since $P * m(z) = 1$ for every z (see § 2.1), the result is correct for $\mu = m$. With the replacement – if necessary – of μ by $\mu - cm$ ($c \in \mathbb{C}$) and with the use of a rotation, it suffices to examine the case $\mu(\mathbb{T}) = \hat{\mu}(0) = 0$ and $\zeta = 1$. Let F be a primitive of μ, i.e. a function on $[-\pi, \pi]$, left-continuous and with bounded variation, such that $\mu[e^{i\alpha}, e^{i\beta}) = F(\beta) - F(\alpha)$, $F(-\pi) = F(\pi)$. As F is defined up to a constant, we can assume that $F(0) = 0$. Integration by parts in the integral

$$P * \mu(z) = \int_{-\pi}^{\pi} P(ze^{-is})\, dF(s), \quad z \in \mathbb{D},$$

gives

$$P * \mu(z) = -\int_{-\pi}^{\pi} \frac{dP(ze^{-is})}{ds} F(s)\, ds = \int_{-\pi}^{\pi} E_z(s) \frac{F(s)}{s}\, ds,$$

where

$$E_z(s) = -s\frac{dP(ze^{-is})}{ds}.$$

We denote $z = re^{i\theta}$ where $|\theta| \leq \pi, 0 \leq r < 1$, and calculate E_z:

$$E_z(s) = -s\frac{d}{ds} \cdot \frac{1 - r^2}{|1 - re^{i(\theta-s)}|^2} = -s\frac{d}{ds} \cdot \frac{1 - r^2}{1 + r^2 - 2r\cos(\theta - s)}$$

$$= -\frac{(1 - r^2)s\sin(\theta - s)}{|1 - re^{i(\theta-s)}|^4} = -\frac{s\sin(\theta - s)}{(1 - r)^2 + 4r\sin^2(\theta - s)/2}P(ze^{-is}).$$

Let us show that the family $\{E_z : z \in S_1\}$ satisfies conditions (i)–(iii) for an approximate identity, given in Lemma 2.1.1(4).

(i) For every $z \in S_1$,

$$\|E_z\|_1 = \int_{-\pi}^{\pi} \left| s\frac{dP(ze^{-is})}{ds} \right| \frac{ds}{2\pi} \leq A \int_{-\pi}^{\pi} P(ze^{-is})\frac{ds}{2\pi} = A,$$

where

$$A = \sup\left\{ \frac{|s\sin(\theta - s)|}{(1 - r)^2 + 4\sin^2(\theta - s)/2} : s \in [-\pi, \pi], z \in S_1 \right\}.$$

It remains to show that $A < \infty$. Let $C > 0$ be a number such that $|\theta| \leq C(1 - r)$ for any $z = re^{i\theta} \in S_1$ (the existence of such a C can be verified as an exercise).

(a) If $|s| \le 2C(1 - r)$, then

$$\frac{|s\sin(\theta - s)|}{(1 - r)^2 + 4\sin^2(\theta - s)/2} \le \frac{4C(1 - r)|\sin(\theta - s)/2|}{(1 - r)^2 + 4\sin^2(\theta - s)/2} \le C.$$

(b) If $|s| > 2C(1 - r)$, then $|s| > 2|\theta|$, and we have

$$\frac{|s\sin(\theta - s)|}{(1 - r)^2 + 4\sin^2(\theta - s)/2} \le \frac{|s| \cdot (|s| + |\theta|)}{4\sin^2(\theta - s)/2} \le \frac{|s| \cdot (|s| + |\theta|)}{4(|\theta - s|/\pi)^2}$$

$$\le \frac{|s| \cdot (|s| + |s|/2)}{4(|s| - |\theta|/\pi)^2} \le \frac{|s|^2 \cdot (3/2)}{4(|s| - |s/2|/\pi)^2} = (3/2)\pi^2.$$

Therefore $A \le \max(C, 3\pi^2/2)$.

(ii) Integration by parts gives

$$\lim_{z \to 1, z \in S_1} \int_{-\pi}^{\pi} E_z(s)\frac{ds}{2\pi} = \lim_{z \to 1, z \in S_1} (1 - P(-z)) = 1.$$

(iii) Let $\delta \le |s| \le \pi$. Then, for $z \in S_1$ sufficiently close to 1, we have $|\theta| < C(1 - r) < \delta/2$, and hence

$$|E_z(s)| = \left|\frac{(1 - r^2)s\sin(\theta - s)}{|1 - re^{i(\theta - s)}|^4}\right| \le \frac{(1 - r^2)\pi}{|1 - re^{i\delta/2}|^4},$$

which tends to 0 when $z \to 1$, $z \in S_1$.

These properties of E_z and the evident relation

$$\lim_{s \to 0} \frac{F(s)}{s} = \frac{1}{2\pi}\frac{d\mu}{dm}(1),$$

as well as (ii) above, imply, when $z \to 1$, $z \in S_1$,

$$P * \mu(z) - \frac{d\mu}{dm}(1) = \int_{-\pi}^{\pi} E_z(s)\left(\frac{F(s)}{s} - \frac{1}{2\pi}\frac{d\mu}{dm}(1)\right)ds + o(1)$$

$$= \int_{-\delta}^{\delta} + \int_{\delta \le |s| \le \pi} + o(1),$$

which tends to 0. Indeed, by (i), for any $\epsilon > 0$ there exists $\delta > 0$ such that

$$\left|\int_{-\delta}^{\delta}\right| \le \max_{|s| \le \delta}\left|\frac{F(s)}{s} - \frac{1}{2\pi}\frac{d\mu}{dm}(1)\right|\int_{-\pi}^{\pi}|E_z(s)|\,ds < \epsilon 2\pi A,$$

and thus, given (iii) above,

$$\varlimsup_{z \to 1, z \in S_1}\left|P * \mu(z) - \frac{d\mu}{dm}(1)\right| \le \epsilon 2\pi A,$$

and the result follows. ∎

Corollary 2.5.2 *Let* $f \in H^p(\mathbb{D})$, $p \geq 1$. *Then there exist a.e. on* $\mathbb{T}$ *non-tangential limits* $\lim_{z \to \zeta} f(z) = f(\zeta) \, (= (bf)(\zeta))$, $\zeta \in \mathbb{T}$.

Indeed, by Theorem 2.2.2, $f(z) = bf * P_z$, $z \in \mathbb{D}$, and it remains only to apply Fatou's theorem. ∎

2.6 The Smirnov Canonical Factorization

The next two theorems are the last steps needed for the factorization theorem (Theorem 2.6.5 below), but they are also important in themselves.

Theorem 2.6.1 (Szegő, 1921: Szegő "maximal functions") *Let* $f \in L^p(\mathbb{T})$, $p > 0$, *be a function such that* $\log |f| \in L^1(\mathbb{T})$. *The function* $[f]$ *is defined by*

$$[f](z) = \exp\left(\int_{\mathbb{T}} \frac{\zeta + z}{\zeta - z} \log |f(\zeta)| \, dm(\zeta) \right), \quad z \in \mathbb{D}.$$

Then

(1) $[f] \in H^p$ *and* $|[f]| = |f|$ *a.e. on* $\mathbb{T}$,
(2) *if* $g \in H^q(\mathbb{D})$ $(q \geq 1)$ *and* $|g| \leq |[f]| = |f|$ *a.e. on* $\mathbb{T}$, *then* $g \in H^p(\mathbb{D})$ *and* $|g(z)| \leq |[f](z)|$ *in* $\mathbb{D}$,
(3) *for every function* $g \in L^p(\mathbb{T})$, *with* $\log |g| \in L^1(\mathbb{T})$, $[fg] = [f] \cdot [g]$, $[f/g] = [f]/[g]$,
(4) $[f](z) \neq 0$ *in* $\mathbb{D}$, *and for any* $\alpha > 0$, $[|f|^\alpha] = [f]^\alpha$.

Proof (1) Evidently, $[f] \in \mathrm{Hol}(\mathbb{D})$. Let $z \in \mathbb{D}$. Using the Jensen convexity inequality (see Appendix A, or Remark 2.3.4 above) with the measure

$$d\mu(\zeta) = \frac{1 - |z|^2}{|\zeta - z|^2} \, dm(\zeta),$$

we obtain

$$|[f](z)|^p = \exp\left(\int_{\mathbb{T}} \frac{1 - |z|^2}{|\zeta - z|^2} \log |f(\zeta)|^p \, dm(\zeta) \right) \leq \int_{\mathbb{T}} \frac{1 - |z|^2}{|\zeta - z|^2} |f(\zeta)|^p \, dm(\zeta).$$

Hence, by Fubini,

$$\int_{\mathbb{T}} |[f](rt)|^p \, dm(t) \leq \int_{\mathbb{T}} |f(\zeta)|^p \left(\int_{\mathbb{T}} \frac{1 - r^2}{|\zeta - rt|^2} \, dm(t) \right) dm(\zeta)$$

$$= \int_{\mathbb{T}} |f(\zeta)|^p \, dm(\zeta) = \|f\|_p^p.$$

This implies $[f] \in H^p$; by Fatou's Theorem 2.5.1,

$$\lim_{r \to 1} \log |[f](rt)| = \log |f(t)| \quad \text{a.e. on } \mathbb{T},$$

and the result follows (a modification for the case $p = \infty$ is immediate).

    (2) This is a direct application of Jensen's inequality Corollary 2.3.3 to g.

(3) & (4) Both are immediate by the definition.   ■

Corollary 2.6.2 (moduli of H^p functions) *Let $w \in L^1(\mathbb{T})$, $w \geq 0$, $w \neq 0$, and $p > 0$. The following assertions are equivalent.*

(1) *There exists a function $f \in H^p$ such that $|f|^p = w$ a.e. $\mathbb{T}$.*
(2) $\log(w) \in L^1(\mathbb{T})$.

Indeed, (1) $\Rightarrow$ (2) by Jensen's inequality Lemma 2.3.1.

(2) $\Rightarrow$ (1) by Theorem 2.6.1 (with $f = [w^{1/p}]$).   ■

Theorem 2.6.3 (Herglotz, 1911) *Let u be a harmonic function in $\mathbb{D}$, $u \geq 0$. Then there exists a unique Borel measure μ, $\mu \geq 0$, such that $u = P * \mu$, i.e.*

$$u(z) = \int_{\mathbb{T}} \frac{1 - |z|^2}{|\zeta - z|^2}\, d\mu(\zeta), \quad z \in \mathbb{D}.$$

Proof Let $0 < r < 1$, $u_r(z) = u(rz)$ and $\mu_r = u_r m$. Because $\text{Var}(\mu_r) = \mu_r(\mathbb{T}) = u_r(0) = u(0) < \infty$ and because the ball in the space $\mathcal{M}(\mathbb{T})$ is weakly compact (Appendix D), there exists a sequence $(\mu_{r_n})_{n \geq 1}$ weakly convergent to a measure $\mu \in \mathcal{M}(\mathbb{T})$. Clearly $\mu \geq 0$ (as $f \in C(\mathbb{T})$ and $f \geq 0$ imply $\int f\, d\mu = \lim_n \int f u_{r_n} \times dm \geq 0$), and since the Poisson kernel is continuous, then

$$u(z) = \lim_n u(r_n z) = \lim_n \int_{\mathbb{T}} \frac{1 - |z|^2}{|\zeta - z|^2}\, d\mu_{r_n}(\zeta) = \int_{\mathbb{T}} \frac{1 - |z|^2}{|\zeta - z|^2}\, d\mu(\zeta)$$

for every $z \in \mathbb{D}$. The uniqueness of μ follows from the Fourier representation of $P * \mu$: $(P * \mu)(r\zeta) = \sum_{n \in \mathbb{Z}} r^{|n|} \hat{\mu}(n) \zeta^n$, and thus if $P * \mu = P * \nu$, $\hat{\mu}(n) = \hat{\nu}(n)$ for every $n \in \mathbb{Z}$, hence $\mu = \nu$.   ■

Corollary 2.6.4 (singular inner functions) *Let $V \in \text{Hol}(\mathbb{D})$. The following assertions are equivalent.*

(1) $0 < |V(z)| \leq 1$ *in* $\mathbb{D}$, $V(0) > 0$ *and* $|V(\zeta)| = 1$ *a.e. on* $\mathbb{T}$.

(2) *There exists a unique measure $\mu \geq 0$ on $\mathbb{T}$, singular with respect to m ($\mu \perp m$), such that*

$$V(z) = V_\mu(z) := \exp\left(-\int_\mathbb{T} \frac{\zeta + z}{\zeta - z}\, d\mu(\zeta)\right), \quad z \in \mathbb{D}.$$

The functions satisfying condition (1) or (2) are called "singular inner functions."

Indeed, $(2) \Rightarrow (1)$, since

$$|V(z)| = \exp\left(-\int_\mathbb{T} \frac{1 - |z|^2}{|\zeta - z|^2}\, d\mu(\zeta)\right)$$

and $d\mu/dm = 0$ a.e. on $\mathbb{T}$, and the rest follows from Fatou's Theorem 2.5.1.

$(1) \Rightarrow (2)$, since applying Theorem 2.6.3 and Fatou's theorem to $u = \log(1/|V|)$ we obtain a unique measure μ such that $d\mu/dm = 0$ a.e. on $\mathbb{T}$ (and hence $\mu \perp m$), and $|V(z)| = |V_\mu(z)|$ in $\mathbb{D}$, thus $V = \lambda V_\mu$ with $|\lambda| = 1$. However, since $V(0) > 0$ and $V_\mu(0) > 0$, we have $V = V_\mu$. $\blacksquare$

The following theorem is the principal result of Chapter 2.

Theorem 2.6.5 (Smirnov, 1928a,b: canonical factorization) *Let $f \in H^p(\mathbb{D})$, $f \neq 0$, $p > 0$. Then there exists a unique representation of f as the product*

$$f = \lambda BV[f],$$

with $\lambda \in \mathbb{T}$, $B = B_k$ the Blaschke product for the divisor $k = k_f$ of the zeros of f, V a singular inner function, and $[f]$ a maximal function of Theorem 2.6.1.

Proof Set $g = f/B_k$, $\lambda = g(0)/|g(0)|$. Then $|g| = |f|$ a.e., $\log|g| \in L^1$ and $[g] = [f]$. By Theorem 2.6.1 and Corollary 2.6.4, the function $V = g/\lambda[g]$ is a singular inner function, which proves the existence of the representation.

For the uniqueness, λ is uniquely determined because the functions B, V, $[f]$ are strictly positive at $z = 0$, and B and $[f]$ are also completely determined by f. $\blacksquare$

Corollary 2.6.6 *Let $f \in H^p(\mathbb{D})$, $f \neq 0$, $p > 0$. Then a.e. on $\mathbb{T}$ there exist non-tangential boundary limits $bf(\zeta) = \lim_{z \to \zeta, z \in S_\zeta} f(z)$, and $bf \in L^p(\mathbb{T})$.*

Indeed, using the canonical factorization $f = \lambda BV[f]$, we have $[f]^p \in H^1(\mathbb{D})$ (Theorem 2.6.1), hence the two functions λBV and $[f]^p$ (and thus $[f]$ and f) admit non-tangential limits a.e., and $([f]^p|\mathbb{T}) \in L^1(\mathbb{T})$; the result follows. $\blacksquare$

We will undertake a thorough study of canonical factorization in Chapter 3. However, one thing is required right now: the identification of the functions

$\lambda[f]$ with the outer functions (Definition 1.4.2) and of the functions λBV with the inner functions (Definition 1.7.1).

Theorem 2.6.7 (Szegő maximal functions, part 2) *Let p, q, $r \geq 1$ and $f \in H^p$. The following assertions are equivalent.*

(1) *There exists a number $\lambda \in \mathbb{T}$ such that $f = \lambda[f]$.*
(2) *For every $\lambda \in \mathbb{D}$, Jensen's inequality Corollary 2.3.2 becomes an equality:*

$$\log|f(\lambda)| = \int_{\mathbb{T}} \frac{1 - |\lambda|^2}{|\lambda - t|^2} \log|f(t)|\, dm(t).$$

(3) $\log|f(0)| = \int_{\mathbb{T}} \log|f(t)|\, dm(t)$.
(4) *There exists a $\lambda \in \mathbb{D}$ verifying the equality (2).*
(5) *If $g \in H^q$ and $|g| \leq |f|$ a.e. on $\mathbb{T}$, then $|g| \leq |f|$ in $\mathbb{D}$.*
(6) *If $g \in H^q$ and $|g| \leq |f|$ a.e. on $\mathbb{T}$, then $|g(0)| \leq |f(0)|$.*
(7) *If $g \in H^q$ and $g/f \in L^r$, then $g/f \in H^r$.*

If $p = 2$, then (1)–(7) are equivalent to:
(8) *f is a (Beurling) outer function.*

Proof Clearly (1) $\Rightarrow$ (2) $\Rightarrow$ (3) $\Rightarrow$ (4).

To show (4) $\Rightarrow$ (1), let $f = \lambda BV[f]$ be the canonical factorization of f; then (4) implies $|B(\lambda)V(\lambda)| = 1$, and hence $|B(\lambda)| = 1$ and $|V(\lambda)| = 1$; if we assume $k_f \neq 0$ (see Remark 2.4.4 for the notation) we would obtain the contradiction $1 = |B(\lambda)| = \prod_{\zeta \in D} |b_\zeta(\lambda)^{k_f(\zeta)}| < 1$ (there is at least one factor < 1), and if we suppose $V \neq 1$, we would obtain $V = V_\mu$, where $\mu \geq 0$, $\mu(\mathbb{T}) > 0$, hence

$$1 = |V(\lambda)| = \exp\left(-\int_{\mathbb{T}} \frac{1 - |\lambda|^2}{|\lambda - t|^2}\, d\mu(t)\right) < 1,$$

which is a contradiction. Thus, $f = \lambda[f]$, and (4) $\Rightarrow$ (1).

Jensen's inequality gives (1) $\Rightarrow$ (5) $\Rightarrow$ (6).

To show (6) $\Rightarrow$ (3), suppose that on the contrary, $|f(0)| < [f](0)$. Set $h_n(\zeta) = \min(|f(\zeta)|, n)$ on $\mathbb{T}$ and $g_n = [h_n]$. Then, $g_n(0) \nearrow [f](0)$ (dominated convergence theorem) and hence, for a large enough index, we have $|f(0)| < g_n(0)$, which contradicts (6) because $g_n \in H^\infty \subset H^q$ and $|g_n| \leq |f|$ a.e. on $\mathbb{T}$. Hence, (6) $\Rightarrow$ (3).

(1) $\Rightarrow$ (7), since if $g = \lambda_1 BV[g] \in H^q$ and $g/f \in L^r$, then, by Theorem 2.6.1, $g/f = (\lambda_1/\lambda)BV[g/f] \in H^r$.

To show (7) $\Rightarrow$ (1), we use the same argument as for (6) $\Rightarrow$ (3): let $L = \min(|f|, 1)$ and $g = [L]$; then, $g \in H^\infty$ and $g/f \in L^\infty$, thus by (7), $g/f = h \in H^r$.

By using the canonical factorization $f = \lambda BV[f]$, $h = \lambda_1 B_1 V_1[h]$, we have $[L] = g = fh = \lambda BV[f] \cdot \lambda_1 B_1 V_1[h]$; by the uniqueness of the factorization, $\lambda BV \cdot \lambda_1 B_1 V_1 = 1$, hence $BV = \text{constant} = \mu$, which implies (1) $(f = \lambda\mu[f])$.

Let $p = 2$. Given the above, the property $(7) = (7)_{p,q,r}$ does not depend on q nor r. Hence, it suffices to see that $(8) \Leftrightarrow (7)_{2,2,2}$: but this is Theorem 1.7.4. ∎

Corollary 2.6.8 (inner–outer factorization versus canonical factorization) *Let $f \in H^2$. Then, f is outer if and only if $f = \lambda[f]$ where $\lambda \in \mathbb{T}$, and f is inner if and only if $f = \lambda BV$ (with the notation of Theorem 2.6.5).*
In the general case where $f = \lambda BV[f]$, we have $f_{out} = \lambda[f]$, $f_{in} = BV$.

Terminology 2.6.9 For a function $f = \lambda BV[f]$, $f_{out} = \lambda[f]$ is said to be the *outer part of f* and $f_{in} = BV$ the *inner part of f*.

2.7 Applications: Szegő Infimum, Weighted Polynomial Approximations, Invariant Subspaces of $L^p(\mathbb{T})$

With the techniques developed in this chapter, we can make a final conclusion on the Szegő problem (see § 1.7 and particularly Theorem 1.7.6), and in addition treat the questions of determination of stationary processes, of cyclic vectors (see below), of the completeness of analytic polynomials $\mathcal{P}_a$ and of invariant subspaces of L^p.

Recall the problem of the Szegő infimum: given a measure μ on $\mathbb{T}$, calculate the quantity

$$d^2(\mu) = \text{dist}^2_{L^2(\mu)}(1, H_0^2(\mu)) = \inf_{p \in z\mathcal{P}_a} \int_{\mathbb{T}} |1 - p|^2 \, d\mu.$$

The following remarkable formula was found by Szegő (1920) for μ absolutely continuous with respect to m $(\mu = wm)$ and then by Verblunsky (1936) for the general case.

Theorem 2.7.1 (Szegő, 1920; Verblunsky, 1936; Kolmogorov, 1941) *Let μ be a Borel measure on $\mathbb{T}$, and $\mu = wm + \mu_s$, $w = d\mu/dm$, its Radon–Nikodym decomposition. Then,*

$$d^2(\mu) = \exp\left(\int_{\mathbb{T}} \log\left|\frac{d\mu}{dm}\right| dm\right) \quad (Szegő–Verblunsky–Kolmogorov\ formula).$$

Proof According to Theorem 1.7.6, either there exists an (outer) function $f \in H^2$ such that $|f|^2 = w$ and then $d(\mu) = |f(0)|$, or $d(\mu) = 0$. By Corollary 2.6.2, the first possibility occurs if and only if $\log(w) \in L^1$. Moreover, if the last condition is verified, then the outer function we seek is $[w^{1/2}] = [w]^{1/2}$ (see Theorem 2.6.1); the result follows. ∎

Corollary 2.7.2 (optimal prediction of stationary processes) *Let $(x_n)_{n\in\mathbb{Z}}$ be a stationary random process and μ its spectral measure (see Theorem 1.6.2).*

(1) *$(x_n)_{n\in\mathbb{Z}}$ is non-deterministic if and only if*

$$\log\left|\frac{d\mu}{dm}\right| \in L^1(\mathbb{T}).$$

(2) *The optimal prediction for 1 step ahead (in quadratic mean and with respect to its past) is*

$$d^2(\mu) = \exp\left(\int_{\mathbb{T}} \log\left|\frac{d\mu}{dm}\right| dm\right).$$

Indeed, by the definitions of 1.6.1, the corollary is simply a probabilistic reformulation of Theorem 2.7.1. ∎

2.7.1 Cyclic Vectors of the Shift Operator M_z

Given a bounded operator $T\colon X \to X$, an important step in the study of the lattice $\mathrm{Lat}(T)$ consists of a description of the T-invariant subspaces E generated by a single element,

$$E = E_x := \mathrm{span}_X(T^n x\colon n \geq 0).$$

In particular, it is interesting to describe the *cyclic vectors* $x \in X$ (if such exist), i.e. x satisfying

$$E_x = X.$$

To explain the terminology, recall that a cyclic vector of a matrix (or linear mapping) $A\colon \mathbb{C}^n \to \mathbb{C}^n$ is an element $x \in \mathbb{C}^n$ such that $\mathrm{span}(A^j x\colon j \geq 0) = \mathrm{span}(A^j x\colon 0 \leq j < n) = \mathbb{C}^n$.

We begin with the shift operator M_z on H^2 where, of course, the question of cyclic elements $f \in H^2$ does not reduce to the simple replacement of the word "cyclic" with "outer" (see Definition 1.7.1). Indeed, it involves finding practical criteria to know whether a given function is outer or not. In Chapter 3 we will study a certain number of tools to recognize outer functions, but for now, we limit ourselves to the following classical criterion (Smirnov showed the principal implication (2) $\Rightarrow$ (1) and Beurling independently showed the equivalence (2) $\Leftrightarrow$ (1)).

Theorem 2.7.3 (Smirnov, 1928a,b; Beurling, 1949) *For a function $f \in H^2$, $f \neq 0$, the following assertions are equivalent.*

(1) *f is cyclic for M_z, i.e.* $\mathrm{span}_{H^2}(z^n f : n \geq 0) = H^2$.
(2) $\log|f(0)| = \int_{\mathbb{T}} \log|f|\, dm$.

Proof This is part of Theorem 2.6.7. ∎

The question of the cyclicity of a function $f \in L^2(\mathbb{T})$ in the space $L^2(\mathbb{T})$, and more generally in $L^2(\mathbb{T}, \mu)$, is also basically resolved, as is shown in the following theorem. Here, it is useful to consider the result of Exercise 1.8.2(d).

Theorem 2.7.4 (Kolmogorov, 1941) *Let $f \in L^2(\mathbb{T}, \mu)$ where $\mu = wm + \mu_s$ is a finite Borel measure on $\mathbb{T}$ (with its Radon–Nikodym decomposition).*

(1) *If f is M_z-cyclic in $L^2(\mathbb{T}, \mu)$ (i.e., $\mathrm{span}_{L^2(\mu)}(z^n f : n \geq 0) = L^2(\mathbb{T}, \mu)$), then $f \neq 0$ μ-a.e. on $\mathbb{T}$.*
(2) *If $f \neq 0$ μ-a.e. on $\mathbb{T}$ and $\mu(\sigma) = 0$ for a set $\sigma \subset \mathbb{T}$, $m(\sigma) > 0$, then f is cyclic in $L^2(\mathbb{T}, \mu)$.*
(3) *If $f \neq 0$ μ-a.e. on $\mathbb{T}$ and $m \ll \mu$ ($\Leftrightarrow w > 0$ m-a.e.), then f is cyclic in $L^2(\mathbb{T}, \mu)$ if and only if $\log(|f|^2 w) \notin L^1(\mathbb{T})$.*

Proof (1) Clear.
(2) Thanks to Exercise 1.8.2(d), the subspace $E_f = \mathrm{span}_{L^2(\mu)}(z^n f : n \geq 0)$ is reducing for M_z, hence $E_f = \chi_A L^2(\mathbb{T}, \mu)$ (see Theorem 1.2.1), and clearly $\chi_A = 1$ μ-a.e., and the result follows.
(3) Let $f = f_s + f_a$ be the decomposition of f corresponding to $L^2(\mathbb{T}, \mu) = L^2(\mathbb{T}, \mu_s) \oplus L^2(\mathbb{T}, wm)$. By Theorem 1.3.5, f is cyclic if and only if f_s and f_a are respectively cyclic in $L^2(\mathbb{T}, \mu_s)$ and $L^2(\mathbb{T}, wm)$. Hence, the condition for f_s is $f_s \neq 0$ μ_s-a.e. on $\mathbb{T}$. For f_a, by using an isometric mapping $L^2(\mathbb{T}, wm) \to L^2(\mathbb{T}, m)$ defined by $h \mapsto hw^{1/2}$, we reduce the question to the cyclicity of $f_a w^{1/2}$ in $L^2(\mathbb{T}, m)$. By Corollary 1.4.1, $f_a w^{1/2}$ is not cyclic in $L^2(\mathbb{T}, m)$ if and only if there exists a unimodular function q such that $\overline{q} f_a w^{1/2} \in H^2(\mathbb{T})$, which is equivalent to $\log(|f|w^{1/2}) \in L^1(\mathbb{T})$ (by Corollary 2.6.2). The result follows. ∎

2.7.2 Weighted Density of Polynomials $\mathcal{P}_a$

Our purpose is to provide a density criterion

$$\mathrm{clos}_{L^2(\mu)} \mathcal{P}_a = L^2(\mathbb{T}, \mu),$$

or, slightly more generally, a description of the spaces

$$H^2(\mu) = \operatorname{span}_{L^2(\mu)}(z^n : n \geq 0) = \operatorname{clos}_{L^2(\mu)} \mathcal{P}_a$$

for a finite Borel measure μ on $\mathbb{T}$. The results have essentially already been obtained. However, we give a last word on this subject.

Theorem 2.7.5 (description of the spaces $H^2(\mu)$) *Let $\mu = wm + \mu_s$ be a finite Borel measure on $\mathbb{T}$ (with its Radon–Nikodym decomposition).*

(1) $\operatorname{clos}_{L^2(\mu)} \mathcal{P}_a = L^2(\mathbb{T}, \mu) \Leftrightarrow \log(w) \notin L^1(\mathbb{T})$.
(2) *If* $\log(w) \in L^1(\mathbb{T})$, *then* $H^2(\mu) = L^2(\mathbb{T}, \mu_s) \oplus ([w^{1/2}]^{-1} H^2)$.

Proof (1) This is an immediate consequence of Theorem 2.7.4(3) with $f = 1$.
(2) We know that $H^2(\mu) = L^2(\mathbb{T}, \mu_s) \oplus H^2(wm)$. As in Theorem 2.7.4(3), by using the isometry $f \mapsto [w^{1/2}]f$ of $L^2(\mathbb{T}, wm)$ in $L^2(\mathbb{T}, m)$, we obtain that the image of $H^2(wm) = \operatorname{clos}_{L^2(wm)} \mathcal{P}_a$ is $H^2 = \operatorname{clos}_{L^2(m)}([w^{1/2}]\mathcal{P}_a)$, and the result follows. $\blacksquare$

Remark 2.7.6 (analytic nature of the spaces $H^2(\mu)$)   By interpreting H^2 as the set of boundary limits of the functions $H^2(\mathbb{D})$, we can read the assertion (2) of the theorem as follows: if $\log(w) \in L^1(\mathbb{T})$, then the convergence of a sequence of polynomials in the space $H^2(\mathbb{T}, wm)$ implies its convergence in the space $\operatorname{Hol}(\mathbb{D})$, and the limit is a function h such that $h[w^{1/2}] \in H^2(\mathbb{D})$.

2.8 Exercises

2.8.1 Invariant Subspaces of $L^p(\mathbb{T}, \mu)$

Here we propose to prove an $L^p(\mathbb{T}, \mu) = L^p(\mu)$ analog of Helson's Theorem 1.3.5. The principal case $\mu = m$ is a theorem of Srinivasan (1963): see (b) below. Since several details of the reasoning are quite similar to those in Theorem 1.3.5 (and in its preparatory lemmas), the hints and solutions will be somewhat brief.

Let $\mu = \mu_s + \mu_a = \mu_s + wm$ be the Radon–Nikodym decomposition of μ. As in the case $p = 2$, the equality

$$\int_{\mathbb{T}} |f|^p \, d\mu = \int_{\mathbb{T}} |f|^p \, d\mu_s + \int_{\mathbb{T}} |f|^p \, d\mu_a$$

can be seen as a direct decomposition

$$L^p(\mathbb{T}, \mu) = L^p(\mathbb{T}, \mu_s) \oplus_p L^p(\mathbb{T}, \mu_a),$$

with

$$\|f\|^p_{L^p(\mu)} = \|f\|^p_{L^p(\mu_s)} + \|f\|^p_{L^p(\mu_a)}$$

to justify the notation $\oplus_p$.

We will use the standard duality $(L^p(\mu))^* = L^{p'}(\mu)$, $1 \le p < \infty$, $1/p + 1/p' = 1$ (see Appendix D) with respect to the bilinear form $\langle f, g \rangle = \int_\mathbb{T} fg \, d\mu$.

Considering $M_z \colon L^p(\mathbb{T}, \mu) \to L^p(\mathbb{T}, \mu)$, we denote $\mathrm{Lat}(M_z)$ the set of closed invariant subspaces $E \subset L^p(\mu)$,

$$M_z E \subset E$$

(when $p = \infty$, for the weak-$*$ topology $\sigma(L^\infty(\mu), L^1(\mu))$, see Appendix D). In this subsection, p is always arbitrary, with $1 \le p \le \infty$, unless otherwise specified.

(a) *Let $M_z \colon L^p(\mathbb{T}, \mu) \to L^p(\mathbb{T}, \mu)$ and $E \in \mathrm{Lat}(M_z, M_z^{-1})$, i.e. $M_z E = E$. Show that there exists a Borel subset $A \subset \mathbb{T}$ (unique modulo μ) such that $E = \chi_A L^p(\mathbb{T}, \mu)$.*

SOLUTION: Repeat literally the arguments of Wiener's Theorem 1.2.1, therefore

$$g \in E^\perp = \{h \in L^{p'}(\mathbb{T}, \mu) : \langle f, h \rangle = 0 \text{ for every function } f \in E\}$$

if and only if $fg = 0$ μ-a.e. for every function $f \in E$. If the space $L^{p'}(\mathbb{T}, \mu)$ is separable (hence, if $p > 1$) the reasoning is completed as in Theorem 1.2.1. If $p = 1$, use the separability of $L^1(\mathbb{T}, \mu)$ to obtain similarly that $E^\perp = \chi_B L^\infty(\mathbb{T}, \mu)$ with a Borel subset $B \subset \mathbb{T}$. Clearly $E = \chi_A L^1(\mathbb{T}, \mu)$ with $A = \mathbb{T} \setminus B$. $\blacksquare$

(b) *Let $M_z \colon L^p(\mathbb{T}, \mu) \to L^p(\mathbb{T}, \mu)$ and $E \in \mathrm{Lat}(M_z)$. Show that $H^\infty(\mathbb{T}) \cdot E \subset E$.*

SOLUTION: A similar property has been seen in part (ii) of Exercise 1.8.3(a). Here is a different argument: if $f \in E$ and $g \in H^\infty(\mathbb{T})$, then for any $0 < r < 1$, $fg_r \in E$ (since the series $g_r = \sum_{k \ge 0} \hat{g}(k) r^k z^k$ converges in norm in $H^\infty(\mathbb{T})$ and $z^k f \in E$) and $\|fg - fg_r\|_p^p = \int_\mathbb{T} |f|^p |g - g_r|^p \, d\mu \to 0$ when $r \to 1$ by Fatou's theorem (and the dominated convergence theorem. For $p = \infty$ and the weak-$*$ convergence, reason similarly with $\langle fg - fg_r, h \rangle = \int_\mathbb{T} fh(g - g_r) \, d\mu$). Thus $fg \in E$, and the result follows. $\blacksquare$

(c) (Srinivasan, 1963). *Let $E \subset L^p(\mathbb{T})$, $E \in \mathrm{Lat}(M_z)$ and $M_z E \ne E$. Show that there exists a unimodular inner function Θ, $|\Theta| = 1$ a.e. (uniquely defined up to a unimodular constant), such that $E = \Theta H^p(\mathbb{T})$.*

SOLUTION: First show that $E = \mathrm{clos}_{L^p}(E \cap L^\infty(\mathbb{T}))$, the closure for the weak-$*$ topology if $p = \infty$. Indeed, if $f \in E$, define $h_n = \min(1, n/|f|)$. Then $0 \le h_n \le 1$ and $\log(h_n) \in L^1(\mathbb{T})$, hence $[h_n]$ exists and $[h_n] \in H^\infty$. By (b) above, $f[h_n] \in E$.

Show that a subsequence $([h_{n_k}])$ converges to 1 m-a.e. on $\mathbb{T}$. By construction, $\lim_n |[h_n]| = 1$ a.e. and $|[h_n]| \leq 1$, hence $\lim_n \|[h_n]\|_2 = 1$ and then $\|[h_n] - 1\|_2^2 = \|[h_n]\|^2 + 1 - 2\operatorname{Re}([h_n](0)) = \|[h_n]\|^2 + 1 - 2\exp \int_{\mathbb{T}} \log |h_n| \, dm$, which tends to 0 (since $\log(h_n) \nearrow 0$ and $\log(h_1) \in L^1(\mathbb{T})$), hence the property. By dominated convergence, $\|f[h_n] - f\|_p \to 0$ (for $p = \infty$ with weak-$*$ convergence, the modification is evident). For $p \leq 2$, complete the reasoning by observing that $E \cap L^\infty(\mathbb{T}) \subset L^2(\mathbb{T}) \subset L^p(\mathbb{T})$, and hence $E = \operatorname{clos}_{L^p} \operatorname{clos}_{L^2}(E \cap L^\infty(\mathbb{T})) = \operatorname{clos}_{L^p} \Theta H^2(\mathbb{T}) = \Theta H^p(\mathbb{T})$ where Θ comes from Helson's Theorem 1.3.5. For $p > 2$, use duality: the subspace $E^\perp = \{h \in L^{p'}(\mathbb{T}) : \langle f, h \rangle = 0$ for every function $f \in E\}$ is M_z-invariant, also $M_z E^\perp \neq E^\perp$ and $p' < 2$, thus $E^\perp = \theta H^{p'}(\mathbb{T})$ and $E = (z\theta) H^p(\mathbb{T})$, by the above. ∎

(d) *Let $E \subset L^p(\mathbb{T}, wm)$, $E \in \operatorname{Lat}(M_z)$, $M_z E \neq E$. Show that $w > 0$ m-a.e. and that there exists a Borel function q (unique up to a multiplicative constant) such that $|q|^p w = 1$ and $E = q H^p(\mathbb{T})$.*

SOLUTION: First, $E \subset M_z^{-1} E$ and $E \neq M_z^{-1} E$: thus there exists a function $g \in L^{p'}(\mathbb{T}, \mu)$ such that $g \in E^\perp$, $g \notin (M_z^{-1} E)^\perp$. Let $f \in E$ with $\langle M_z^{-1} f, g \rangle \neq 0$: then $0 = \langle z^n f, g \rangle = \int_{\mathbb{T}} z^n f g w \, dm$ for $n \geq 0$. But $fgw \neq 0$, and the Riesz brothers' Theorem 1.5.4 guarantees that $fgw \in H^1(\mathbb{T})$, hence $\log(|fg|w) \in L^1(\mathbb{T})$, so that $w > 0$ m-a.e.

Next, clearly the mapping $J \colon L^p(\mathbb{T}, wm) \to L^p(\mathbb{T})$, $Jf = w^{1/p} f$, is an isometric isomorphism between the indicated spaces and $JM_z = M_z J$, hence JE is an M_z-invariant subspace of $L^p(\mathbb{T})$, but not M_z, M_z^{-1}-invariant. By (c), $JE = \Theta H^p(\mathbb{T})$ where $|\Theta| = 1$ a.e. on $\mathbb{T}$, thus $E = w^{-1/p} \Theta H^p(\mathbb{T})$: the proof is complete. ∎

(e) An analog of Helson's Theorem 1.3.5. *Let $M_z \colon L^p(\mathbb{T}, \mu) \to L^p(\mathbb{T}, \mu)$ and $E \in \operatorname{Lat}(M_z)$. Show that*

(i) *if $M_z E = E$, then there exists a Borel set $A \subset \mathbb{T}$ (unique modulo μ) such that $E = \chi_A L^p(\mathbb{T}, \mu)$;*

(ii) *if $M_z E \neq E$, then there exist a Borel set $A \subset \mathbb{T}$ (unique modulo μ_s) and a Borel function q such that $E = \chi_A L^p(\mathbb{T}, \mu_s) \oplus_p q H^p(\mathbb{T})$ and $|q|^p w = 1$ m-a.e. on $\mathbb{T}$.*

A is uniquely defined modulo μ, and so is q, up to a unimodular constant.

In particular, $E \in \operatorname{Lat}(M_z) \Rightarrow M_z E = E$, hence $E = \chi_A L^p(\mathbb{T}, \mu)$ for some Borel set A, if and only if $m \not\ll \mu$, i.e. if and only if there exists $A \subset \mathbb{T}$ such that $mA > 0$, $\mu A = 0$.

SOLUTION: Part (i) is proved in (a). For (ii), let $f \in L^p(\mu) = L^p(\mu_s) \oplus_p L^p(\mu_a)$ with its decomposition $f = f_s + f_a$. Then $f \in E \Leftrightarrow \langle z^n f, g \rangle = 0, \forall n \geq 0$ for every function $g \in E^\perp$, i.e. $\int_{\mathbb{T}} z^n fg \, d\mu = 0$ for every $n \geq 0$ and $g \in E^\perp$. The Riesz brothers' theorem 1.5.4 gives $fg\mu \ll m$, hence $fg = 0$ μ_s-a.e., or $f_s g = 0$ for every $g \in E^\perp$.

Conversely, such a function f_s is in E, thus $E = E_s \oplus_p E_a$ where $E_s = \chi_A L^p(\mu_s)$ (by (a)) and $E_a \subset L^p(\mu_a)$ is an invariant subspace satisfying $M_z E_a \neq E_a$; (d) implies that $E_a = qH^p(\mathbb{T})$ with a function q such that $|q|^p w = 1$ m-a.e. on $\mathbb{T}$. The uniqueness is clear from the above. $\blacksquare$

(f) Invariant subspaces of H^p, $1 \leq p \leq \infty$.

(i) *Let $E \subset H^p$, $E \in \mathrm{Lat}(M_z)$, $E \neq \{0\}$. Show that there exists an inner function Θ (unique up to a unimodular constant) such that $E = \Theta H^p$.*

(ii) *Let $f \in H^p$, $f \neq 0$. Then $E_f := \mathrm{span}_{H^p}(z^n f : n \geq 0) = f_{in} H^p$.*

(iii) **Cyclic functions of H^p.** *Let $f \in H^p$; show that $E_f = H^p$ if and only if $f = \lambda[f]$ with $[f]$ the Szegő maximal function, see Theorems 2.6.1 and (2.6.7).*

SOLUTION: Part (i) follows from (c) above because $M_z^{-1} E \subset E$ is impossible ($M_z^{-n} f \in H^p, \forall n \geq 0 \Rightarrow f = 0$) and the function Θ of (c) is clearly in H^p.

For (ii), let Θ be an inner function such that $E_f = \Theta H^p$. As $f \in \Theta H^p$, Θ divides f_{in}, but Θ is also an H^p limit of a sequence $f g_n = f_{in} f_{out} g_n$, $g_n \in \mathcal{P}_a$. As H^p is complete (it is a closed subspace of L^p), the Cauchy sequence $(f_{out} g_n)$ converges in H^p to a limit h, hence $\Theta = f_{in} h$, and thus f_{in} divides Θ. Consequently, $\Theta = \lambda f_{in}$, $\lambda \in \mathbb{C}$.

Part (iii) is clear by (ii). $\blacksquare$

2.8.2 Factorization on the H^p Scale, $0 < p < \infty$

(a) *Let $r > 0$, $s > 0$, $t > 0$ be such that $1/r = 1/s + 1/t$. Show that $H^r = H^s \cdot H^t$ and moreover that $\|f\|_r = \min\{\|g\|_s \|h\|_t : g \in H^s, h \in H^t$ such that $f = gh\}$.*

SOLUTION: By Hölder's inequality (Appendix A), if $g \in H^s(\mathbb{D})$, $h \in H^t(\mathbb{D})$ then $f = gh \in \mathrm{Hol}(\mathbb{D})$ and for every ρ, $0 < \rho < 1$, we have $\|f_\rho\|_r \leq \|g_\rho\|_s \|h_\rho\|_t$, which implies $f \in H^r(\mathbb{D})$ and $\|f\|_r \leq \|g\|_s \|h\|_t$. Conversely, if $f \in H^r(\mathbb{D})$, with $f = \lambda BV[f]$ its canonical factorization, then by setting $g = \lambda BV[f]^{r/s}$, $h = [f]^{r/t}$, we obtain $f = gh$ and $\|f\|_r = \|g\|_s \|h\|_t$. $\blacksquare$

(b) *Let $\lambda \in \mathbb{D}$ and let φ_λ be an evaluation functional on H^p, $1 \leq p \leq \infty$, i.e.*

$$\varphi_\lambda(f) = f(\lambda), \quad f \in H^p.$$

Show that $\|\varphi_\lambda\| = (1 - |\lambda|^2)^{-1/p}$.

SOLUTION: When $p = 2$, $\varphi_\lambda(f) = f(\lambda) = \sum_{k \geq 0} \hat{f}(k) \lambda^k = (f, k_\lambda)_{H^2}$, where

$$k_\lambda(z) = \sum_{k \geq 0} \overline{\lambda}^k z^k \quad (z \in \mathbb{D})$$

is the *Szegő reproducing kernel* of H^2, hence $\|\varphi_\lambda\| = \|k_\lambda\|_2 = (1 - |\lambda|^2)^{-1/2}$. When p is arbitrary, recall that, for every function f, $|f(\lambda)| \le |[f](\lambda)|$ and $\|f\|_p = \||f|\|_p = \|[f]^{p/2}\|_2^{2/p}$, which leads to

$$\|\varphi_\lambda\| = \sup\{|f(\lambda)| : f \in H^p, \|f\|_p \le 1\} = \sup\{|[f]^{p/2}(\lambda)|^{2/p} : \|[f]^{p/2}\|_2 \le 1\}$$
$$= ((1 - |\lambda|^2)^{-1/2})^{2/p}. \qquad \blacksquare$$

(c) (Neuwirth and Newman, 1967). *Let $f \in H^p(\mathbb{D})$, $p > 0$. Show that $f = $ constant if one of the following hypotheses is verified:*

(i) *$p \ge 1$ and $f(\zeta)$ is real a.e. $\zeta \in \mathbb{T}$,*
(ii) *$p \ge 1/2$ and $f(\zeta) \ge 0$ a.e. $\zeta \in \mathbb{T}$.*

Show that the conclusion no longer holds if $p < 1$ (respectively, $p < 1/2$).

SOLUTION: Case (i) is evident, because in this case, $f, \overline{f} \in H^1(\mathbb{T})$, which implies $f = $ constant.

Case (ii) and Theorem (2.6.5) imply $f = Bg^2$ with $g \in H^1(\mathbb{T})$, which can be written $f = |f| = |g^2|$; hence $Bg = \overline{g}$ a.e. on $\mathbb{T}$. By the last equation, the function $Bg = \overline{g}$ and its complex conjugate are in $H^1(\mathbb{T})$, thus $g = $ constant, $Bg = $ constant, and hence $f = $ constant.

For the last assertion, consider the functions $f_1 = i(1 + z)/(1 - z)$ (respectively $f_2 = f_1^2$): it is easy to see that $f_1 \in H^p(\mathbb{D})$ for any $p < 1$, and $f_2 \in H^p(\mathbb{D})$ for $p < 1/2$. $\qquad \blacksquare$

Donald Newman (1930–2007), an American mathematician, was one of the great experts in "hard analysis" of the second half of the twentieth century. After submitting his thesis in 1958 at Harvard, he worked with Leon Ehrenpreis, John Nash (1994 Nobel Prize for game theory), Harold Shapiro, and many others. Several pages of the book *A Beautiful Mind* by Sylvia Nasar are devoted to Newman's friendship with Nash. Newman published more than 180 articles and five books, among which is a gem of number theory, *Analytic Number Theory* (1998). As one of the best problem solvers, he also published more than 190 notes and articles in the section "Problems and Solutions" in *American Mathematical Monthly*. He was an unequaled master of elegant and refined proofs, with a knack for clear thinking. To mention only two of his masterworks, let us cite his proof of the *Wiener $1/f$ theorem* (a model of clarity and transparency), or his celebrated proof of the theorem of distribution of prime numbers (even an expository article – by Don Zagier – explaining the latter proof to the general public received a prestigious prize: the 2000 MAA Writing Awards). Certain of Newman's discoveries (sometimes without signature or precise attribution) revolutionized the disciplines concerned: his theorem of the rational approximation of $|x|$ renewed approximation theory, as did his contribution to the famous "Corona theorem" for complex analysis (see the striking testimony of his friend Leon Ehrenpreis in an "In Memoriam" article published in *Journal of Approximation Theory*, vol. 154 (2008), p. 39). To sum up his philosophy of mathematics, let us quote his words: "Some of us are old enough to remember ... the days when Math was fun ...," and his daily greeting to his friends, "I have a problem ..." instead of "Hello."

According to one of his friends, "Newman is the Vivaldi of Mathematics."

2.8.3 The Hilbert and Hardy Inequalities

(a) *Let $f, g \in H^2$ and $h = fg$. Show that $|\hat{h}(n)| \leq \sum_{k+j=n} |\hat{f}(k)| \cdot |\hat{g}(j)|$.*

SOLUTION: The Fourier series $g = \sum_{j \in \mathbb{Z}} \hat{g}(j) z^j$ converges in $L^2(\mathbb{T})$, hence by Cauchy's inequality (Appendix A), the series $h = fg = \sum_{j \in \mathbb{Z}} \hat{g}(j) f z^j$ converges in $L^1(\mathbb{T})$, and by continuity of $h \mapsto \hat{h}(n)$, we obtain $\hat{h}(n) = \sum_{j \in \mathbb{Z}} \hat{f}(n-j) \hat{g}(j)$: the result follows. ∎

(b) *Let $\varphi(e^{it}) = i(t - \pi)$ for $0 < t < 2\pi$. Find the Fourier coefficients of φ.*

SOLUTION: $\hat{\varphi}(0) = 0$, and for $k \neq 0$,

$$\hat{\varphi}(k) = \int_0^{2\pi} i(t - \pi) e^{-ikt} dt/2\pi$$

$$= [-(t - \pi) e^{-ikt}/2\pi k]_{t=0}^{t=2\pi} + \int_0^{2\pi} e^{-ikt} dt/2\pi k = -1/k.$$ ∎

(c) The Hilbert inequality, 1908. *Let $f, g \in H^2$. Show that*

$$\left| \sum_{k,j \geq 0} \frac{\hat{f}(k)\hat{g}(j)}{k + j + 1} \right| \leq \pi \|f\|_2 \|g\|_2.$$

Hint The double sum converges absolutely and coincides with $(\varphi f, \bar{z} \cdot \bar{g})$, where φ is defined in (b).

SOLUTION: For $F, G \in L^2(\mathbb{T})$ and $\Phi \in L^\infty(\mathbb{T})$, just as in (a) above, we have $(\Phi F, \bar{G}) = \sum_{i+j+k=0} \widehat{\Phi}(i) \widehat{F}(k) \widehat{G}(j)$, which gives

$$(\varphi f, \overline{zg}) = -\sum_{k,j \geq 0} \frac{\hat{f}(k)\hat{g}(j)}{k + j + 1}.$$

Then the result follows from

$$|(\varphi f, \overline{zg})| \le \|\varphi f\|_2 \, \|\overline{zg}\|_2 \le \|\varphi\|_\infty \, \|f\|_2 \, \|g\|_2 = \pi \, \|f\|_2 \, \|g\|_2.$$

∎

(d) The Hardy inequality, 1926. *Deduce from (a), (c) and § 2.8.2(a) that for every function $h \in H^1$,*

$$\sum_{k \ge 0} \frac{|\hat{h}(k)|}{k + 1} \le \pi \|h\|_1.$$

SOLUTION: By § 2.8.2(a), $h = fg$ with $f, g \in H^2$ and $\|f\|_2^2 = \|g\|_2^2 = \|h\|_1$, and by § 2.8.3(a, c),

$$\sum_{k \ge 0} \frac{|\hat{h}(k)|}{k + 1} \le \sum_{k \ge 0} \frac{\sum_{i+j=k} |\hat{f}(i)| \cdot |\hat{g}(j)|}{k + 1} \le \pi \, \|f\|_2 \, \|g\|_2 = \pi\|h\|_1.$$

∎

(e) *Show that the constant π is sharp in (c) and (d).*

Hint Consider $h_\epsilon(z) = (1 - z)^{-1+\epsilon}$, where $\epsilon > 0$.

SOLUTION: It suffices to verify that $\sup\{\sigma(h) : \|h\|_1 \le 1\} \ge \pi$ with

$$\sigma(h) = \sum_{k \ge 0} \frac{|\hat{h}(k)|}{k + 1}.$$

For $\epsilon > 0$, set

$$h_\epsilon(z) = (1 - z)^{-1+\epsilon} = \sum_{k \ge 0} \binom{k}{-1 + \epsilon}(-z)^k \quad (z \in \mathbb{D}),$$

where

$$\binom{k}{\alpha} = \frac{\alpha(\alpha - 1)\dots(\alpha - k + 1)}{k!}$$

is a binomial coefficient. By a familiar asymptotic estimation of $\binom{k}{-1+\epsilon}$, when $\epsilon \to 0$, uniformly in k, we obtain

$$\log\left|\binom{k}{-1 + \epsilon}\right| = \sum_{j=1}^{k} \log\left|1 - \frac{\epsilon}{j}\right| = \sum_{j=1}^{k} \frac{-\epsilon}{j} + O(\epsilon^2)$$

$$= -\epsilon(\log(k + 1) + O(1)) + O(\epsilon^2),$$

hence $\left|\binom{k}{-1+\epsilon}\right| = (k + 1)^{-\epsilon} e^{O(\epsilon)}$. When $\epsilon \to 0$, this leads to

$$\sigma(h_\epsilon) = e^{O(\epsilon)} \sum_{k \ge 0} \frac{(k + 1)^{-\epsilon}}{k + 1} \sim \int_1^\infty \frac{dx}{x^{1+\epsilon}} = \frac{1}{\epsilon}.$$

Moreover, it is easy to see that

$$\|h_\epsilon\|_1 = \int_{\mathbb{T}} |1 - z|^{-1+\epsilon} \, dm(z) \sim \int_{-\pi}^{\pi} |t|^{-1+\epsilon} \, dt/2\pi = \pi^\epsilon/\epsilon\pi,$$

and hence $\overline{\lim}_{\epsilon \to 0} \, \sigma(h_\epsilon)/\|h_\epsilon\|_1 \ge \pi$.

∎

(f) *Show that there exists a function $f \in L^1(\mathbb{T})$ such that*

$$\sum_{k \geq 0} \frac{|\hat{f}(k)|}{k+1} = \infty.$$

Hint For every positive convex sequence $(a_n)_{n \geq 0}$ tending to zero, there exists a function $f \in L^1(\mathbb{T})$ such that $\hat{f}(n) = a_{|n|}$ for every $n \in \mathbb{Z}$: see W. H. Young, 1913, and Zygmund (1959, Ch. 5, § 1).

SOLUTION: It suffices to set $a_n = 1/\log(e(n+1))$ in Young's theorem. ∎

(g) *Show that the Riesz projection defined on the trigonometric polynomials by the formula*

$$P_+ \left(\sum_{k \in \mathbb{Z}} a_k z^k \right) = \sum_{k \geq 0} a_k z^k$$

is bounded neither in the space $L^1(\mathbb{T})$ nor in $L^\infty(\mathbb{T})$.

SOLUTION: For $L^1(\mathbb{T})$, it suffices to compare (f) and (d) above; for $L^\infty(\mathbb{T})$, it is easy to see that P_+ is *self-adjoint* with respect to the sesquilinear duality $(L^1(\mathbb{T}))^* = L^\infty(\mathbb{T})$, $\langle f, g \rangle = \int_{\mathbb{T}} f\bar{g}\, dm$, i.e. $(P_+|L^1(\mathbb{T}))^* = P_+|L^\infty(\mathbb{T})$. The result follows. ∎

David Hilbert (1862–1943) was a German mathematician whose works had a decisive influence on all the mathematics of the twentieth century, from the choice of problems to be tackled, to the development of the technical tools. His contribution to the theories of invariants, geometry, integral equations and functional analysis, mathematical physics, the calculus of variations, and mathematical logic was fundamental and profoundly innovative. His plenary conference at the second International Congress of

Mathematicians (Paris, 1900) contained 23 unsolved problems across all these domains, and was decisive for the development of mathematics for decades to come. Several disciplines were quite simply created by David Hilbert, including integral equations and functional analysis. In particular, the founding course on integral equations that he gave at the University of Göttingen in 1905–1908 (first published by his student Hermann Weyl (1908) and then by Hilbert himself (Hilbert, 1912)) contained what is known today as *Hilbert spaces*, the *Hilbert transform*, and the *Hilbert inequality* (theorem of § 2.8.2(c) in this text). His 69 students, i.e. those who submitted a thesis under his supervision, include stars of twentieth-century mathematics such as Bernstein, Blumenthal, Courant, Haar, Hecke, Hellinger, Schmidt, and Steinhaus. Many others, without having formally been his doctoral students, spent long periods at Hilbert's seminars in Göttingen: these include Harald Bohr, Born, Lasker (chess champion), von Neumann, Toeplitz, Weyl, Zermelo, and dozens of others.

There are innumerable anecdotes about Hilbert. His first article on invariants, which contained among other results the celebrated *Nullstellensatz*, was rejected by Paul Gordan, a referee for the *Mathematischen Annalen*, who said of his proof of the (pure) existence of a finite number of generators: *Das ist nicht Mathematik. Das ist Theologie* ("That is not mathematics. That is theology.") Hilbert's own credo was: "A perfect formulation of a problem is already half of its solution" (reported by Constance Reid in her biography *Hilbert* (Reid, 1970)). Then there is the famous final motto of his retirement ceremony (1930), *Wir müssen wissen, wir werden wissen* ("We must know, we will know"), on the eve (!) of the announcement by Kurt Gödel of his incompleteness theorem concerning the Zermelo–Fraenkel axioms.

2.8.4 Harmonic Conjugates and the Riesz Projection (1927), Following Calderón (1950)

(a) *Let $1 \le p \le \infty$ and set*

$$h^p(\mathbb{D}) = \{u : u \text{ harmonic function on } \mathbb{D}, \text{ such that } \sup_{0 < r < 1} \|u_r\|_p < \infty\},$$

with $u_r(\zeta) = u(r\zeta)$. Show that the mapping $h \mapsto h * P_z$ is an isometric isomorphism of $L^p(\mathbb{T})$ onto $h^p(\mathbb{D})$, with $h * P_z = h * P_r(\zeta)$ for $z = r\zeta \in \mathbb{D}$, $\zeta \in \mathbb{T}$.

SOLUTION: Use a slight modification of the reasoning of Theorem 2.2.2. ∎

(b) *Let $1 < p \leq 2$. Show that*

$$1 \leq \beta(\cos(t))^p - \alpha(\cos(pt))$$

for every $t \in [-\pi/2, \pi/2]$, with $\beta = \beta(p) := \alpha^p(1 + \alpha)$, $\alpha = 1/\cos(\delta)$ and $\delta = \pi/(p + 1)$.

SOLUTION: Denote $f(t) = \beta(\cos(t))^p - \alpha(\cos(pt))$. Then f is an even function and, for $0 \leq t \leq \delta$, we have $f(t) \geq \beta(\cos(\delta))^p - \alpha = 1 + \alpha - \alpha = 1$. For $\delta \leq t \leq \pi/2$ we have $f(t) \geq -\alpha \cdot \cos(pt) \geq -\alpha \cdot \cos(p\delta) = -\alpha \cdot \cos(\pi - \delta) = \alpha \cdot \cos(\delta) = 1$. ∎

(c) *Let $1 < p < \infty$, $h \in L^p(\mathbb{T})$ and*

$$\Gamma h(z) = \int_{\mathbb{T}} \frac{\zeta + z}{\zeta - z} h(\zeta)\, dm(\zeta), \quad z \in \mathbb{D},$$

the Herglotz transform of h. Show that

(i) *if $h \geq 0$ a.e. on $\mathbb{T}$, then $\mathrm{Re}(\Gamma h(z)) \geq 0$,*
(ii) *for $1 < p \leq 2$, $0 < r < 1$ and $h \geq 0$ a.e. on $\mathbb{T}$,*
$$|(\Gamma h)_r|^p \leq \beta(\mathrm{Re}((\Gamma h)_r))^p - \alpha\,\mathrm{Re}((\Gamma h)_r^p) \text{ (α and β are defined in (b) above),}$$
(iii) *for every function $h \in L^p(\mathbb{T})$, we have $\Gamma h \in H^p(\mathbb{D})$ and*

$$\|\Gamma h\|_p \leq A_p \|h\|_p,$$

with $A_p = 4\beta(p)^{1/p}$ for $1 < p \leq 2$ and $A_p = 4\beta(p')^{1/p'}$, $1/p + 1/p' = 1$ for $2 < p < \infty$.

SOLUTION: (i) We have

$$\mathrm{Re}(\Gamma h(z)) = \int_{\mathbb{T}} \frac{1 - |z|^2}{|\zeta - z|^2} h(\zeta)\, dm(\zeta) = h * P_z \geq 0$$

since $h \geq 0$.

(ii) For $\zeta \in \mathbb{T}$ and $h \geq 0$ a.e. on $\mathbb{T}$, we have $(\Gamma h)_r(\zeta) = |(\Gamma h)_r(\zeta)|e^{it}$ with $|t| \leq \pi/2$. Hence inequality (b) gives

$$|(\Gamma h)_r(\zeta)|^p \leq |(\Gamma h)_r(\zeta)|^p(\beta(\cos(t))^p - \alpha(\cos(pt)))$$
$$= \beta(\mathrm{Re}((\Gamma h)_r(\zeta)))^p - \alpha\,\mathrm{Re}((\Gamma h)_r^p(\zeta)).$$

(iii) For $1 < p \leq 2$ and $h \geq 0$, by integrating the last inequality and using

$$\|\mathrm{Re}((\Gamma h)_r)\|_p = \|h * P_r\|_p \leq \|h\|_p,$$

we obtain $\|(\Gamma h)_r\|_p^p \leq \beta\|\mathrm{Re}((\Gamma h)_r)\|_p^p - \alpha\|h\|_1^p \leq \beta\|\mathrm{Re}((\Gamma h)_r)\|_p^p \leq \beta\|h\|_p^p$; a generic function $h \in L^p(\mathbb{T})$ is a combination $h = h_1 - h_2 + ih_3 - ih_4$ with $0 \leq h_j \leq |h|$, thus $\|\Gamma h\|_p \leq 4\beta^{1/p}\|h\|_p$.

The case $2 < p < \infty$ follows from the duality $(L^p)^* = L^{p'}$ (see Appendix D): indeed, it is easy to verify that for every $0 < r < 1$ and $g \in L^{p'}(\mathbb{T})$, we have

$$\int_{\mathbb{T}} (\Gamma h)_r g(\overline{z}) \, dm(z) = \int_{\mathbb{T}} (\Gamma g)_r h(\overline{z}) \, dm(z).$$

Therefore, by Hölder and $1 < p' < 2$,

$$\left| \int_{\mathbb{T}} (\Gamma h)_r g(\overline{z}) \, dm(z) \right| \leq \|(\Gamma g)_r\|_{p'} \|h\|_p \leq 4\beta(p')^{1/p'} \|g\|_{p'} \|h\|_p.$$

Then, by passing to the sup on $\|g\|_{p'} \leq 1$, we obtain $\|(\Gamma h)_r\|_p \leq 4\beta(p')^{1/p'} \|h\|_p$.   ∎

(d) Harmonic conjugate according to Marcel Riesz (1927). *Let $1 < p < \infty$. Show that for every real function $u \in L^p(\mathbb{T})$, there exists one and only one real function $Hu \in L^p(\mathbb{T})$ such that $\widehat{Hu}(0) = 0$ and $u + iHu \in H^p$. Moreover, $\|Hu\|_p \leq A_p\|u\|_p$, where A_p is defined in part (iii) of (c) above.*

SOLUTION: The uniqueness follows from § 2.8.2(c). Then, in (c) above, set $h = u$ and $Hu = \text{Im}(\Gamma u)$. This completes the proof.   ∎

(e) The Riesz projection on H^p. *Let $1 < p < \infty$. Show that the mapping*

$$P_+\left(\sum_{k \in \mathbb{Z}} a_k z^k \right) = \sum_{k \geq 0} a_k z^k$$

is well-defined and bounded $L^p(\mathbb{T}) \to L^p(\mathbb{T})$ with $\|P_+\| \leq A_p$ where A_p is defined in part (iii) of (c) above. Thus

$$P_+ L^p(\mathbb{T}) = H^p(\mathbb{T}) \quad and \quad L^p(\mathbb{T}) = H^p(\mathbb{T}) + H^p_-(\mathbb{T}),$$

where $H^p_-(\mathbb{T}) = \{ f \in L^p(\mathbb{T}) : \hat{f}(k) = 0 \text{ for } k \geq 0 \}$.

SOLUTION: For $z \in \mathbb{D}$ and $h \in L^p(\mathbb{T})$, we have

$$\Gamma h(z) = \int_{\mathbb{T}} \left(1 + \frac{2z}{\zeta - z} \right) h(\zeta) \, dm(\zeta) = \hat{h}(0) + 2z \int_{\mathbb{T}} \frac{\overline{\zeta}}{1 - \overline{\zeta}z} h(\zeta) \, dm(\zeta)$$

$$= \hat{h}(0) + 2z \sum_{k \geq 0} z^k \int_{\mathbb{T}} \overline{\zeta}^{k+1} h(\zeta) \, dm(\zeta)$$

$$= \hat{h}(0) + 2z \sum_{k \geq 0} z^k \hat{h}(k+1) = 2P_+ h(z) - \hat{h}(0),$$

which implies $P_+ h \in H^p$ and $\|P_+ h\|_p \leq (\|\Gamma h\|_p + \|h\|_p)/2 \leq (1 + A_p/2)\|h\|_p \leq A_p\|h\|_p$. The decomposition $L^p(\mathbb{T}) = H^p(\mathbb{T}) + H^p_-(\mathbb{T})$ is given by $f = P_+ f + (I - P_+)f$.   ∎

(f) *Let $1 < p < \infty$ and $f \in L^p(\mathbb{T})$. Show that the Fourier series $\sum_{k \in \mathbb{Z}} \hat{f}(k)z^k$ converges to f for the $L^p(\mathbb{T})$ norm, and hence that (z^k) is a Schauder basis of $L^p(\mathbb{T})$.*

SOLUTION: Let $P_{m,n}f = \sum_{m \leq k < n} \hat{f}(k)z^k$ and $P_m = z^m P_+ \bar{z}^m$. Then $P_{m,n} = (1 - P_n)P_m$ (to be verified!), $\|P_m\| = \|P_+\|$, and $\|I - P_n\| = \|P_+\|$ (for the last equation use $\|f(z)\|_p = \|f(\bar{z})\|_p$ for every function f). Hence

$$\|P_{m,n}\| \leq \|P_+\|^2.$$

As clearly the Fourier series converges for the trigonometric polynomials $f \in \mathcal{P}$ ($\lim_{m,n \to \infty} \|f - P_{m,n}f\|_p = 0$), the rest follows from the Banach–Steinhaus *uniform boundedness principle*. $\blacksquare$

(g) *Let $p = 1$ or $p = \infty$. Show that P_+ is not bounded on $L^p(\mathbb{T})$ and that (z^k) is not a basis of $L^p(\mathbb{T})$. In particular, $L^p(\mathbb{T}) \neq H^p(\mathbb{T}) + H^p_-(\mathbb{T})$ for these values of p. (For $p = \infty$, see also (i) below for an improvement.)*

SOLUTION: By (e) above, clearly P_+ is bounded if and only if the harmonic conjugate H is bounded. Moreover, setting $u = \arg(1 - z)$, we obtain $Hu = \log|1 - z|$, and hence $u \in L^\infty(\mathbb{T})$ while $Hu \notin L^\infty(\mathbb{T})$, and the result follows for $L^\infty(\mathbb{T})$.

For $p = 1$, P_+ is not bounded by duality. The inequality $L^p(\mathbb{T}) \neq H^p(\mathbb{T}) + H^p_-(\mathbb{T})$ ensues from the closed graph theorem: if we had $L^p(\mathbb{T}) = H^p(\mathbb{T}) + H^p_-(\mathbb{T})$, then P_+ would be well-defined and closed (to be verified), and hence bounded, which is not the case. $\blacksquare$

(h) *Show that the norm $\|P_+\|_r$, $0 < r < \infty$, has linear growth as $r \to 1$ or $r \to \infty$, i.e. that there exist constants $a, b > 0$ such that*

$$\frac{a}{r - 1} \leq \|P_+\|_r = \|P_+\|_{r'} \leq \frac{b}{r - 1}$$

when $r \to 1$ and $ar \leq \|P_+\|_r = \|P_+\|_{r'} \leq br$ when $r \to \infty$.

SOLUTION: By the duality $\|P_+\|_r = \|P_+\|_{r'}$, and by (e) and (b) above, clearly for $r \to 1$ we have

$$\|P_+\|_r \leq A_r \sim \frac{4}{\pi(r - 1)}.$$

Moreover, for $r \to \infty$, using the example of (g), we have

$$\|P_+\|_r \geq c\big\|\log|1 - z|\big\|_r \sim d\left(\int_0^1 (\log(1/t))^r \, dt\right)^{1/r}$$

$$= d\left(\int_0^\infty s^r e^{-s} \, ds\right)^{1/r} = d(\Gamma(r + 1))^{1/r}$$

where $c, d > 0$ are constants and Γ is the Euler Γ function. Then, by Stirling's formula, we have $(\Gamma(r + 1))^{1/r} \sim ((r/e)^r \sqrt{2\pi r})^{1/r} \sim r/e$, and the result follows. $\blacksquare$

(i) (Newman, 1962). *Show that $L^\infty(\mathbb{T}) \neq \operatorname{clos}_{L^\infty}(H^\infty(\mathbb{T}) + H^\infty_-(\mathbb{T}))$.*

SOLUTION: First show that there exists $C > 0$ such that, for every function $f \in L^\infty(\mathbb{T})$, we have

$$|\Gamma f(r)| \leq C\|f\|_\infty \log(1-r)^{-1}, \quad 0,5 < r < 1,$$

where Γ is the Herglotz operator from (c) above: indeed, by the definition,

$$|\Gamma f(r)| \leq \|f\|_\infty \int_\mathbb{T} |(\zeta + r)/(\zeta - r)|\, dm(\zeta)$$

$$\leq 4\|f\|_\infty \int_0^\pi |e^{it} - r|^{-1} dt/2\pi = (2\|f\|_\infty/\pi)\left(\int_0^{1-r} + \int_{1-r}^\pi\right).$$

For the integrals, we have

$$\int_0^{1-r} |e^{it} - r|^{-1} dt \leq \int_0^{1-r} (1-r)^{-1}\, dt = 1,$$

$$\int_{1-r}^\pi |e^{it} - r|^{-1}\, dt = \int_{1-r}^\pi (1 + r^2 - 2r\cos(t))^{-1/2}\, dt$$

$$= \int_{1-r}^\pi ((1-r)^2 + 4r\sin^2(t/2))^{-1/2}\, dt$$

$$\leq \int_{1-r}^\pi (2\sqrt{r}\sin(t/2))^{-1}\, dt$$

$$\leq \int_{1-r}^\pi (2\sqrt{r}t/\pi)^{-1}\, dt$$

$$= (2\sqrt{r}/\pi)^{-1} \log(\pi/1 - r),$$

and thus

$$|\Gamma f(r)| \leq \|f\|_\infty(2/\pi + r^{-1/2}\log(\pi/1 - r)),$$

hence the stated estimate. Consequently, for every function $f \in \operatorname{clos}_{L^\infty}(H^\infty(\mathbb{T}) + H^\infty_-(\mathbb{T}))$, we have $|\Gamma f(r)| = o(\log(1-r)^{-1})$ when $r \to 1$: indeed, for $g \in H^\infty(\mathbb{T})$ we have $\Gamma g = 2g - g(0)$ and for $h \in H^\infty_-(\mathbb{T}) - \Gamma h = 0$, hence for every function g and h we obtain

$$\varlimsup_{r\to 1} |\Gamma f(r)|/\log(1-r)^{-1} = \varlimsup_{r\to 1} |\Gamma(f - g - h)(r)|/\log(1-r)^{-1} \leq C\|f - g - h\|_\infty.$$

Since the last norm can be arbitrarily small, we obtain

$$\varlimsup_{r\to 1} |\Gamma f(r)|/\log(1-r)^{-1} = 0.$$

Moreover, clearly there exists $u \in L^\infty(\mathbb{T})$ satisfying $\varlimsup_{r\to 1} |\Gamma u(r)|/\log(1-r)^{-1} > 0$: for example, $u = \arg(1 - z)$ for which $\Gamma u = \log(1 - z)$ (because $\operatorname{Re}(\log(1-z)) = u$ and $\Gamma u(0) = 0$), and the result follows. ∎

(j) (Kolmogorov, 1925). *Let $u \in L^\infty(\mathbb{T})$ be a real function. Show that $Hu \in L^p(\mathbb{T})$ for every $p < \infty$, and even $e^{tHu} \in L^1(\mathbb{T})$ if $t\|u\|_\infty < \pi/2$.*

Hint Use Smirnov's theorem from § 3.1.1(d): no vicious circle is lurking, because this § 2.8.4(j) will be used only in § 4.6.

SOLUTION: Set $f = \exp(-i(u + iHu))$. Then, by § 3.1.1(d) we have $f \in H^t$ for every t such that $0 < t < \pi/(2\,\|u\|_\infty)$, hence $|f|^t = e^{tHu} \in L^1(\mathbb{T})$. ∎

2.8.5 The Kolmogorov Weak Type Inequality

Another important property of the harmonic conjugate transform H (and of the Riesz projection P_+) is an inequality due to Kolmogorov (1925) for the distribution function of Hf, known as a weak type inequality (in contrast with strong type inequalities concerning the norm of Hf). The proof below, of incomparable transparency and elegance, is attributed to Lennart Carleson and Yitzhak Katznelson (1968) and to Otar Tsereteli (1976).

(a) H and P_+ are weak type operators $L^1 \longrightarrow L^1$ (Kolmogorov, 1925). *Let* $u \in L^1(\mathbb{T})$ *and*

$$\lambda_{Hu}(t) = m\left(\{\zeta \in \mathbb{T} : |Hu(\zeta)| \geq t\}\right), \quad t > 0.$$

Show that

$$\lambda_{Hu}(t) \leq \frac{8\sqrt{2}}{t}\|u\|_{L^1(\mathbb{T})}$$

(respectively $\leq (4/t)\|u\|_{L^1(\mathbb{T})}$ if u is real, and $\leq (2/t)\|u\|_{L^1(\mathbb{T})}$ if $u \geq 0$).

SOLUTION: First, suppose that $u \geq 0$ and $\|u\|_1 := \|u\|_{L^1(\mathbb{T})} = \int_{\mathbb{T}} u\,dm = 1$. Then $Hu = \operatorname{Im}(\Gamma u)$ is well-defined a.e. on $\mathbb{T}$ (see for example § 3.1.1(d)) and $\operatorname{Re}(\Gamma u) \geq 0$. For any $t > 0$, set $\varphi_t = 1 + (\Gamma u - t)/(\Gamma u + t)$. Note that $\operatorname{Re}(\varphi_t) \geq 0$ a.e. on $\mathbb{T}$ and that

$$|Hu(\zeta)| = |\operatorname{Im}(\Gamma u(\zeta))| \geq t \Rightarrow \operatorname{Re}(\varphi_t(\zeta)) = 1 + \frac{|\Gamma u(\zeta)|^2 - t^2}{|\Gamma u(\zeta) + t|^2} \geq 1,$$

hence

$$\lambda_{Hu}(t) \leq \lambda_{\operatorname{Re}(\varphi_t)}(1) \leq \int_{\mathbb{T}} \operatorname{Re}(\varphi_t)\,dm = \operatorname{Re}(\varphi_t(0)) = \frac{2}{1+t} \leq \frac{2}{t}.$$

By homogeneity, we have $\lambda_{Hu}(t) \leq (2/t)\|u\|_1$ for every $u \geq 0$. A real function u can be written $u = u_1 - u_2$ where $u_j \geq 0$ and $\|u_1\|_1 + \|u_2\|_1 = \|u\|_1$. Moreover, we have

$$\{\zeta \in \mathbb{T} : |Hu(\zeta)| \geq t\} \subset \{\zeta \in \mathbb{T} : |Hu_1(\zeta)| \geq t/2\} \cup \{\zeta \in \mathbb{T} : |Hu_2(\zeta)| \geq t/2\},$$

hence $\lambda_{Hu}(t) \leq (4/t)\|u\|_1$. Now, a general function $u \in L^1(\mathbb{T})$ can be written $u = u_1 + iu_2$ with u_j real, thus $\|u_1\|_1 + \|u_2\|_1 \leq \sqrt{2}\|u\|_1$, and the result follows. ∎

(b) *Deduce that $H(L^1(\mathbb{T})) \subset \bigcap_{0<p<1} L^p(\mathbb{T})$.*

> Solution: This is immediate by
>
> $$\int_{\{|Hu|>1\}} |Hu|^p \, dm = p \int_1^\infty t^{p-1} \lambda_{Hu}(t) \, dt$$
>
> (see Appendix A). ∎

2.8.6 The Littlewood Subordination Principle (1925)

Let $\omega \in H^\infty$, $\|\omega\|_\infty \le 1$, and let $C_\omega f = f \circ \omega$ be a *composition operator*.

(a) *Let $f \in \mathrm{Hol}(\mathbb{D})$ and let S^* be the backward shift, $S^* f = (f - f(0))/z$. Show that $f = \hat{f}(0) + z S^* f$.*

> Solution: This is trivial. ∎

(b) *Suppose $\omega(0) = 0$, and let f be a polynomial $f \in \mathcal{P}_a$, $\deg(f) \le n$. Show that $\|C_\omega f\|_2^2 \le |\hat{f}(0)|^2 + \|C_\omega(S^* f)\|_2^2$, and then – by induction – that $\|C_\omega f\|_2^2 \le \sum_{k=0}^n |\hat{f}(k)|^2 = \|f\|_2^2$.*

> Solution: By (a), $C_\omega f = C_\omega(\hat{f}(0)) + C_\omega(z S^* f) = \hat{f}(0) + \omega C_\omega(S^* f)$, thus
>
> $$\|C_\omega f\|_2^2 = |\hat{f}(0)|^2 + \|\omega C_\omega(S^* f)\|_2^2 \le |\hat{f}(0)|^2 + \|C_\omega(S^* f)\|_2^2.$$
>
> By applying the inequality to $S^* f$, we obtain
>
> $$\|C_\omega f\|_2^2 \le |\hat{f}(0)|^2 + |\widehat{S^* f}(0)|^2 + \|C_\omega(S^{*2} f)\|_2^2 \le |\hat{f}(0)|^2 + |\hat{f}(1)|^2 + \|C_\omega(S^{*2} f)\|_2^2,$$
>
> and hence by iteration,
>
> $$\|C_\omega f\|_2^2 \le \sum_{k=0}^n |\hat{f}(k)|^2 + \|C_\omega(S^{*(n+1)} f)\|_2^2 = \sum_{k=0}^n |\hat{f}(k)|^2 = \|f\|_2^2. \qquad \blacksquare$$

(c) *Using (b) and the canonical factorization, show that if $\omega(0) = 0$ then $C_\omega(H^p) \subset H^p$ and $\|C_\omega : H^p \to H^p\| = 1$ for every p, $1 \le p \le \infty$.*

> Solution: Clearly $\|C_\omega : H^p \to H^p\| \ge 1$ because $C_\omega 1 = 1$. Moreover, the result of (b) and the density of $\mathcal{P}_a$ in H^2 show that $C_\omega(H^2) \subset H^2$ and $\|C_\omega : H^2 \to H^2\| \le 1$. Then, if $f \in H^p$, we have $f = f_{in} f_{out}$ and hence
>
> $$\|C_\omega f\|_p = \|(f_{in} \circ \omega)(f_{out} \circ \omega)\|_p \le \|f_{out} \circ \omega\|_p = \left(\int_\mathbb{T} |(f_{out} \circ \omega)^{p/2}|^2 \, dm\right)^{1/p}$$
>
> $$\le \left(\int_\mathbb{T} |f_{out}^{p/2}|^2 \, dm\right)^{1/p} = \|f\|_p. \qquad \blacksquare$$

(d) *Let $b_\lambda = (\lambda - z)/(1 - \bar\lambda z)$, where $\lambda \in \mathbb{D}$. Show that*

$$\|C_{b_\lambda}\colon H^p \to H^p\| = \|b'_\lambda\|_\infty^{1/p} = \left(\frac{1+|\lambda|}{1-|\lambda|}\right)^{1/p}.$$

SOLUTION: We have $b_\lambda \circ b_\lambda = \mathrm{id}$ (to be verified!). Thus the mapping $b_\lambda\colon \mathbb{T} \to \mathbb{T}$ is bijective, with the Jacobian

$$|b'_\lambda(\zeta)| = \frac{1 - |\lambda|^2}{|1 - \bar\lambda\zeta|^2}.$$

Hence

$$\|f \circ b_\lambda\|_p^p = \int_\mathbb{T} |f \circ b_\lambda|^p \, dm = \int_\mathbb{T} |f|^p |b'_\lambda| \, dm \leq \|b'_\lambda\|_\infty \|f\|_p^p,$$

therefore $\|C_{b_\lambda}\colon H^p \to H^p\| \leq \|b'_\lambda\|_\infty^{1/p}$. Then

$$\|C_{b_\lambda}\colon H^p \to H^p\|^p = \sup\left\{\int_\mathbb{T} |f|^p |b'_\lambda| \, dm\colon f \in H^p, \int_\mathbb{T} |f|^p \, dm \leq 1\right\}$$

$$\geq \sup\left\{\int_\mathbb{T} h|b'_\lambda| \, dm\colon h \geq 0, \inf_\mathbb{T} h > 0, \int_\mathbb{T} h \, dm \leq 1\right\}$$

(the last inequality follows from the existence, for any such function h, of $f \in H^p$, with $|f|^p = h$: see Corollary 2.6.2). It is easy to see that the last "sup" equals $\|b'_\lambda\|_\infty$. $\blacksquare$

(e) Littlewood subordination principle (1925). *Show that for every $\omega \in H^\infty$, $\|\omega\|_\infty \leq 1$, we have*

$$\sup_{\lambda\in\mathbb{D}} \left(\frac{1-|\lambda|^2}{1-|\omega(\lambda)|^2}\right)^{1/p} \leq \|C_\omega\colon H^p \to H^p\| \leq \left(\frac{1+|\omega(0)|}{1-|\omega(0)|}\right)^{1/p}.$$

In the cases $\omega(0) = 0$ or $\omega = b_\lambda$ ($\lambda \in \mathbb{D}$), the two inequalities become equalities.

Hint For the upper bound, use $\omega = b_\lambda \circ b_\lambda \circ \omega$ with $\lambda = \omega(0)$. For the lower bound, use $C_\omega^* \varphi_\lambda = \varphi_{\omega(\lambda)}$ with φ_λ an evaluation functional of § 2.8.2(b).

SOLUTION: Let $\varphi = b_\lambda \circ \omega$. Then $\varphi(0) = 0$, and by (c) and (d), we have

$$\|C_\omega\colon H^p \to H^p\| = \|C_{b_\lambda} C_\varphi\| \leq \|C_{b_\lambda}\| \cdot \|C_\varphi\| = \|C_{b_\lambda}\| = \left(\frac{1+|\omega(0)|}{1-|\omega(0)|}\right)^{1/p}.$$

For the lower bound, we calculate the adjoint operator C_ω^* on the evaluation functionals $\varphi_\lambda \in (H^p)^*$: for every $f \in H^p$, $\langle C_\omega f, \varphi_\lambda \rangle = \langle f \circ \omega, \varphi_\lambda \rangle = f(\omega(\lambda)) = \langle f, \varphi_{\omega(\lambda)} \rangle$, thus $C_\omega^* \varphi_\lambda = \varphi_{\omega(\lambda)}$. Therefore

$$\|C_\omega\colon H^p \to H^p\| \geq \sup_{\lambda\in\mathbb{D}} \frac{\|\varphi_{\omega(\lambda)}\|}{\|\varphi_\lambda\|},$$

and an application of § 2.8.2(b) completes the verification. $\blacksquare$

John Edensor (J. E.) Littlewood (1885–1977), a British mathematician, played a key role in the development of analysis in Great Britain in the first half of the twentieth century. The principal collaborator of G. H. Hardy for more than three decades (with at least 100 articles published together!), he focused on the elaboration of the basic techniques of modern harmonic analysis. It suffices to recall the *Hardy–Littlewood maximal function*, the *Littlewood–Paley decomposition theory*, the *Littlewood subordination principle* (§ 2.8.6(e)), the Littlewood and Hardy–Littlewood Tauberian theorems, etc., not to mention the fundamental impact of Hardy and Littlewood in number theory (on the distribution of prime numbers and of the zeros of the zeta function).

In the 1930s, Hilbert created a ranking of British mathematicians: the first was Hardy, the second was Littlewood, and the third was Hardy–Littlewood.

Littlewood spent most of his life at the University of Cambridge, where he was unrivaled for the elegance and importance of his work, and his sharp sense of humor. There are hundreds of mathematical jokes and anecdotes attributed to Littlewood; part of this folklore is captured in his famous popular science book *A Mathematician's Miscellany* (1953). Here is an example. At the end of an article published in the *Comptes Rendus*, Littlewood added three notes at the bottom of the page (in French): " [1]I am greatly indebted to Prof. Riesz for translating the present paper. [2]I am indebted to Prof. Riesz for translating the preceding footnote. [3]I am indebted to Prof. Riesz for translating the preceding footnote." In *A Mathematician's Miscellany*, he added: "Actually I stop legitimately at number 3: however little French I know I am capable of *copying* a French sentence."

Littlewood's research students included Chowla, Collingwood, Davenport, Ingham, Ramanujan, and Spencer.

He is also known for his creative longevity: in an article published at the age of 85, Littlewood wrote that he had solved a problem that "raised difficulties which defeated me for some time. I have now overcome them" (Littlewood, 1970, p. 239). Here is another quote from Littlewood: "Try a hard problem. You may not solve it, but you will prove something else" (www-groups.dcs.st-and.ac.uk/history/Quotations/Littlewood.html).

2.9 Notes and Remarks

Definition 2.2.1 is the original definition of G. H. Hardy (1915), whereas (almost) all the beginning of the theory can be credited to Frigyes Riesz (1923), in particular Theorem 2.2.2. The immediate consequences of the identification of Convention 2.2.3 are not all mentioned explicitly in the text; for example, we could have added that $H^p(\mathbb{D})$ is separable if $p < \infty$ (as a subspace of a separable space).

Jensen's inequality of Lemma 2.3.1 and Corollary 2.3.2, proved in Jensen (1899), admits a conformal invariant form: if Ω is a Jordan domain, then

$$\log |f(z)| \le \int_{\partial\Omega} \log |f(t)| \, d\omega_z(t),$$

where ω_z is the harmonic measure of Ω at the point $z \in \Omega$, i.e. the inverse image $\omega_z = \varphi^{-1} \circ m$ of the Lebesgue measure on $\mathbb{T}$ for the conformal mapping $\varphi \colon \mathbb{D} \to \Omega$, $\varphi(0) = z$. An equivalent description: $z \longmapsto \omega_z(I)$ is a harmonic function in Ω having boundary limits equal to χ_I, for every arc I of the boundary $\partial\Omega$).

The uniqueness theorem Corollary 2.3.3 was found by the Riesz brothers (Riesz and Riesz, 1916). It was reinforced as follows by Luzin and Privalov (1930, published in Privalov (1941)): if a function $f \in \mathrm{Hol}(\mathbb{D})$ (without any growth condition!) admits zero non-tangential limits on a set $A \subset \mathbb{T}$, with $m(A) > 0$, then $f = 0$. Here, "non-tangential" cannot be replaced by "radial": for every sequence $a_n \searrow 0$ and pair of real measurable functions u and v, there exists a function $f \in \mathrm{Hol}(\mathbb{D})$, $f = \sum_{n \ge 0} \hat{f}(n) z^n$, $\sum_{n \ge 0} |\hat{f}(n)|^2 a_n < \infty$ such that $\lim_{r \to 1} \mathrm{Re}(f(r\zeta)) = u(\zeta)$, $\lim_{r \to 1} \mathrm{Im}(f(r\zeta)) = v(\zeta)$ a.e. on $\mathbb{T}$ (Kahane and Katznelson, 1971).

The Blaschke products in Theorem 2.4.2 originated in Blaschke (1915), as well as Theorem 2.4.3 in the case $p = \infty$; the general case was treated by

Frigyes Riesz (1923). Fatou's Theorem 2.5.1 (Fatou, 1906) and its Corollary 2.5.2 (proved by Fatou for $p = \infty$) were generalized by Plessner (1927): for every holomorphic (or even meromorphic) function f in $\mathbb{D}$, the circle $\mathbb{T}$ can be written as $\mathbb{T} = N \cup L \cup C$, where

(1) $m(N) = 0$,
(2) at every point of L, there exists a finite non-tangential limit of f,
(3) at every point ζ of C, the set of partial non-tangential limits $\lim_{z_n \to \zeta} f(z_n)$
     (*cluster set at ζ*) is the entire plane $\mathbb{C}$.

The Szegő maximal functions were introduced in Szegő (1921). The same terminology is used by Smirnov (1928a,b). The term "outer" comes from Beurling (1949), while Wiener and Masani (1958), in their theory of filtering and stationary processes, speak of "optimal" functions. The description in Corollary 2.6.2 of the moduli of functions H^p is of vital importance for the weighted approximation of § 2.7, as well as for the prediction of the processes (see Chapter 5 on filtering).

   For Theorem 2.6.3 see Herglotz (1911), and for Corollary 2.6.4 and Theorem 2.6.5 see Smirnov (1928b). Only half in jest, we could say that the singular inner functions V_μ corresponding to a continuous measure μ (without point masses) finally provide a positive answer to a fairly silly question: "Have analysts invented a single new function unknown to Weierstrass?" (this question made the rounds in the years 1960–1970; of course, the Blaschke products are modified Weierstrass products, not to mention Cauchy-type integrals known long before Weierstrass).

   The characterizations of Theorem 2.6.7 are from Smirnov (1932), Szegő (1921), and Beurling (1949).

   The formula of the prediction distance of Theorem 2.7.1 and Corollary 2.7.2 was proved by Szegő (1920) for $\mu \ll m$ and by Verblunsky (1936) in the general case. Theorem 2.7.4 was essentially obtained by Smirnov (1928a ((2) $\Rightarrow$ (1)), 1932 ((1) $\Leftrightarrow$ (2)), and then reinvented by Beurling (1949). The cyclic functions of $L^2(\mu)$ (Theorem 2.7.5) were characterized by Kolmogorov (1941).

   The description in Theorem 2.7.5 of the spaces $H^2(\mu)$ is a very particular case of the following approximation problem: given a Borel measure μ in $\mathbb{C}$ such that $\mathcal{P}_a \subset L^p(\mu)$, find a description of the closure

$$H^p(\mu) := \operatorname{clos}_{L^p(\mu)} \mathcal{P}_a,$$

and in particular, find a criterion for the occurrence of the equality $H^p(\mu) = L^p(\mu)$ (this is the problem of completeness of the polynomials). Theorem 2.7.5 corresponds to the case where $\operatorname{supp}(\mu) \subset \mathbb{T}$, $p = 2$. The case $p = \infty$,

$\mathrm{supp}(\mu) \subset \mathbb{R}$ is a classical problem of Bernstein (1924): with a slight modification of the notation in the case of a Borel measure μ absolutely continuous with respect to the Lebesgue measure λ, $\mu = w\lambda$, we redefine $L^p(\mathbb{R}, \lambda) = \{f : fw \in L^p(\mathbb{R}, \lambda)\}$; hence for $p = \infty$, the norm becomes $\|f\|_{w,\infty} = \|fw\|_{L^\infty(\mathbb{R})}$. After a long evolution of the subject, the solutions were provided by Akhiezer (1956) and Mergelyan (1956) (in the case of considerable "regularity" of w, the criterion of the completeness of the polynomials is $\int_{\mathbb{R}} (1 + t^2)^{-1} \log w(t)\, dt = -\infty$).

The case of the measures $\mu = \sum_n w_n \delta_{t_n}$ where $w_n > 0$ and $t_n \in \mathbb{R}$ was treated by Koosis (1966), de Branges (1959), Borichev (2001) and others. For an overview of the results in the case $\mathrm{supp}(\mu) \subset \mathbb{R}$ see Borichev and Sodin (2001). Good progress has also been made in the special case of the measures μ with compact $\mathrm{supp}(\mu)$. In particular, the result of Thomson (1991) indicates that $H^p(\mu) \neq L^p(\mu)$ if and only if there exists $z \in \mathbb{C}$ such that the functional $p \longmapsto p(z)$ is continuous on $\mathcal{P}_a$ equipped with the norm $L^p(\mu)$ (the previous results of Akhiezer and Mergelyan provided the same answer, along with algorithms to decide if this is really the case or not).

The description in § 2.8.1(c) is due to Srinivasan (Srinivasan, 1963; Helson, 1964); a generalization to the case where the space $L^p(\mathbb{T}, wm)$ is replaced by an arbitrary lattice $X \subset L^1(\mathbb{T})$ can be found in Nikolski (1986). The theorem in § 2.8.1(f) was proved by Beurling (1949).

For § 2.8.2(c) see Neuwirth and Newman (1967). Hilbert's inequality of § 2.8.3(c) appeared in his course on integral equations given in 1905–1908 at Göttingen and published in Weyl's (1908) notes (with the constant 2π), as well as in Hilbert (1912), while Hardy's inequality of § 2.8.3(d) appeared in Hardy and Littlewood (1926). In fact, there exist important improvements of these two inequalities, specifically

$$\sum_{k,j\geq 0} \frac{|\widehat{f}(k)\widehat{g}(j)|}{k + j + 1/2} \leq \pi \|f\|_2 \|g\|_2,$$

$$\sum_{k\geq 0} \frac{|\widehat{h}(k)|}{k + 1/2} \leq \pi \|h\|_1,$$

which ensue from an inequality of Hardy (1913) and Fejér and Riesz (1921) (later generalized by Gabriel in 1932 and Zygmund in 1934: see Zygmund (1959, Ch. 4 (6.27))) for their proof and some references):

$$I := \int_{-1}^{1} |f(x)|^2\, dx \leq \pi \int_{\mathbb{T}} |f|^2\, dm \quad (\forall f \in H^2).$$

The proof of Hardy–Fejér–Riesz is simpler. If the coefficients $\widehat{f}(k)$, $k \geq 0$, are real, then for $-1 < x < 1$, we have $|f(x)|^2 = f(x)^2$. Hence by the Cauchy integral theorem (Appendix B) for the semicircle $C_+ = \overline{\mathbb{D}} \cap \overline{\mathbb{C}}_+$, where $\mathbb{C}_+ = \{z \in \mathbb{C}: \operatorname{Im}(z) > 0\}$ is the upper half-plane, we have

$$0 = \int_{C_+} f(z)^2 \, dz = \int_{-1}^{1} f(x)^2 \, dx + \int_{\mathbb{T}_+} f(z)^2 \, dz,$$

and thus $I \leq 2\pi \int_{\mathbb{T}_+} |f|^2 \, dm$. With the same inequality for $\mathbb{T}_-$ we obtain $I \leq \pi \int_{\mathbb{T}} |f|^2 \, dm$.

In the general case $f \in H^2$, we write $f = g + ih$ where the coefficients of g and h are real. On $]{-1}, 1[$ we have $|f(x)|^2 = g(x)^2 + h(x)^2$, and on $\mathbb{T}$, we have

$$|f(z)|^2 = |g(z)|^2 + |h(z)|^2 + 2 \operatorname{Re}(\overline{g(z)}h(z)i).$$

However,

$$\operatorname{Re}(\overline{g(z)}h(z)i) = (i/2)(\overline{g(z)}h(z) - g(z)\overline{h(z)}) = (i/2)(g(\bar{z})h(z) - g(z)h(\bar{z})) := F(z),$$

and thus $F(\bar{z}) = -F(z)$ and $\int_{\mathbb{T}} F \, dm = 0$. Finally,

$$\int_{-1}^{1} |f(x)|^2 dx = \int_{-1}^{1} (g^2 + h^2) \, dx \leq \pi \int_{\mathbb{T}} (|g|^2 + |h|^2) \, dm = \pi \int_{\mathbb{T}} |f|^2 \, dm. \qquad \blacksquare$$

Once the Hardy–Fejér–Riesz inequality is established, we use it for $f(z^2)$ in place of f to obtain

$$\sum_{k,j \geq 0} \frac{2 |\widehat{f}(k)| \cdot |\widehat{f}(j)|}{2k + 2j + 1} \leq \pi \|f\|^2;$$

the rest of the argument is identical to that of § 2.8.3(d).

In fact, Ingham (1936) proved an even stronger inequality: for every λ, $0 < \lambda < 1$, there exists a constant $M(\lambda) > 0$ such that

$$\sum_{k,j \geq 0} \frac{|\widehat{f}(k)\widehat{g}(j)|}{k + j + \lambda} \leq M(\lambda) \|f\|_2 \|g\|_2 \quad (\forall f, g \in H^2).$$

There exist also a number of generalizations of these important inequalities: with the group $\mathbb{R}$ instead of $\mathbb{Z}$, with multipliers λ_k other than the $\lambda_k = (k + 1/2)^{-1}$, etc. The simplest way is to note that the same proof leads to

$$\sum_{k \geq 0} \lambda_k |\widehat{h}(k)| \leq \|h\|_1 \left\| \sum_{k \geq 0} \lambda_k z^k \right\|_{BMOA}$$

where $BMOA = L^\infty / H^\infty_-$ (the inequality is the best possible). Much more profound is the generalization given by McGehee, Pigno, and Smith (1981): for every monotone sequence of integers $(n_k)_{k \geq 1}$ and every function $f \in L^1(\mathbb{T})$

with $\mathrm{supp}(\hat{f}) \subset (n_k)$ we have $\sum_{k \geq 1} |\hat{f}(n_k)|/k \leq 30\,\|f\|_1$. For a more complete discussion see DeVore and Lorentz (1993) or Steele (2004).

Theorems § 2.8.4(d)–(f) are attributed to Marcel Riesz (1927); the proof in the text follows Calderón (1950) (and is the same as those of Rudin (1998) and Zygmund (1959)). In Chapter 4, the question will be treated again in the framework of the weighted spaces $L^p(\mathbb{T}, wm)$. The proof of the Newman § 2.8.4(i) property is published in Hoffman (1962). With the identification $H^p = H^p(\mathbb{D})$, it is natural to examine the harmonic conjugate mapping $u \mapsto Hu$ in the disk $\mathbb{D}$. In fact, this is exactly the basic approach initially developed in complex analysis: $f = u + iHu$ (u real) means that in the disk, $f(z) = f * P_z = u * P_z + i(Hu) * P_z = u(z) + iv(z)$, $v(0) = 0$; hence u and v satisfy the system of Cauchy–Riemann (C-R) equations,

$$\partial_x u = \partial_y v, \quad \partial_x v = -\partial_y u, \tag{C-R}$$

which of course implies that u is *harmonic*: $\Delta u := \partial_x^2 u + \partial_y^2 u = 0$. Hence the differential form $\alpha = (\partial_x u)\,dy - (\partial_y u)\,dx$ is exact in $\mathbb{D}$, which immediately implies that the function v defined by

$$v(x, y) = -\int_0^x \partial_y u(t, 0)\,dt + \int_0^y \partial_x u(x, t)\,dt = \int_0^y \partial_x u(0, t)\,dt - \int_0^x \partial_y u(t, y)\,dt$$

satisfies the C-R system and $v(0, 0) = 0$, hence $v = Hu$.

The subordination principle § 2.8.6 is from Littlewood (1925). It is easy to see that § 2.8.6(b) remains correct (with its proof) in any weighted space

$$l_a^2(w_n) = \left\{ f = \sum_{k \geq 0} c_k z^k : \|f\|^2 = \sum_{k \geq 0} |c_k|^2 w_k < \infty \right\},$$

for which $w_k \searrow$; in particular, for the Bergman spaces ($w_k = (k+1)^{-\alpha}$, $\alpha > 0$). Indeed, clearly for the shift operator M_z on $l_a^2(w_n)$ we have $\|M_z\| \leq 1$ and hence for every $\omega \in H^\infty$, $\|\omega(M_z)\| \leq \|\omega\|_\infty$ (by the von Neumann inequality: see Nikolski (2002)). Under the hypothesis $\omega(0) = 0$, this leads to $\|C_\omega f\|^2 \leq |\hat{f}(0)|^2 + \|\omega\|_\infty \|C_\omega(S^* f)\|^2 \leq |\hat{f}(0)|^2 + \|C_\omega(S^* f)\|^2$, which already implies $\|C_\omega f\| \leq \|f\|$ (exactly as in the text of § 2.8.6). The Littlewood principle and its generalizations play an important role in the theory of conformal mappings (see Goluzin, 1966), and in the dynamics of composition operators (see Shapiro, 1993).

3

The Smirnov Class $\mathcal{D}$ and the Maximum Principle

Topics. Admissible operations on the outer functions, spectrum of an inner function, GCD and LCM of a family of inner functions, analytic extension and spectrum, classes of Nevanlinna and of Smirnov, the conformally invariant framework, the generalized Phragmén–Lindelöf principle.

This chapter develops the applied potential of the techniques seen in the preceding chapters, in particular the "maximal" property of outer functions (and from there to a very general maximum principle), as well as the rules for finding the "characteristic" function Θ of an invariant subspace $E = \Theta H^p$. The latter leads to a study of the arithmetic of the inner functions and of their analytic behavior at the boundary. All these properties will be useful for applications to filtering theory (Chapter 5) and to the study of the distribution of the zeros of the Euler ζ function (Chapter 6).

3.1 Calculus of Outer Functions

Recall that in the terminology of § 2.6.9 and Theorem 2.6.1, an outer function f was defined as an element of the space H^p, $p > 0$, satisfying $f = \lambda[h]$. In fact, to define a function $[h]$, the integrability of h is not required, but only that of $\log|h|$.

Definition 3.1.1 (general outer functions) *Let h be a measurable function on $\mathbb{T}$ with $\log|h| \in L^1(\mathbb{T})$. An outer function (of absolute value $|h|$) is a function $f = \lambda[h]$ with $|\lambda| = 1$ and, as in Theorem 2.6.1,*

$$[h](z) = \exp\left(\int_{\mathbb{T}} \frac{\zeta + z}{\zeta - z} \log|h(\zeta)|\, dm(\zeta) \right), \quad z \in \mathbb{D}.$$

82

The following list of properties of outer functions is not exhaustive; however, to recapitulate, a few facts already mentioned are nonetheless included.

3.1.1 Properties of Outer Functions

(a) An outer function f admits non-tangential boundary limits bf and $f \in H^p(\mathbb{D}) \Leftrightarrow bf \in L^p(\mathbb{T})$. (See Theorem 2.5.1, Corollary 2.6.2.)

(b) Let $f \in H^p$, $p \geq 1$. Then, f is outer if and only if $E_f = \mathrm{clos}_{H^p}(f\mathcal{P}_a) = H^p$ ($\Leftrightarrow f$ is a cyclic function in H^p). (See Exercise 2.8.1(f).)

(c) If $f \in H^p$ and $1/f \in H^q$, with $p > 0$ and $q > 0$, then f is outer.

Indeed, by the canonical factorization of Theorem 2.6.5, $f = \lambda_1 B_1 V_1[f]$ and $1/f = \lambda_2 B_2 V_2[1/f]$, hence $1 = \lambda BV$, and by the uniqueness of the factorization $B = B_1 B_2 = $ constant, $V = V_1 V_2 = $ constant, hence $f = \lambda_1[f]$. ∎

(d) Theorem (Smirnov, 1928a,b)

(1) *Let $f \in \mathrm{Hol}(\mathbb{D})$ with $\mathrm{Re}(f) \geq 0$ in $\mathbb{D}$. Then f is outer and*

$$f \in H^p(\mathbb{D}) \text{ for every } 0 < p < 1$$

(but perhaps $f \notin H^1(\mathbb{D})$.
(2) *More generally, if $f \in \mathrm{Hol}(\mathbb{D})$, $f(z) \neq 0$ and $\alpha := \sup_{z\in\mathbb{D}} |\arg(f(z))| < \infty$, then f is outer and*

$$f \in H^p(\mathbb{D}) \text{ for every } 0 < p < \pi/2\alpha$$

(but perhaps $f \notin H^{\pi/2\alpha}(\mathbb{D})$).
(3) *For every $h \in L^1(\mathbb{T})$, $\Gamma h \in \bigcap_{0<p<1} H^p(\mathbb{D})$ where*

$$\Gamma h(z) = \int_{\mathbb{T}} \frac{\zeta + z}{\zeta - z} h(\zeta)\, dm(\zeta)$$

is the Herglotz operator of Exercise 2.8.4(c).

Proof (1) As the values of f are in the half-plane

$$\overline{\mathbb{C}}^{+} = \{z \in \mathbb{C}: \ \mathrm{Re}(z) \geq 0\},$$

the function $z \longmapsto f(z)^p$ is well-defined and holomorphic in $\mathbb{D}$, and

$$|\arg(f(z)^p)| \leq \pi p/2.$$

Thus, for any $z \in \mathbb{D}$ and $0 < p < 1$,

$$|f(z)^p| = \mathrm{Re}(f(z)^p)/\cos(\arg(f(z)^p)) \leq \mathrm{Re}(f(z)^p)/\cos(\pi p/2),$$

and consequently, for $0 < r < 1$, by the mean value theorem,

$$\int_{\mathbb{T}} |f(r\zeta)|^p \, dm(\zeta) \leq \int_{\mathbb{T}} \mathrm{Re}(f(r\zeta)^p)/\cos(\pi p/2) \, dm(\zeta) = \mathrm{Re}(f(0)^p)/\cos(\pi p/2).$$

Hence $f \in H^p(\mathbb{D})$.

Since $\mathrm{Re}(1/f(z)) \geq 0$ in $\mathbb{D}$, we have $1/f \in H^p(\mathbb{D})$ for $0 < p < 1$, thus f is outer by (c) above.

(2) It suffices to apply (1) to $g = f^{\pi/2\alpha}$.

(3) If $h \geq 0$ then $\mathrm{Re}(\Gamma h(z)) \geq 0$ in $\mathbb{D}$, hence $\Gamma h \in \bigcap_{0<p<1} H^p(\mathbb{D})$. The general case follows from $h = h_1 - h_2 + i h_3 - i h_4$ where $0 \leq h_j \leq |h|$.   ■

Remark By Herglotz's Theorem 2.6.3, the general form of a function $f \in \mathrm{Hol}(\mathbb{D})$ with $\mathrm{Re}(f) \geq 0$ is

$$\Gamma\mu(z) = \int_{\mathbb{T}} \frac{\zeta + z}{\zeta - z} \, d\mu(\zeta) + ic,$$

where μ is a positive measure on $\mathbb{T}$ and $c \in \mathbb{R}$.

(e) If $f \in H^\infty$ and $\|f\|_\infty \leq 1$, then $1 + f$ is outer. (See Exercise 1.8.3(c), or (d) above.)

(f) The set of outer functions is a commutative group for standard point-by-point multiplication.

(Clear by Definition 3.1.1 and Theorem 2.6.1.)   ■

(g) *Let $f, g \in H^p(\mathbb{D})$, $p > 0$. Then,*

  (i) *fg is outer if and only if f and g are outer,*
 (ii) *if f is outer and $|f| \leq |g|$ in $\mathbb{D}$, then g is outer,*
(iii) *if f is outer and $|f| \geq |g|$ on $\mathbb{T}$, then $f + g$ is outer.*

Indeed, for (i), we write the canonical factorizations $f = \lambda_1 B_1 V_1[f]$ and $g = \lambda_2 B_2 V_2[g]$, hence $fg = (\lambda_1\lambda_2)B_1 B_2 V_1 V_2[fg]$, and then we use the uniqueness part of Smirnov's Theorem 2.6.5.

For (ii), we do the same, but for the product $f = gh$ where $h = f/g \in H^\infty$.

For (iii) (which is a sort of Rouché's theorem), we write $f + g = f(1 + h)$ where $h = g/f \in H^\infty$ (by Theorem 2.6.7, $|g| \leq |f|$ in $\mathbb{D}$), $\|h\|_\infty \leq 1$ and use (e) and (f) above.   ■

(h) A function $f \in \mathrm{Hol}(\mathbb{D})$ ($f \neq 0$) is outer if and only if it is a decreasing limit of outer functions separated from zero:

$$f(z) = \lim_n f_n(z), \quad \inf_{z\in\mathbb{D}} |f_n(z)| > 0 \quad \forall n, \quad and \quad |f_n(z)| \searrow |f(z)| \quad (z \in \mathbb{D}).$$

If $f \in H^p$, $p > 0$, we can select $f_n \in H^p$, $n \geq 1$.

Indeed, if $f = \lambda[f]$ and $f_n = \lambda[\max(1/n, |f|)]$, then all the properties mentioned are clear, including the passage to the limit by the monotone convergence

$$\lim_n \int_{\mathbb{T}} \frac{\zeta + z}{\zeta - z} \log |f_n(\zeta)| \, dm(\zeta) = \int_{\mathbb{T}} \frac{\zeta + z}{\zeta - z} \log |f(\zeta)| \, dm(\zeta), \quad z \in \mathbb{D}.$$

Conversely, let f_n be functions satisfying the conditions of the theorem. By Fatou's Theorem 2.5.1, there exist limits at the boundary (denoted by the same name), hence $f_n = \lambda_n[f_n]$, $|\lambda_n| = 1$. Clearly

$$(|f_n(z)| \geq |f_{n+1}(z)|, \ z \in \mathbb{D}) \Leftrightarrow (|f_n| \geq |f_{n+1}| \text{ a.e. } \mathbb{T}),$$

also the limit $\lim_n |f_n(t)| := h(t)$ exists a.e. $\mathbb{T}$, and $\log |h| \in L^1(\mathbb{T})$. For the last property, when $z \in \mathbb{D}$ with $f(z) \neq 0$, the Beppo Levi theorem can be applied (see Appendix A), by separately considering the positive part $(\log |f_n|)^+$ and negative part $(\log |f_n|)^-$, with the measure $P_z(\zeta) \, dm(\zeta))$, to obtain

$$0 < |f(z)| = \lim_n |[f_n](z)| = \lim_n \exp\left(\int_{\mathbb{T}} P_z(\zeta) \log |f_n(\zeta)| \, dm(\zeta) \right)$$
$$= \exp\left(\int_{\mathbb{T}} P_z(\zeta) \log |h(\zeta)| \, dm(\zeta) \right), \quad z \in \mathbb{D}.$$

Again applying the Beppo Levi theorem: for any $z \in \mathbb{D}$, the limit $\lim_n [f_n](z) = [h](z)$ exists, and hence so does $\lim_n \lambda_n = \lambda$. Clearly $f = \lambda[h]$. ∎

(i) A function $f \in \mathrm{Hol}((1 + \epsilon)\mathbb{D})$ is outer if and only if $f(z) \neq 0$ for every $z \in \mathbb{D}$. There exist outer functions $f \in H^\infty$ where a neighborhood in H^∞ contains only outer functions (for example, the inner functions f with $\inf_{\mathbb{D}} |f| > 0$), but this does not hold for certain other outer functions $f \in H^\infty$.

Indeed, if $f \in \mathrm{Hol}((1 + \epsilon)\mathbb{D})$ without zeros in $\mathbb{D}$, then $f = pg$ where p is a polynomial without zeros in $\mathbb{D}$ and $g^{\pm 1} \in H^\infty(\mathbb{D})$. Hence, g is outer (see (c) above) and $p = A \prod_{k=1}^n (1 - z/\lambda_k)$, $|\lambda_k| \geq 1$, which leads to $\mathrm{Re}(1 - z/\lambda_k) \geq 0$ for $z \in \mathbb{D}$. By (e) and (f), f is outer.

If $\epsilon := \inf_{\mathbb{D}} |f| > 0$ and $\|g\|_\infty < \epsilon$, then $f + g$ is outer by part (iii) of (g) above.

To justify the last assertion, let $f = 1 - z$ and $f_\epsilon = (1 - z)V_\epsilon$ where

$$V_\epsilon = \exp\left(-\epsilon \frac{1 + z}{1 - z} \right).$$

It is easy to see that $\lim_{\epsilon \to 0} \|f - f_\epsilon\|_\infty = 0$, that f is outer, but not the functions f_ϵ. ∎

(j) Here we use certain properties of § 3.3.1, more precisely those of § 3.3.1(d) and § 3.3.1(f); as can be easily assured, the proofs of § 3.3.1(d) and § 3.3.1(f) are independent of § 3.1.1(j). The function e^f is outer for any $f \in H^1(\mathbb{D})$.

Indeed, when $h := \mathrm{Re}(f) \in L^1(\mathbb{T})$, then $h(z) = h * P_z$ and

$$f = ic + \int_{\mathbb{T}} \frac{\zeta + z}{\zeta - z} h(\zeta)\, dm(\zeta),$$

where $c \in \mathbb{R}$ (see also Exercise 2.8.4(c)). Set $h_n(\zeta) = \min(h(\zeta), n)$ and

$$f_n(z) = ic + \int_{\mathbb{T}} \frac{\zeta + z}{\zeta - z} h_n(\zeta)\, dm(\zeta), \quad F_n = \exp(f_n).$$

Then, $f_n \in H^p$, $0 < p < 1$ (see (d) above) and $\mathrm{Re}(f_n(\zeta)) = h_n(\zeta) \nearrow h(\zeta) = \mathrm{Re}(f(\zeta))$ a.e. on $\mathbb{T}$. Hence, $|F_n(z)| = \exp(\mathrm{Re}(f_n(z))) \le e^n$ and

$$|F_n(z)| = \exp(\mathrm{Re}(f_n(z))) \nearrow \exp(\mathrm{Re}(f(z))) = |e^{f(z)}|.$$

By § 3.3.1(f), we obtain $e^f \in \mathcal{D}$.

With the same argument but for $h_n(\zeta) = \max(h(\zeta), -n)$ we obtain $e^{-f} \in \mathcal{D}$. Thus $e^{\pm f} \in \mathcal{D}$, which implies that e^f is an outer function (see § 3.3.1(d)).   ∎

Remark In a way, this result is optimal, as $\exp(-(1 + z)/(1 - z))$ is not outer.

3.2 Calculus of Inner Functions: The Spectrum

The first goal of this section is to study the links between the set operations on the M_z-invariant subspaces of H^p and the arithmetic of inner functions, having in mind a bijection $\Theta \longmapsto \Theta H^p$ between these two sets. The other is to study the spectrum of the inner part of a function of H^p and its links with the analytic properties of this function. For simplicity, we formulate them in the framework of the space H^2, but they hold without any changes for any arbitrary p.

We begin with the arithmetic.

Definition 3.2.1 *Let Θ, θ be two inner functions and τ a family of inner functions.*

(1) *The function Θ is said to divide θ (notation: $\Theta \mid \theta$) if $\theta = \Theta\theta'$ where θ' is an inner function.*

(2) *The function Θ is said to be the greatest common divisor of τ (notation: $\Theta = \mathrm{GCD}(\tau)$) if $\forall \theta \in \tau$, $\Theta \mid \theta$ and if for every inner function Θ' such that $\forall \theta \in \tau$, $\Theta' \mid \theta$ then $\Theta' \mid \Theta$.*

(3) *The function Θ is the least common multiple of τ (notation: $\Theta = \mathrm{LCM}(\tau)$) if $\forall \theta \in \tau, \theta \mid \Theta$ and if for every inner function Θ' such that $\forall \theta \in \tau, \theta \mid \Theta'$ then $\Theta \mid \Theta'$.*

By convention, the constant 1 is an inner function, and $\mathrm{LCM}(\tau) = 0$ if there does not exist an inner function divisible by all the inner functions $\theta \in \tau$.

3.2.1 Properties of the Divisors, GCDs, and LCMs

Let Θ_1, Θ_2 be two inner functions and τ a family of inner functions.

(a) $\Theta_1 \mid \Theta_2 \Leftrightarrow \Theta_1 H^2 \supset \Theta_2 H^2$.

Proof If $\Theta_2 = \Theta_1 \Theta_3$, then $\Theta_2 H^2 = \Theta_1 \Theta_3 H^2 \subset \Theta_1 H^2$. Conversely, if $\Theta_2 H^2 \subset \Theta_1 H^2$, then $\Theta_2 \in \Theta_1 H^2$, hence $\Theta_2 = \Theta_1 f$ where $f \in H^2$. Clearly f is inner. ∎

(b) $\bigcap_{\theta \in \tau}(\theta H^2) = \Theta_1 H^2$ where $\Theta_1 = \mathrm{LCM}(\tau)$, and $\mathrm{span}(\theta H^2 : \theta \in \tau) = \Theta_2 H^2$ where $\Theta_2 = \mathrm{GCD}(\tau)$.

Proof $E := \bigcap_{\theta \in \tau}(\theta H^2)$ is an invariant subspace and hence, if $E \neq \{0\}$, there exists an inner function Θ such that $E = \Theta H^2$. By definition, for every inner function $\theta \in \tau$, $\theta \mid \Theta_1$ so that $\Theta_1 H^2 \subset \theta H^2$, hence $\Theta_1 H^2 \subset E$. Moreover, for every inner function $\theta \in \tau$, $\theta H^2 \supset \Theta H^2$, so that $\theta \mid \Theta$, thus $\Theta_1 \mid \Theta$, giving $\Theta_1 H^2 \supset \Theta H^2 = E$. Hence, $\Theta_1 H^2 = E$.

If $E = \{0\}$, the formula remains valid by the convention of Definition 3.2.1. Similar reasoning proves the second formula. ∎

(c) An invariant subspace generated by a family of functions. Let $\mathcal{F} \subset H^p$, $p \geq 1$, and let $E_{\mathcal{F}} = \mathrm{span}_{H^p}(z^n f : n \geq 0, f \in \mathcal{F})$ be the subspace of H^p generated by $\mathcal{F}$. Then,

$$E = \Theta H^p \text{ where } \Theta = \mathrm{GCD}(f_{in} : f \in \mathcal{F}).$$

Proof For every function $f \in \mathcal{F}$, $E_f = f_{in} H^p$, and the rest follows from (b). ∎

To find the *explicit expressions* of the GCD and LCM of a family of inner functions, we use the notation introduced in Remark 2.4.4 and Corollary 2.6.4 for an inner function Θ:

$$\Theta(z) = B_k(z) V_\mu(z) = \left(\prod_{\lambda \in \mathbb{D}} b_\lambda(z)^{k(\lambda)} \right) \exp\left(- \int_{\mathbb{T}} \frac{\zeta + z}{\zeta - z} \, d\mu(\zeta) \right), \quad z \in \mathbb{D},$$

where $k = k_\Theta$ is a zero divisor satisfying the Blaschke condition $\sum_{\lambda \in \mathbb{D}} k_\Theta(\lambda)(1 - |\lambda|) < \infty$ and $\mu = \mu_\Theta$ a measure on $\mathbb{T}$ singular with respect to m. In property (d) below, we also permit *divisors k_1 that do not satisfy the Blaschke condition*; in this case, we set $B_{k_1} = 0$. Similarly *for the measures*: if $\mu_1(\mathbb{T}) = \infty$, we set $V_{\mu_1} = 0$.

(d) For any divisors k_1, k_2, we have $B_{k_1} B_{k_2} = B_{k_1+k_2}$, and for any singular measures μ_1, μ_2, we have $V_{\mu_1} V_{\mu_2} = V_{\mu_1+\mu_2}$. Moreover, if $\Theta_1 = B_{k_1} V_{\mu_1}$ and $\Theta_2 = B_{k_2} V_{\mu_2}$, then $\Theta_1 \mid \Theta_2$ if and only if $B_{k_1} \mid B_{k_2}$ and $V_{\mu_1} \mid V_{\mu_2}$, and if and only if $k_1 \leq k_2$ and $\mu_1 \leq \mu_2$ (meaning $\mu_2 - \mu_1 \geq 0$).

Proof Clear. ∎

(e) Let $\Theta_1 = \mathrm{LCM}(\tau)$ and $\Theta_2 = \mathrm{GCD}(\tau)$. Then, $\Theta_1 = B_{k_1} V_{\mu_1}$ and $\Theta_2 = B_{k_2} V_{\mu_2}$, where

$$k_1(\lambda) = \sup_{\theta \in \tau} k_\theta(\lambda), \qquad\qquad k_2(\lambda) = \inf_{\theta \in \tau} k_\theta(\lambda) \quad (\lambda \in \mathbb{D}),$$

$$\mu_1(A) = \sup \sum_{\theta \in \tau} \mu_\theta(A_\theta), \qquad \mu_1(A) = \inf \sum_{\theta \in \tau} \mu_\theta(A_\theta),$$

where sup and inf are taken over all finite Borel partitions of A (A an arbitrary Borel subset of $\mathbb{T}$), i.e. $\cup_{\theta \in \tau} A_\theta = A$, $A_\theta \cap A_{\theta'} = \emptyset$ for $\theta \neq \theta'$ and $A_\theta = \emptyset$ for all but a finite number $\theta \in \tau$.

Proof The expressions for k_1 and k_2 are clear by the division property (d), as are the formulas for μ_1 and μ_2, since (again by (d)) they correspond, respectively, to sup (the upper bound) and inf (the lower bound) in the set of positive measures ordered by the relation $\mu \leq \nu$ ($\Leftrightarrow \nu - \mu \geq 0$); see Appendix A. ∎

In the the rest of this section, we study the relationship between the holomorphic extension of a function of H^p and the "size" of its inner factor. We know from complex analysis that an analytic function is "well-defined" by its behavior on neighborhoods of the points where it loses analyticity, hence at its singularities. By applying this maxim to $1/f$ where f is an inner function, we obtain the notion of the spectrum, i.e. the set of singularities, of an inner function.

Definition 3.2.2 *Let $\Theta = BV$ be an inner function with its canonical factorization, $B = B_k$ (k its zero divisor: see Remark 2.4.4) and $V = V_\mu$ (μ is a singular measure on $\mathbb{T}$: see Corollary 2.6.4). The spectrum of Θ is defined by*

$$\sigma(\Theta) = \mathrm{supp}(k) \cup \mathrm{supp}(\mu),$$

where $\mathrm{supp}(k) = \mathrm{clos}\{z \in \mathbb{D} : k(z) > 0\}$ *(closure of the zero set of Θ).*

Note immediately that $\sigma(\Theta) = \emptyset \Leftrightarrow \Theta = $ constant, and that $\sigma(\Theta_1\Theta_2) = \sigma(\Theta_1) \cup \sigma(\Theta_2)$ where the Θ_j are inner functions. Both are immediate consequences of the uniqueness of the factorization of Θ_j: see Remark 2.4.4 and Corollary 2.6.4.

Theorem 3.2.3 (spectrum of an inner function) *Let Θ be an inner function and $\zeta \in \mathbb{T}$. The following assertions are equivalent.*

(1) $\zeta \notin \sigma(\Theta)$.
(2) Θ *admits an analytic extension in a neighborhood of ζ.*
(3) *There exists a neighborhood U_ζ of ζ such that*

$$\epsilon := \inf\left\{|\Theta(z)| : z \in U_\zeta \bigcap \mathbb{D}\right\} > 0.$$

Proof (1) $\Rightarrow$ (2) Let $\Theta = B_k V_\mu$ be the canonical factorization of Θ. Then have $\sigma(\Theta) = \sigma(B_k) \cup \sigma(V_\mu)$, hence $\zeta \in \mathbb{T} \setminus \sigma(B_k)$ and thus B can be analytically extended in a neighborhood of ζ (since B_k exists and is holomorphic in $\mathbb{C} \setminus \text{clos}\{1/\bar{\lambda} : k(\lambda) > 0\}$, see Theorem 2.4.2). The same is evident for V_μ because $\zeta \notin \sigma(V_\mu) = \text{supp}(\mu)$ and

$$V_\mu(z) = \exp\left(-\int_{\text{supp}(\mu)} \frac{\zeta + z}{\zeta - z} \, d\mu(\zeta)\right).$$

This proves property (2).

(2) $\Rightarrow$ (3) Clear: as a consequence of (2), Θ is continuous on a neighborhood $U_\zeta \cap \overline{\mathbb{D}}$ of ζ and $|\Theta| = 1$ on $U_\zeta \cap \mathbb{T}$.

(3) $\Rightarrow$ (1) It suffices to show that $\zeta \notin \sigma(B_k)$ and $\zeta \notin \sigma(V_\mu)$. Given the hypothesis, the first non-inclusion is evident. For the second, suppose that U_ζ is an open set satisfying (3) and $\Delta \subset \mathbb{T}$ a closed arc such that $\Delta \subset U_\zeta$ (see diagram overleaf). Set

$$V_{\mu|\Delta}(z) = \exp\left(-\int_\Delta \frac{\zeta + z}{\zeta - z} \, d\mu(\zeta)\right), \quad z \in \mathbb{D}.$$

Then, by the hypothesis, $|V_{\mu|\Delta}(z)| \geq |V_\mu(z)| \geq |\Theta(z)| \geq \epsilon$ for $z \in U_\zeta \cap \mathbb{D}$ (). Also $\inf_{z \in \mathbb{D} \setminus U_\zeta} |V_{\mu|\Delta}(z)| > 0$ (since $\text{dist}(\mathbb{D} \setminus U_\zeta, \Delta) = \inf\{|\zeta - z| : \zeta \in \Delta, z \in U_\zeta \cap \mathbb{D}\} > 0$), and hence $1/V_{\mu|\Delta} \in H^\infty$, which implies $V_{\mu|\Delta} = $ constant. By the uniqueness of the Herglotz representation (see Theorem 2.6.3 and Corollary 2.6.4), we obtain $\mu \mid \Delta = 0$, thus $\zeta \notin \sigma(V_\mu)$. $\blacksquare$

Corollary 3.2.4 *Let Θ be an inner function. Then*

$$\sigma(\Theta) = \left\{\zeta \in \overline{\mathbb{D}} : \lim_{\epsilon \to 0} \inf_{|z|<1, |z-\zeta|<\epsilon} |\Theta(z)| = 0\right\} = \bigcap_{\epsilon>0} \text{clos}(z : |\Theta(z)| \leq \epsilon).$$

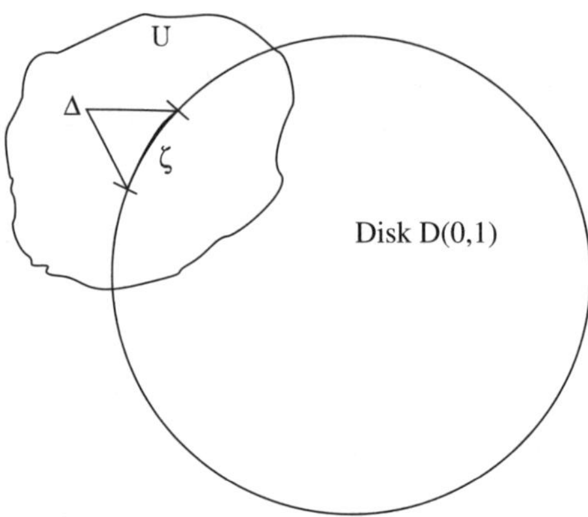

The arc Δ in the set U_ζ, where Θ is separated from zero.

Theorem 3.2.5 *Let $f \in H^p$ ($f \neq 0$), $p > 0$, and $f = f_{in}f_{out}$ its inner–outer factorization. Then, f admits an analytic extension at a point $\zeta \in \mathbb{T}$ if and only if so do the functions f_{in}, f_{out}.*

Proof The sufficiency is evident, so we turn to the necessity. Suppose that f is analytically extendable at a neighborhood of a point $\zeta \in \mathbb{T}$. It suffices to show that f_{in} is analytically extendable at a neighborhood of ζ. Use the same letter f to denote the analytic extension of f in an open neighborhood U_ζ of ζ and by k the multiplicity of the zero of f at ζ. Then, $f(z) = (z - \zeta)^k g(z)$ where g is holomorphic in $\mathbb{D}$ and in the same neighborhood of ζ, and $g(\zeta) \neq 0$. By Theorem 2.6.7(f), $g \in H^p$ and since $(z - \zeta)^k$ is outer (see § 3.1.1(i)), we obtain $f_{in} = g_{in}$. Consider a sequence of open neighborhoods $U_\zeta'' \subset \overline{U}_\zeta'' \subset U_\zeta' \subset \overline{U}_\zeta' \subset U_\zeta$. Then the function g is bounded and separated from zero on $\overline{U}_\zeta'$, say $0 < \epsilon \leq |g| \leq C$ on $\mathbb{T} \cap \overline{U}_\zeta'$, and hence for $z \in \mathbb{D} \cap U_\zeta''$,

$$\log|g_{out}(z)| = \int_{\mathbb{T}} P_z(t) \log|g(t)|\, dm(t) = \int_{\mathbb{T} \cap \overline{U}_\zeta'} + \int_{\mathbb{T} \setminus \overline{U}_\zeta'}$$

$$\leq \log C \int_{\mathbb{T} \cap \overline{U}_\zeta'} P_z(t)\, dm(t) + \int_{\mathbb{T} \setminus \overline{U}_\zeta'} P_z(t) \log|g(t)|\, dm(t)$$

$$\leq \log C + \int_{\mathbb{T} \setminus \overline{U}_\zeta'} P_z(t) \log|g(t)|\, dm(t) \leq \log C + C_1,$$

since the last integral is well-defined and continuous on $\mathbb{C} \setminus (\mathbb{T} \setminus U_\zeta')$, and in particular, bounded on $\mathrm{clos}(\mathbb{D} \cap U'') \subset \mathbb{C} \setminus (\mathbb{T} \setminus U_\zeta')$. Conclusion: g_{out} (as well as g) is bounded on $\mathbb{D} \cap U_\zeta''$. Consequently,

$$|g_{in}(z)| = \frac{|g(z)|}{|g_{out}(z)|} \geq \frac{\epsilon}{Ce^{C_1}} \text{ for } z \in \mathbb{D} \cap U_\zeta'',$$

and Theorem 3.2.3 implies that $g_{in} = f_{in}$ can be analytically extended on U_ζ''. $\blacksquare$

3.2.2 Logarithmic Residues

Given a function $f \in H^p$, $p > 0$, there exists a simple method to find the *discrete part of the singular measure* $\mu_{f_{in}}$ directly as a function of the values of f, without having to find the canonical factorization of f. In fact, for a Borel measure μ on $\mathbb{T}$, we can detail the Radon–Nikodym decomposition $\mu = \mu_s + \mu_a$ by separating in μ_s the point masses and the continuous part: $\mu_s = \mu_d + \mu_{sc}$, where $\mu_d = \sum_{t \in \mathbb{T}} \mu(\{t\}) \delta_t$ (a discrete part of μ_s, or indeed of μ; in fact, $\mu(\{t\}) = 0$ for every $t \in \mathbb{T}$ with the exception of a set at most countable in size) and μ_{sc} is a continuous singular measure, i.e. $\mu_{sc}(\{t\}) = 0$, $\forall t \in \mathbb{T}$. To calculate $\mu_{f_{in}}(\{t\})$ we use the asymptotic behavior of $f(z)$ as z approaches t. In the following theorem, we first consider the case of an isolated singularity $t \in \mathbb{T}$ (to be used in the next step), and then the general case, where we rely on a corollary of the generalized maximum principle of § 3.4 presented in Exercise 3.5.3. There is no risk of a vicious circle since this part of Theorem 3.2.6 will never be used before § 3.5.4).

Theorem 3.2.6 (logarithmic residue) *Let $f \in H^p$ ($f \neq 0$), $p > 0$, with the canonical factorization $f = \lambda BV[f]$, and $\zeta \in \mathbb{T}$.*

(1) *If $\zeta \notin \sigma(B)$ then*

$$\mu_V(\{\zeta\}) = -(1/2) \lim_{r \to 1}(1 - r) \log |f(r\zeta)|.$$

(2) *In the general case,*

$$\mu_V(\{\zeta\}) = -(1/2) \overline{\lim_{r \to 1}}(1 - r) \log |f(r\zeta)|.$$

This limit is called the logarithmic residue of f at the point ζ.

Proof Without loss of generality we can suppose $\zeta = 1$. Let $\mu = \log |f| \cdot m - \mu_V$, a real-valued measure (finite, since $\log |f| \in L^1(\mathbb{T})$). Then

$$(1 - r) \log |f(r)| = (1 - r) \log |B(r)| + \int_{\mathbb{T}} (1 - r)\frac{1 - r^2}{|1 - rt|^2} \, d\mu(t),$$

where

$$(1 - r)\frac{1 - r^2}{|1 - rt|^2} \le 1 + r \le 2 \quad \text{and} \quad \lim_{r \to 1}(1 - r)\frac{1 - r^2}{|1 - rt|^2} = 2\chi_{\{1\}}(t)$$

for every $t \in \mathbb{T}$. By the dominated convergence theorem,

$$\lim_{r \to 1} \int_{\mathbb{T}} (1 - r)\frac{1 - r^2}{|1 - rt|^2} \, d\mu(t) = -2\mu_V(\{1\}).$$

(1) In the case where $1 \notin \sigma(B)$, clearly $\lim_{r \to 1} \log |B(r)| = 0$, and hence the formula is established.

(2) In the general case, it only remains to show that $\overline{\lim}_{r \to 1}(1 - r) \log |B(r)| = 0$. Suppose $\overline{\lim}_{r \to 1}(1 - r) \log |B(r)| < 0$. In this case, there exists $\alpha > 0$ such that $(1 - r) \log |B(r)| < -\alpha$ for $0 < r_0 < r < 1$, and hence the quotient B/S is bounded on the boundary $\partial \mathbb{D}^+$ where

$$S = \exp\left(-\frac{\alpha}{2} \cdot \frac{1 + z}{1 - z}\right)$$

and $\mathbb{D}^+ = \{z \in \mathbb{D}: \ \text{Im}(z) > 0\}$. By Exercise 3.5.3 (a corollary of the Phragmén–Lindelöf principle), B/S is bounded in $\mathbb{D}^+$. The same argument for $\mathbb{D}^- = \{z \in \mathbb{D}: \ \text{Im}(z) < 0\}$ shows that B/S is bounded on the disk $\mathbb{D}$, hence $S|B$, which is absurd. ∎

Example 3.2.7   For every $0 < \alpha < 1$ and $A > 0$, the function

$$f = f_\alpha(z) = \exp\left(-A\left(\frac{1 + z}{1 - z}\right)^\alpha\right), \quad z \in \mathbb{D},$$

is outer (and $f \in H^\infty$) because it can be extended analytically on $\mathbb{C} \setminus [0, \infty)$, and hence $\sigma(f_{in}) \subset \{1\}$ and $\mu_{f_{in}} = \mu_{f_{in}}(\{1\})\delta_1$. Moreover, $\lim_{r \to 1}(1 - r) \log |f(r)| = 0$ and Theorem 3.2.6(1) give $\mu_{f_{in}} = 0$, and the result follows.

3.3 The Nevanlinna ($\mathcal{N}$) and Smirnov ($\mathcal{D}$) Classes

We introduce here two spaces of holomorphic functions, $\mathcal{N}$ and $\mathcal{D}$, respectively bearing the names of Rolf Nevanlinna and Vladimir Smirnov, which provide the most general and natural framework for a theory of boundary behavior: in a way, ($\mathcal{N}$) is "maximal" for the existence of non-tangential boundary limits, and ($\mathcal{D}$) is "maximal" for having a maximum principle.

Definition 3.3.1 *The Nevanlinna class $\mathcal{N}$ and the Smirnov class $\mathcal{D} = \mathcal{N}_+$ are defined as follows:*

$$\mathcal{N} = \left\{ f \in \mathrm{Hol}(\mathbb{D}) : f = f_1/f_2 \text{ where } f_1, f_2 \in \bigcup_{p>0} H^p \right\},$$

$$\mathcal{D} = \left\{ f \in \mathrm{Hol}(\mathbb{D}) : f = f_1/f_2 \text{ where } f_1, f_2 \in \bigcup_{p>0} H^p \text{ and } f_2 \text{ is outer} \right\}.$$

Rolf Nevanlinna (1895–1980), a Finnish mathematician, was one of the key figures in complex analysis of the twentieth century. After defending his thesis in Helsinki in 1919 (under the supervision of Ernst Lindelöf, professor at the University and a cousin of his father), he became famous as the author of the value distribution theory of meromorphic functions (1925), culminating in his influential monograph *Eindeutige analytische Funktionen* (1936). President of the International Mathematical Union from 1959 to 1963, throughout his career he received numerous signs of professional recognition (he was an honorary professor of several universities, member of various academies, etc.).

The Nevanlinna/Neovius family produced professional mathematicians over at least five generations. Rolf Nevanlinna's grandfather was a major-general in the army of the Russian Empire (Finland was a province of Russia until 1917) and a professor of mathematics at the military academy. Rolf's brother Frithiof was also a renowned mathematician, as were two of his descendants (son and grandson). His father, Otto Neovius, was professor of astronomy at the Pulkovo observatory (Saint Petersburg) and a player in the patriotic movement for the liberation of Finland. In 1906 Neovius changed his family name from Swedish to its Finnish translation (Nevanlinna, which can be translated as "Neva river"). In keeping with the same tradition, Nevanlinna was involved in many social movements during his career, often under the colors of the extreme right: a member of the People's Patriotic Movement, from 1942 to 1943 he presided over the Committee of Volunteers of the Waffen SS. Finally, in 1946, his Germanophile position cost him his position as President of the University of Helsinki. One of his five children (the renowned architect Arne Nevanlinna, born in 1925) gave an unflattering portrait of his father in his book *Isan maa* (*The Land of the Fathers*).

Several mathematical objects are named after Nevanlinna: the Nevanlinna *value distribution theory*, *Nevanlinna–Pick interpolation*, *Nevanlinna meromorphic functions*, the *Nevanlinna characteristic*, etc., and also the Nevanlinna Prize awarded every four years at the International Congress of Mathematicians (since 1982). He supervised 28 doctoral theses; among his students were Ahlfors (Fields Medal 1936, the very first after its creation), Karhunen, and Lehto.

3.3.1 A Few Properties of $\mathcal{N}$ and $\mathcal{D}$, by Smirnov (1932)

(a) $\mathcal{D} \subset \mathcal{N}$, and both $\mathcal{N}$ and $\mathcal{D}$ are sub-algebras of the algebra $\mathrm{Hol}(\mathbb{D})$; $H^p(\mathbb{D}) \subset \mathcal{D}$ ($\forall p > 0$); the functions of $\mathcal{N}$ admit non-tangential boundary limits a.e. on $\mathbb{T}$; if $f \in \mathcal{N}$ and $f \neq 0$ then $\log|f| \in L^1(\mathbb{T})$.

(b) $\mathcal{N} = \{f \in \mathrm{Hol}(\mathbb{D}): f = f_1/f_2 \text{ where } f_1, f_2 \in H^\infty\}$,

　　$\mathcal{D} = \{f \in \mathrm{Hol}(\mathbb{D}): f = f_1/f_2 \text{ where } f_1, f_2 \in H^\infty \text{ and } f_2 \text{ is outer}\}$.

Proof Let $f \in \mathcal{N}$, $f = f_1/f_2$ where $f_1, f_2 \in H^p$, $p > 0$, let $f_1 = \lambda_1 B_1 V_1[f_1]$, $f_2 = \lambda_2 B_2 V_2[f_2]$ be their canonical factorizations, and let $B = B_1/B_2$. Setting $g_1 = [\min(1, |f_1/f_2|)]$ and $g_2 = [\min(1, |f_2/f_1|)]$, we obtain $g_1, g_2 \in H^\infty$ and $f = \lambda B V_1 g_1/V_2 g_2$ where $\lambda = \lambda_1/\lambda_2$, which justifies the first formula, as well as the second: if $f \in \mathcal{D}$, we can suppose that $V_2 = 1$. ∎

(c) Every outer function (see Definition 3.1.1) is in $\mathcal{D}$. Moreover,

$$\mathcal{D} = \{f: f = f_{in}f_{out}, f_{in} \text{ and } f_{out} \text{ inner and outer functions, resp.}\}.$$

A function $f \in \mathcal{D}$ is outer if and only if

$$\log|f(0)| = \int_{\mathbb{T}} \log|f(t)| \, dm(t).$$

Proof For the outer functions, we use the same verification as before: if $f = [h]$ and $\log|h| \in L^1(\mathbb{T})$, then $f = g_1/g_2$ where $g_1 = [\min(1,|h|)]$ and $g_2 = [\min(1,|1/h|)]$.

The representation of a function of $\mathcal{D}$ as the product of the inner and outer parts follows from this and the second formula in (b) above.

The criterion with Jensen's identity follows from (b) and the canonical factorization (see Theorem 2.6.7 for details). $\blacksquare$

(d) If $f_1, f_2 \in \mathcal{D}$ and $f_1 f_2$ is outer, then so are f_1 and f_2. ($\mathcal{D}$ cannot be replaced by N: $f_1 = \exp(-(1+z)/(1-z))$, $f_2 = 1/f_1$.) In particular, if both f and $1/f \in \mathcal{D}$, the function f is outer.

Proof As functions in $\mathcal{D}$, f_1 and f_2 can be uniquely written in the form $f_1 = \lambda_1 B_1 V_1[h_1]$, $f_2 = \lambda_2 B_2 V_2[h_2]$ where $\log|h_k| \in L^1(\mathbb{T})$ ($k = 1,2$), hence $f_1 f_2 = \lambda_1 \lambda_2 B_1 B_2 V_1 V_2 [h_1 h_2]$ and $B_1 B_2 V_1 V_2 = \text{constant}$. By uniqueness, B_1, B_2, V_1, V_2 must be constants. $\blacksquare$

(e) Let $f \in \text{Hol}(\mathbb{D})$, $g \in \mathcal{D}$ and $|f| \le |g|$ in $\mathbb{D}$. Then, $f \in \mathcal{D}$.

Proof Indeed, $g = g_1/g_2$ where $g_k \in H^\infty$ and g_2 is outer. By the hypothesis, $|fg_2| \le |g_1|$ in $\mathbb{D}$, hence $fg_2 \in H^\infty$ and $f = fg_2/g_2 \in \mathcal{D}$, since g_2 is outer. $\blacksquare$

(f) Let $f \in \text{Hol}(\mathbb{D})$. Then, $f \in \mathcal{D}$ if and only if there exists a sequence (f_n), $f_n \in H^\infty(\mathbb{D})$ such that $f(z) = \lim_n f_n(z)$ and $|f_n(z)| \nearrow |f(z)|$ for every $z \in \mathbb{D}$.

Proof If $f \in \mathcal{D}$, then $f = g/h$ where $g, h \in H^\infty$ and h is outer (see (b) above). By § 3.1.1(h), there exists a sequence (h_n) of outer functions such that $\inf_{z \in \mathbb{D}} |h_n(z)| > 0$ for every n and $h(z) = \lim_n h_n(z)$, $|h_n(z)| \searrow |h(z)|$ for every $z \in \mathbb{D}$. Then the sequence $f_n = g/h_n$ verifies all the required properties.

Conversely, let $f_n \in H^\infty(\mathbb{D})$ and $f(z) = \lim_n f_n(z)$, $|f_n(z)| \nearrow |f(z)|$ for every $z \in \mathbb{D}$, and let $f_n = \lambda_n B_n V_n[f_n]$ be the canonical factorizations. We rewrite the condition $|f_n(z)| \le |f_{n+1}(z)|$ in the form

$$\left| \frac{B_n(z) V_n(z)}{B_{n+1}(z) V_{n+1}(z)} \right| \le |[f_{n+1}/f_n](z)|, \quad z \in \mathbb{D},$$

which shows that $(B_n V_n/B_{n+1} V_{n+1}) \in \text{Hol}(\mathbb{D})$ and, by (e), $(B_n V_n/B_{n+1} V_{n+1}) \in \mathcal{D}$. By the uniqueness of the canonical factorization, we obtain $B_{n+1} \mid B_n$ and $V_{n+1} \mid V_n$, hence $k_{n+1} \le k_n$ (the corresponding divisors) and $\mu_{n+1} \le \mu_n$ (the corresponding singular measures). This implies the monotone convergence $B(z) := \lim_n B_n(z)$, $|B_n(z)| \nearrow |B(z)|$ and $V(z) := \lim_n V_n(z)$, $|V_n(z)| \nearrow |V(z)|$

for every $z \in \mathbb{D}$, and similarly for $[f_n]$. In particular, $|f_n| \nearrow$ on $\mathbb{T}$. Let h be the limit and $z \in \mathbb{D}$ such that $B_1(z) \neq 0$. Then

$$|B_1(z)V_1(z)[f_n](z)| \leq |B_n(z)V_n(z)[f_n](z)| \leq |f(z)| < \infty,$$

hence $\int_{\mathbb{T}} P_z(t) \log |f_n(t)| \, dm(t) < |f(z)/ B_1(z)V_1(z)|$ for every $n = 1, 2, \ldots,$ which implies $\log(h) \in L^1(\mathbb{T})$. The properties of the B_n, V_n, and $[f_n]$ lead to a representation $f = \lambda BV[h]$, thus $f \in \mathcal{D}$. ∎

(g) Generalized maximum principle (Smirnov, 1932). Let $f \in \mathcal{D}$ and let g be an outer function. If $|f| \leq |g|$ on $\mathbb{T}$, then $|f| \leq |g|$ in $\mathbb{D}$. In particular,

$$H^p(\mathbb{D}) = \mathcal{D} \cap L^p(\mathbb{T}), \quad 0 < p \leq \infty.$$

Proof Let $f_1, f_2, g_1, g_2 \in H^\infty$ where f_2, g_1, g_2 are outer and such that $f = f_1/f_2$, $g = g_1/g_2$. By the hypothesis, we have $|f_1 g_2| \leq |g_1 f_2|$ on $\mathbb{T}$, where $g_1 f_2$ is an outer function. Applying part (5) of Theorem 2.6.7, we obtain $|f_1 g_2| \leq |g_1 f_2|$ in $\mathbb{D}$, and the result follows.

For the formula ("integral maximum principle"), the inclusion $H^p(\mathbb{D}) \subset \mathcal{D} \cap L^p(\mathbb{T})$ is evident, and the converse follows from what has already been proved by setting $g = [f]$. ∎

3.4 The Generalized Phragmén–Lindelöf Principle

In this section we show that the Smirnov maximum principle (§ 3.3.1(g)) contains as special cases a variety of very useful propositions, known collectively as the *Phragmén–Lindelöf principle*. We begin with the definition and a few properties of the spaces $\mathcal{N}$ and $\mathcal{D}$ in domains of the complex plane $\mathbb{C}$ different from $\mathbb{D}$.

3.4.1 The Spaces $\mathcal{N}$ and $\mathcal{D}$: Conformally Invariant Versions

We study the *Jordan domains* Ω in the extended complex plane $\overline{\mathbb{C}} = \mathbb{C} \cup \{\infty\}$ defined as the conformal images $\Omega = \omega(\mathbb{D})$ of the disk $\mathbb{D}$ by a bi-holomorphic mapping

$$\omega : \mathbb{D} \to \overline{\mathbb{C}},$$

continuous up to the boundary and bijective in $\overline{\mathbb{D}} = \mathrm{clos}\, \mathbb{D}$; see Appendix B for some references. In § 3.4, Ω will always denote a Jordan domain.

(a) Definition

$$H^\infty(\Omega) = \{f \in \mathrm{Hol}(\Omega) \colon \|f\|_\infty = \sup_{z \in \Omega} |f(z)| < \infty\},$$

$$\mathcal{N}(\Omega) = \{f \in \mathrm{Hol}(\Omega) \colon f = g/h; g, h \in H^\infty(\Omega)\},$$

$$\mathcal{D}(\Omega) = \{f \in \mathrm{Hol}(\Omega) \colon f = g/h; g, h \in H^\infty(\Omega), h \text{ is outer}\},$$

where g *outer in* Ω means that $g \circ \omega$ is outer in $\mathbb{D}$.

The following properties are immediate by the definitions and § 3.3.

(b) Properties

(i) $\mathcal{N}(\Omega) = \{f \colon f \circ \omega \in \mathcal{N}(\mathbb{D})\}$, $\mathcal{D}(\Omega) = \{f \colon f \circ \omega \in \mathcal{D}(\mathbb{D})\}$.

(ii) $f \in \mathcal{D}(\Omega) \Leftrightarrow f(z) = \lim_n f_n(z)$ *where* $f_n \in H^\infty(\Omega)$, $|f_n(z)| \nearrow |f(z)|$ ($\forall z \in \Omega$).

(iii) *A function* $f \in H^\infty(\Omega)$ *is outer if and only if* $f(z) = \lim_n f_n(z)$ *where* $f_n \in H^\infty(\Omega)$, $\inf_{z \in \Omega} |f_n(z)| > 0$ ($\forall n$) *and* $|f_n(z)| \searrow |f(z)|$ ($\forall z \in \Omega$).

(iv) *Let* $f \in \mathrm{Hol}(\Omega)$ *and* $g \in \mathcal{D}(\Omega)$, $|f| \le |g|$ *in* Ω. *Then* $f \in \mathcal{D}(\Omega)$ *(evident by* § *3.3.1(e)).*

(v) *Let* $\Omega_1 \subset \Omega_2$ *be two Jordan domains and* $f \in \mathrm{Hol}(\Omega_2)$. *If* f *is outer in* Ω_2, *then* $f \mid \Omega_1$ *is outer in* Ω_1 *(evident by (iii)). Similarly,* $f \in \mathcal{D}(\Omega_2) \Rightarrow f|\Omega_1 \in \mathcal{D}(\Omega_1)$ *(evident by (ii)).*

(c) Generalized maximum principle

Let $\lambda \in \partial\Omega$, $f \in \mathcal{D}(\Omega) \cap C(\overline{\Omega} \setminus \{\lambda\})$ and let $g \in C(\overline{\Omega} \setminus \{\lambda\})$ be an outer function in Ω such that $|f| \le |g|$ on $\partial\Omega \setminus \{\lambda\}$. Then, $|f| \le |g|$ in Ω.

Proof Evident by the Smirnov theorem § 3.3.1(g). ∎

3.4.2 Generalized Phragmén–Lindelöf Principle

In fact, the propositions of Theorem 2.6.1, § 3.3.1(g) and § 3.4.1(c) are already versions of the Phragmén–Lindelöf principle, but in applications, the condition $f \in \mathcal{D}(\Omega)$ is often replaced by an upper estimate $|f(z)| \le M(z)$ of "outer type," where M is not necessarily holomorphic but is bounded above by an outer function $M(z) \le |g(z)|$, $z \in \Omega$ (which already implies $f \in \mathcal{D}(\Omega)$ by part (iv) of § 3.4.1(b)). To formalize this passage we introduce a definition.

(a) **Definition.** Let M and M_* be two non-negative functions on Ω, and $w \in C(\partial\Omega \setminus \{\lambda\})$ where $\lambda \in \partial\Omega$, $w > 0$. The function M_* is called a *Phragmén–Lindelöf majorant* for a pair M, w if the conditions $f \in \mathrm{Hol}(\Omega) \cap C(\overline{\Omega} \setminus \{\lambda\})$, $|f| \le M$ in Ω and $|f| \le w$ on $\partial\Omega \setminus \{\lambda\}$ imply $|f| \le M_*$ in Ω.

Thus one can say that a Phragmén–Lindelöf principle is established for a given pair (M, w), if we can find at least one Phragmén–Lindelöf majorant M_*. The continuity condition $C(\overline{\Omega} \setminus \{\lambda\})$ is, of course, excessive and simply means that we do not wish to discuss the "boundary values" of the functions in Ω. In applications, often $\lambda = \infty$.

Lars Edvard Phragmén (1863–1937), a Swedish mathematician, is known for important developments, first concerning the classical Liouville theorem (1904), then (with Ernst Lindelöf, 1908) for the maximum modulus principle, and also for his work for Mittag-Leffler's journal *Acta Mathematica*, where he started as a member of staff in 1888. He obtained his doctorate at the University of Uppsala in 1889. He gained a large international recognition by his participation in the discovery of what Jean-Christophe Yoccoz called "a fertile error of Henri Poincaré" (SMF *Gazette*, vol. 107 (January 2006)). This concerns an event that took place in 1889, when Poincaré won a prize awarded by King Oscar II of Sweden and Norway for resolving the question of the stability of the three-body problem in celestial mechanics. The prize was presented, and the article accepted by *Acta Mathematica*, but the young Lars Phragmén, tasked with re-reading the proofs, found a certain number of errors (90 pages of remarks and objections for a manuscript of 160 pages!). Poincaré himself then detected a major gap and withdrew the manuscript. He re-submitted it a year and a half later, significantly enriched and already 270 pages long; it was later to form a major part of his masterpiece *Méthodes nouvelles de la Mécanique céleste*, the origin of the theory of dynamical systems, ergodic theory, and chaos theory. Soon after, Phragmén easily obtained a position as professor at the University of Stockholm, where he stayed until 1903 when he left to work for a private insurance company.

Today, Phragmén is especially known for the *Phragmén–Lindelöf principle*, published in *Acta Mathematica* in 1908 (a version is presented in §§ 3.4 and 3.5).

Ernst Leonard Lindelöf (1870–1946) was a Finnish mathematician known for his work in topology and analysis. He obtained his doctorate at the University of Helsinki (Helsingfors at the time, when Finland was controlled by Russia) in 1895 and became professor at the same university in 1902. He is especially known as a topologist (Lindelöf spaces), but also as a specialist in complex analysis: the Phragmén–Lindelöf Principle, published in 1908 in *Acta Mathematica*, and then the *Lindelöf hypothesis* on the Euler ζ function, which has resisted all efforts and remains open. For decades, from 1907 to 1938, he was one of the editors of *Acta Mathematica*, and was an Honorary Professor of several Scandinavian universities.

(b) Theorem (universal Phragmén–Lindelöf principle). Let M and w be two functions as in (a) above and let $F \in \mathcal{D}(\Omega)$, $G \in \mathcal{N}(\Omega) \cap C(\overline{\Omega} \setminus \{\lambda\})$ be such that $M \leq |F|$ in Ω and $w \leq |G|$ on $\partial\Omega \setminus \{\lambda\}$. Then, either

(i) $(f \in \mathrm{Hol}(\Omega) \cap C(\overline{\Omega} \setminus \{\lambda\}), |f| \leq M$ (in Ω), $|f| \leq w$ (on $\partial\Omega$)) $\Rightarrow f = 0$, and then $M_* = 0$ is a Phragmén–Lindelöf majorant, or

(ii) there exists an outer function $[w \circ \omega]$, and then

$$M_* = [w \circ \omega] \circ \omega^{-1}$$

is a Phragmén–Lindelöf majorant for (M, w).

Proof By part (iv) of § 3.4.1(b), for a holomorphic function f, the estimates $|f| \leq M \leq |F|$ in Ω imply $f \in \mathcal{D}(\Omega)$. If there exists a function $f \neq 0$ such that $f \in \mathrm{Hol}(\Omega) \cap C(\overline{\Omega} \setminus \{\lambda\})$, $|f| \leq M$ (in Ω) and $|f| \leq w$ (on $\partial\Omega$), then $|f \circ \omega| \leq w \circ \omega \leq |G \circ \omega|$ a.e. on $\mathbb{T}$. Since $f \circ \omega, G \circ \omega \in \mathcal{N}(\mathbb{D})$, this implies the existence of an outer function $[w \circ \omega]$. Applying § 3.4.1(c) to $g = [w \circ \omega]$, we obtain (ii). If such a function f does not exist, we set $M_* = 0$. ∎

3.4.3 Classical Examples

In § 3.4.2(b) we spoke of a "universal principle" because all known implementations of the Phragmén–Lindelöf principles, beginning with the original theorems of Phragmén and Lindelöf (1908), Examples 3.4.3(a)–(b) below and the exercises of § 3.5, are special cases of § 3.4.2(b). Moreover, Theorem 3.4.2(b) explains the *true nature of the maximum principles* and gives an indication of how to construct them in the domain of interest: these are exactly the majorizations of the Smirnov class $\mathcal{D}$ that can be extended from an estimate on the boundary $\partial\Omega$ to an inequality over the interior of Ω (using the construction of a Szegő maximal function).

(a) Example (Phragmén and Lindelöf, 1908)

Let $\Omega = \mathbb{C}^+ = \{z \in \mathbb{C}: \mathrm{Re}(z) > 0\}$ and $f \in \mathrm{Hol}(\Omega)$ be such that

$$|f(z)| \le A \exp(B|z|^\alpha), \quad z \in \Omega,$$

where $A, B > 0$ and $0 \le \alpha < 1$. If $f \mid i\mathbb{R}$ is bounded, then $f \in H^\infty(\Omega)$ (and hence, $\|f\|_{H^\infty(\Omega)} = \|f\|_{L^\infty(i\mathbb{R})}$). The condition $\alpha < 1$ is optimal (consider for example $f(z) = e^z$).

Indeed, we apply § 3.4.2(b) with $w = G = 1$ and $F(z) = C \exp(Cz^\alpha)$ and $C > 0$ large enough. For $z \in \Omega$, we have $|z|^\alpha \le \mathrm{Re}(z^\alpha)/\cos(\pi\alpha/2))$ and hence $A \exp(B|z|^\alpha) \le |F(z)|$ if C is sufficiently grand. Moreover, F is an outer function in Ω, because this is the case for its transplantation to the disk (see Example 3.2.7 above):

$$F \circ \omega(\zeta) = \exp\!\left(C\left(\frac{1+\zeta}{1-\zeta}\right)^\alpha\right),$$

where ω is a conformal mapping of $\mathbb{D}$ into $\mathbb{C}^+$ ($\omega(\zeta) = (1+\zeta)/(1-\zeta)$).   ∎

(b) Example (Phragmén and Lindelöf, 1908)

Let $f \in \mathrm{Hol}(\mathbb{C}^+)$, $\mathbb{C}^+ = \{z \in \mathbb{C}: \mathrm{Re}(z) > 0\}$ and let $0 < \alpha, \beta < 1$ be such that

$$|f(z)| \le A \exp(B|z|^\alpha),\ z \in \mathbb{C}^+, \quad \text{and} \quad |f(iy)| \le C \exp(D|iy|^\beta),\ iy \in i\mathbb{R} = \partial\mathbb{C}^+,$$

where $A, B, C, D > 0$. Then,

$$|f(re^{it})| \le C \exp(D' r^\beta \cos(\beta t)), \quad \text{where } D' = \frac{D}{\cos(\pi\beta/2)}, \ r > 0, |t| \le \frac{\pi}{2}.$$

Indeed, we apply § 3.4.2(b) with $F(z) = K \exp(Kz^\alpha)$ and $K > 0$ large enough and

$$G = [w] = C[\exp(D|z|^\beta)] = C \cdot \exp(D'z^\beta), \quad \mathrm{Re}(z) > 0.$$

The fact that F and G are outer has already been mentioned (Example (a)).   ∎

3.5 Exercises

3.5.1 An Improvement of Liouville's Theorem

Let $f \in \mathrm{Hol}(\mathbb{C})$ and let $0 < \alpha < 2$ be such that

$$|f(z)| \le A \exp(B|z|^\alpha), \quad z \in \mathbb{C},$$

where $A, B > 0$, and f is bounded on $\mathbb{R}$ and $i\mathbb{R}$.

(1) *Show that f = constant.*

(2) *Show with an example that the condition on the exponent α is optimal.*

SOLUTION: (1) Indeed, it suffices to verify that the restrictions $f|\mathbb{C}_\pm^\pm$ on each quarter-plane $\mathbb{C}_\pm^\pm = \{z \in \mathbb{C}: \ \mathrm{Re}(\pm z) > 0, \mathrm{Im}(\pm z) > 0\}$ are bounded (and then use Liouville's theorem: every entire bounded holomorphic function is constant). For this, it suffices to note that the exponential $F = \exp(Cz^\alpha)$ is an outer function in $\mathbb{C}_\pm^\pm$, and to use the same reasoning as in Examples 3.4.3(a) and 3.4.3(b). This last property is verified in the same manner as in the examples, with the help of a conformal mapping

$$\omega: \mathbb{D} \to \mathbb{C}_+^+, \quad \omega(\zeta) = \left(i\frac{1+\zeta}{1-\zeta}\right)^{1/2}.$$

(2) $f(z) = \exp(iz^2)$. $\blacksquare$

Joseph Liouville (1809–1882) was a French mathematician, a brilliant student of the École Polytechnique from 1825 to 1827, where he studied under Ampère and Arago, and graduated at the age of 18 (!), with Poisson as examiner. After taking a break for health reasons, he taught in different establishments in Paris (for 35–40 hours per week!) before being named professor at the École Polytechnique (1838), and then at the Collège de France (1850) and the Faculté de Sciences in Paris (1857). While accumulating all these positions, Liouville was also active at the Académie des Sciences (elected in 1838) and at the Bureau des Longitudes (1840). In 1836 he founded the *Journal de Mathématiques Pures et Appliquées* (also known as the *Journal de Liouville*) which played an important role in French mathematical life of the nineteenth and twentieth centuries.

Liouville wrote more than 400 articles in analysis, number theory, mathematical physics, and even astronomy: in analysis, the *Sturm–*

Liouville theory (historically, the foundation of Hilbert's spectral theory) and a Liouville theorem on entire functions (important but today simple: a generalization is presented in § 3.4); in number theory, an explicit construction of transcendental numbers with the aid of continued fractions (*Liouville numbers*) and a fundamental theorem on Diophantine approximation; in mathematical physics, the invariance of phase space volume for Hamiltonian dynamics (hence for Newtonian mechanics). A well-known episode in Liouville's career is linked to the unpublished manuscripts of Évariste Galois (containing a revolutionary idea which led to what is now known as *Galois theory*): it was Liouville who recovered them after the brutal death of the author, interpreted them, and published them in his journal.

3.5.2 The Case of a Strip (Phragmén and Lindelöf, 1908)

Let $\Omega = \{z \in \mathbb{C}: 0 < \mathrm{Im}(z) < \pi\}$ and $f \in \mathrm{Hol}(\Omega)$ be such that

$$|f(x + iy)| \leq A \exp(Be^{\alpha|x|}), \quad x + iy \in \Omega,$$

where $A, B > 0$ and $0 < \alpha < 1$. Show that if f is bounded on $\partial\Omega$, then $f \in H^\infty(\Omega)$ and $\|f\|_{H^\infty(\Omega)} = \|f\|_{L^\infty(\partial\Omega)}$.

SOLUTION: Indeed, if $\omega(z) = e^z$ is a conformal isomorphism $\Omega \to \mathbb{C}_+ = \{z \in \mathbb{C}: \mathrm{Im}(z) > 0\}$ and $g(\zeta) = f \circ \omega^{-1}(\zeta) = f(\log \zeta)$, a pullback of f to $\mathbb{C}_+$, then

$$|g(\zeta)| \leq A \exp(Be^{\alpha|\log|\zeta||}) = A \exp(B \max(|\zeta|^\alpha, 1/|\zeta|^\alpha)), \quad \zeta \in \mathbb{C}_+.$$

We conclude the proof in the same manner as in Example 3.4.3(a) with a modification of the function F: we apply § 3.4.2(b) with $w = G = 1$ and $F(\zeta) = C \exp(C(\zeta^\alpha + \zeta^{-\alpha}))$ and $C > 0$ large enough. Using $|\zeta^\alpha + \zeta^{-\alpha}| \leq 2|\zeta|^\alpha$ for $|\zeta| > 1$ and $(1 - 2^{-2\alpha})|\zeta|^\alpha \leq |\zeta^\alpha + \zeta^{-\alpha}|$ for $|\zeta| < 1/2$, we obtain (as in 3.4.3(a)) that for $C > 0$ large enough:

$$\exp(B \max(|\zeta|^\alpha, 1/|\zeta|^\alpha)) \leq |F(\zeta)|, \quad \zeta \in \mathbb{C}_+.$$

As a product of two outer functions $F = C \exp(C\zeta^\alpha) \exp(C\zeta^{-\alpha})$ (see Example 3.4.3(a)) F is also outer. ∎

3.5.3 An Inner Function Which Becomes Outer on a Subdomain

Let

$$\Omega = \mathbb{D}_+ = \{z \in \mathbb{D}: \mathrm{Im}(z) > 0\} \quad and \quad f_\alpha(z) = \exp\left(A\left(\frac{1 + z}{1 - z}\right)^\alpha\right),$$

$A \in \mathbb{R}$, $0 < \alpha < 2$. *Then, f_α is an outer function in Ω (however f_1 is inner in $\mathbb{D}$).*

SOLUTION: Indeed, by choosing a conformal isomorphism

$$\omega_1 : \mathbb{D}_+ \to \mathbb{C}_+ = \{z \in \mathbb{C} : \operatorname{Im}(z) > 0\}, \quad \omega_1(z) = \left(\frac{1+z}{1-z}\right)^2,$$

we obtain $f \circ \omega_1^{-1}(w) = \exp(Aw^{\alpha/2})$, $w \in \mathbb{C}_+$. Next, with $\omega_2 : \mathbb{C}_+ \to \mathbb{D}$, $\omega_2(w) = (w-i)/(w+i)$ and $\omega = \omega_2 \circ \omega_1$, we have

$$F(\zeta) := f \circ \omega^{-1}(\zeta) = \exp\left(A\left(i\frac{1+\zeta}{1-\zeta}\right)^{\alpha/2}\right), \quad \zeta \in \mathbb{D}.$$

We first examine the easy case when $0 < \alpha \leq 1$. In this case,

$$\operatorname{Re}\left(i\frac{1+\zeta}{1-\zeta}\right)^{\alpha/2} \geq 0$$

for every $\zeta \in \mathbb{D}$, and hence for $A < 0$ the function F is bounded in $\mathbb{D}$, holomorphic in $\mathbb{C} \setminus [1, \infty)$, and thus $\sigma(F_{in}) \subset \{1\}$. With the zero logarithmic residue,

$$\lim_{r \to 1}(1-r)\log|F(r)| = \lim_{r \to 1}(1-r)\left|A\left(i\frac{1+r}{1-r}\right)^{\alpha/2}\right| = 0,$$

we see that F is outer (and hence, so is f). If $A > 0$, we apply the preceding arguments to $1/F$.

In the case where $1 < \alpha < 2$, the function F is no longer bounded, but we can get around this obstacle by, for example, using § 3.1.1(j). Indeed,

$$\varphi := A\left(i\frac{1+z}{1-z}\right)^{\alpha/2} \in H^p(\mathbb{D})$$

for every $p < 2/\alpha$ where $2/\alpha > 1$, hence $F = e^\varphi$ is outer by § 3.1.1(j). ∎

3.5.4 Division by a Singular Function with a Point Measure

Let $f \in H^p(\mathbb{D})$, $p > 0$ and

$$V = \exp\left(-A\frac{1+z}{1-z}\right)$$

where $A > 0$. Show that $V \mid f_{in}$ if and only if

$$|f(r)| \leq C_\epsilon \exp\left(-\frac{A-\epsilon}{1-r}\right)$$

for every $\epsilon > 0$ and a certain $C_\epsilon > 0$.

SOLUTION: Apply Theorem 3.2.6(2) to see that $\mu_{f_{in}}(\{1\}) \geq A$. ∎

3.6 Notes and Remarks

As already explained, the principal goal of this chapter is to show the decisive role of the outer functions and the Smirnov class $\mathcal{D}$ in maximum-principle-type estimations. This role, and the class $\mathcal{D}$ itself, as well as its importance for problems of polynomial approximation, were discovered by Smirnov (1932) when studying the applicability of the classical Cauchy and Green formulas in Jordan domains Ω,

$$f(z) = \frac{1}{2\pi i} \int_{\partial\Omega} \frac{f(t)}{t - z}\, dt,$$
$$f(z) = \frac{1}{2\pi} \int_{\partial\Omega} f(t) \frac{\partial G(t, z)}{\partial n}\, |dt|,$$

where G is the *Green's function* of Ω with a pole at z. The conclusion of Smirnov (1932) is as follows: if the boundary $\partial\Omega$ is rectifiable, the formulas are applicable if and only if *the integrals converge absolutely and $f \in \mathcal{D}(\Omega)$*. Other subjects where the class $\mathcal{D}$ plays an important role include the theory of conformal mappings ($\mathcal{D}$ intervenes in the definition of *Smirnov domains* in this theory; see for example Duren (1970) and Goluzin (1966)), the description of ideals in algebras of holomorphic functions, etc.

Theorem § 3.1.1(d) and its proof appeared in Smirnov (1928b), the important conformally invariant characterization § 3.1.1(h) is from Smirnov (1932). For the definition of the spectrum of an inner function in Theorem 3.2.3, refer to Nikolski (1986), and for the contents of § 3.2 to Hoffman (1962) or Nikolski (1986).

The original definition of the Nevanlinna class $\mathcal{N}$ is different from Definition 3.3.1, namely: $f \in \mathcal{N}$ if and only if f is holomorphic in $\mathbb{D}$ and

$$\sup_{0<r<1} \int_{\mathbb{T}} \log^+ |f(rt)|\, dm(t) < \infty,$$

where $\log^+(x) = \max(0, \log(x))$ for $x > 0$ (see Nevanlinna and Nevanlinna, 1922). The equivalence of the two definitions is not at all obvious; the proof can be found in Nevanlinna and Nevanlinna (1922), Privalov (1941), Duren (1970), and Koosis (1980). Using the same terminology, the class $\mathcal{D}$ can equally be characterized as follows (Tumarkin; see Privalov, 1941): $f \in \mathcal{D}$ if and only if f is holomorphic in $\mathbb{D}$ and the integrals

$$\int_E \log^+ |f(rt)|\, dm(t), \quad E \subset \mathbb{T},$$

are uniformly absolutely continuous, or if and only if

$$\lim_{r \to 1} \int_{\mathbb{T}} \log^+ |f(t) - f(rt)| \, dm(t) = 0$$

(the non-tangential boundary limits $f(t)$ are assumed to exist).

The properties of § 3.3.1 are attributed to Smirnov (1932); among these, (f) is particularly important – it allows $\mathcal{D}$ to be defined in a conformally invariant manner.

The Smirnov maximum principle § 3.3.1(g) is the source and the reason for the various forms of *maximum principles* and *Phragmén–Lindelöf principles*, these latter being generalizations of both the classical maximum principle (Ω is a bounded open set, $f \in \text{Hol}(\Omega) \cap C(\overline{\Omega}) \Rightarrow \sup_\Omega |f| = \sup_{\partial\Omega} |f|$) and Liouville's theorem (a bounded entire function is constant). In our presentation of the subject we have followed Helson (1964). Researchers needing to construct and apply their own Phragmén–Lindelöf principle must know how to distinguish the outer functions in domains Ω that can at times be quite complicated. A few examples are given in § 3.4.3 and § 3.5; others can be found in Boas (1954), Levin (1956), and Pólya and Szegő (1925). Nikolski and Volberg (1990) propose a different method, based on the properties of Green's functions. In particular, they show that a restriction $B \mid \Omega$ of a Blaschke product on any simply connected domain that does not contain any zeros of B is outer in Ω, and the same is true for the restrictions $V \mid \Omega$ of a singular function V on a domain "sufficiently thin" (for example, the restriction of $\exp(-\frac{1+z}{1-z})$ on a disk $D(r, 1-r), 0 < r < 1$, is not outer, but the restriction on a Stolz angle S_1 is; see also § 3.5.3). It should also be mentioned (continuing the discussion of § 2.9 on Jensen's inequality) that a necessary and sufficient condition on the modulus of the boundary values of an outer function in Ω is $\log |f| \in L^1(\partial\Omega, \omega_z)$ where ω_z is the harmonic measure of Ω at a point $z \in \Omega$.

Examples 3.4.3(a,b), as well as the exercises of § 3.5, are taken from Phragmén and Lindelöf (1908); since then, they have been widely used, have become classics, and have given rise to a multitude of variations and generalizations. We refer to the monographs and texts already mentioned above.

4

An Introduction to Weighted Fourier Analysis

Topics. Generalized Fourier series, Schauder bases, skew projections, angle between the past and the future, Hilbert operator (harmonic conjugation in $L^2(\mathbb{T})$), Helson–Szegő theorem, angular operators, Babenko's example, basis constant versus unconditional basis constant, Gram matrices, the McCarthy–Schwartz inequality.

In this chapter the reader will find an introduction to Fourier analysis in the weighted spaces $L^2(\mathbb{T}, \mu)$. The point of origin is the "flat" space $L^2(\mathbb{T}, m)$, where the exponentials $(e^{int})_{n \in \mathbb{Z}}$ form an orthonormal basis, and hence the Fourier developments are automatically convergent, and Parseval's identity establishes a direct link between the values of a function and its Fourier coefficients (see Appendix C). This fundamental fact allows the techniques of Fourier analysis to be widely extended, including applications to integral operators, problems of optimization, approximation theory, etc.

The situation changes completely if a "weight" is introduced, i.e. if we pass to $L^2(\mathbb{T}, wm)$, and then to $L^2(\mathbb{T}, \mu)$, with μ an arbitrary Borel measure. In this chapter we introduce a technique capable of managing such a situation. The first three sections are devoted to a few preliminaries on subspaces and sequences, that will later be used uniquely in the framework of Hilbert spaces. However, their geometrical nature renders these properties more transparent in the framework of Banach spaces. Thus we first place ourselves in this more general context.

106

Joseph Fourier (1768–1830) was a French mathematician and physicist, the founder of Fourier analysis and of the mathematical theory of the propagation of heat, the discoverer of the greenhouse effect in climatology, member of the Académie des Sciences, permanent secretary (section of mathematical sciences, 1826), member of the Académie Française, Prefect of the Department of Isère (1802), Baron of the Empire (1809), chevalier (1804) and then officer of the Legion of Honor, and member of the Swedish Royal Academy (1823). Ninth of a family of twelve children, and orphaned when ten years old, Fourier studied under Lagrange, Laplace, and Monge at the École Normale in "Year III" (1795). In 1793, as a member of the Revolutionary Committee at Auxerre, he had considered resigning after the start of the Terror; when he was arrested in 1794, it was only Robespierre's execution that spared him from the guillotine. He took part in the Egyptian campaign, where he was at the heart of the Institut d'Égypte. In Grenoble, he met Champollion and passed on to him his passion for Egypt. As Prefect of Isère, Fourier oversaw several important projects, including the construction of a road from Grenoble to Turin. At the same time, he formulated his theory of heat (1804–1807), for which, in particular, he discovered the equation of heat propagation, the method of separation of variables, and *Fourier series*, i.e. the expansion of an "arbitrary" function f as a trigonometric series $f = \sum c(n) \exp(inx)$ with a calculation rule for the coefficients $c(n)$ based on an integral formula. For more than 15 years, Fourier fought against the objections of Lagrange, Laplace, Poisson (in particular concerning the use of series whose convergence is not known), and Biot (who claimed to have discovered the "true" heat equation in 1804, but which turned out to be false). Fourier was even obliged to prepare a pamphlet in his defense, *Précis historique*, which he distributed in academic circles but never published. He only succeeded in publishing his work *Théorie analytique de la chaleur* in 1822 (with the quotation *Et ignem regunt numeri*, "The numbers govern even fire," which he attributes to Plato though the source cannot be found). Over the next two centuries, this work remained a cornerstone of harmonic analysis. Among other things, the common notation for the definite integral $\int_a^b \varphi(x)\,dx$ was introduced in this text.

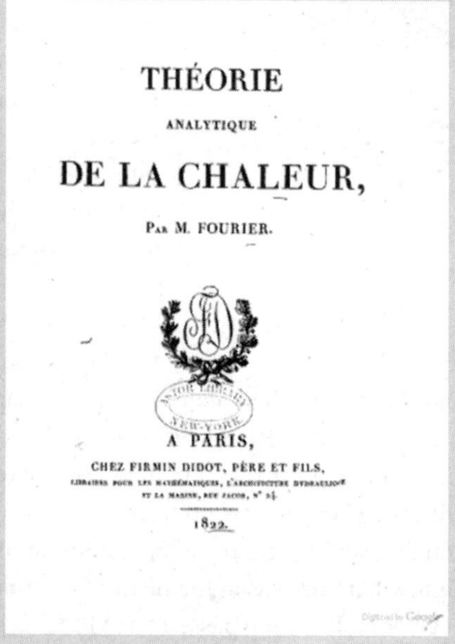

Fourier's *Théorie analytique de la chaleur* (1822), which provided the impetus for the later development of PDEs and harmonic analysis (despite the objections of contemporaries such as Biot and Poisson).

4.1 Generalized Fourier Series

In a Banach space X (always on the field $\mathbb{C}$), we study the sequences x_n indexed by $\mathbb{Z}$, or by $\mathbb{Z}_+ = \{n \in \mathbb{Z}: n \geq 0\}$, or by a finite set of indices. To abbreviate the notation, in each case we write $\mathcal{X} = (x_n)$ for such a sequence, if necessary stating the set of indices. We associate with the space X its dual space X^* and let $\langle \cdot, \cdot \rangle$ denote the bilinear form representing the duality (see also Appendix D).

4.1.1 Minimal Sequences

Minimal sequences are introduced with a list of definitions and short lemmas.

(a) Definition. Let $\mathcal{X} = (x_n)$ be a sequence in X. It is called *minimal* (or topologically free) if, for every n,

$$x_n \notin \mathrm{span}_X(x_k : k \neq n),$$

and *uniformly minimal* if

$$\delta(\mathcal{X}) := \inf_n \mathrm{dist}\left(\frac{x_n}{\|x_n\|}, \mathrm{span}_X(x_k : k \neq n) \right) > 0.$$

(b) Lemma. With the notation of (a), a sequence $\mathcal{X}$ is minimal if and only if, for every k, there exists a functional $x'_k \in X^*$ such that

$$\langle x_n, x'_k \rangle = \delta_{n,k},$$

where $\delta_{n,k} = 1$ if $n = k$ and $\delta_{n,k} = 0$ otherwise (this is the *Kronecker delta*). For every finite linear combination,

$$\left\langle \sum a_n x_n, x'_k \right\rangle = a_k \quad (a_n \in \mathbb{C}),$$

and hence, in the case where $\mathcal{X}$ generates X, $X = \mathrm{span}_X(\mathcal{X})$ ($\mathcal{X}$ is said to be complete in X), the x'_k are uniquely determined by $\mathcal{X}$.

Proof The Hahn–Banach theorem (see Appendix D) implies that for a set $A \subset X$ and an element $x \in X$, $x \in \mathrm{span}_X(A) \Leftrightarrow (f \in X^*, f \mid A = 0 \Rightarrow \langle x, f \rangle = 0)$, and the result follows. $\blacksquare$

(c) Definition. The functional x'_k of (b) is called a *coordinate functional*, and the sequence $\mathcal{X}'$ is said to be the *dual* (to $\mathcal{X}$). A pair $(\mathcal{X}, \mathcal{X}')$, where $\mathcal{X}$ is a minimal sequence and $\mathcal{X}'$ a dual sequence, is called a *biorthogonal pair*.

(d) Lemma. $\mathcal{X}$ is uniformly minimal if and only if there exists a dual sequence such that $\sup_n \|x_n\| \cdot \|x'_n\| < \infty$. If $\mathcal{X}$ is complete in X, then

$$\sup_n \|x_n\| \cdot \|x'_n\| = \frac{1}{\delta(\mathcal{X})}.$$

Proof By a corollary of the Hahn–Banach theorem, for any $x \in X$ and every subspace $E \subset X$, we have

$$\mathrm{dist}_X(x, E) = \max\{|\langle x, f\rangle|: f \in X^*, f \mid E = 0, \|f\| \leq 1\},$$

hence

$$\min\{\|f\|: f \in X^*, \langle x, f\rangle = 1, f \mid E = 0\} = 1/\mathrm{dist}_X(x, E).$$

Thus, by choosing a coordinate functional x'_n of minimal norm, we obtain

$$1/\|x'_n\| = \mathrm{dist}(x_n, \mathrm{span}_X(x_k : k \neq n)),$$

or

$$\mathrm{dist}\left(\frac{x_n}{\|x_n\|}, \mathrm{span}_X(x_k : k \notin n)\right) = \frac{1}{\|x_n\| \cdot \|x'_n\|},$$

and the result follows. $\blacksquare$

(e) Definition. Let $(X, \mathcal{F})$ be a biorthogonal pair in X. With each $x \in X$ we associate a *generalized Fourier series*

$$x \sim \sum_n \langle x, x'_n\rangle x_n,$$

and the *partial sums*

$$P_{k,l}x = \sum_{k \leq n \leq l} \langle x, x'_n\rangle x_n.$$

The following least upper bound (if it is finite) is called the *basis constant of* X (or of $\mathcal{F}$):

$$b(X) := \sup_{k,l} \|P_{k,l}\|.$$

The quantity

$$m(X) := \sup_k \|P_{k,k}\| = \sup_k \|x_k\| \cdot \|x'_k\|$$

is called the *uniform minimality constant of* X. We have $m(X) = 1/\delta(X)$ and $m(X) \leq b(X)$. Given that $P_{k,l}$ is a projection on X (i.e. $P_{k,l}^2 = P_{k,l}$), we have $\|P_{k,l}\| \geq 1$, and if $X = H$ is a Hilbert space the equality $\|P_{k,l}\| = 1$ is equivalent to the fact that $P_{k,l}$ is an orthogonal projection. Hence we always have $b(X) \geq m(X) \geq 1$ and, in a Hilbert space, $m(X) = 1$ (or $b(X) = 1$) if and only if X is an orthogonal basis.

4.1.2 Bases

The notion of a basis of a Banach space was introduced in Banach's founding text of functional analysis, the *Théorie des opérations linéaires* (Banach, 1932).

Stefan Banach (1892–1945), a Polish mathematician, creator of the theory of normed vector spaces, was one of the founders (with Maurice Fréchet) of twentieth century functional analysis. Son of Stefan Greczek, a simple soldier in the service of the Austro-Hungarian Empire, who was married to Katarzyna Banach, young Stefan was abandoned by his mother four days after his birth (presumably, as he was never able to be certain on this subject), and then by his father, who entrusted him to some friends. He studied at the Technical University of Lwów (1910–1914) (Lwów = Lviv = Lvov = Lemberg), and then continued at the Jagiellonian University in Krakow. There, in 1916, in the torment of the First World War, Hugo Steinhaus, who was about to take up a professorship at Lwów, was strolling in Krakow's Planty Park when he heard the words "Lebesgue measure" in a conversation between two youths: at this time and place, almost nobody could be expected to know this combination of words! It was Stefan Banach and his friend Otto Nikodym. This is how the Banach–Steinhaus collaboration began (Banach moved to Lwów); it was highly productive and lasted up to the beginning of the Second World War. The mathematical community of Lwów was substantial: Steinhaus, Banach, Orlicz, Saks, Mazur, Ulam, Schauder, Mark Kac, and others. They gathered together almost every day at the "Scottish Café" to discuss mathematics, pose problems, and work together. The results of each day were written up in a notebook, the "Scottish Book," which became famous because of its character of a daily mathematical gazette and its collection of unsolved problems and their associated prizes. Some of these were quite humorous: for example, for the *basis problem* (find a separable Banach space without a Schauder basis: problem 153 in the book, posed by Mazur in 1936), Per Enflo (who provided an example in 1972) received the prize promised in 1936 – a "live goose." Banach founded the international journal *Studia Mathematica*, and then the series Monografie Matematyczne (the first volume was Banach's celebrated *Théorie des opérations linéaires*, the bible of functional analysis specialists for more than 40 years!). In 1939 he was elected President of the Polish Mathematical Society. After the Nazi occupation of Lwów (in June 1941) Banach was arrested, and then released (his thesis advisor, Professor Lomnicki, was killed). During the Nazi occupation (1941–1944), bereft of any teaching activities, Banach gained a living as a lice feeder in the German Institute for Infectious Diseases. When the war was over, Banach resumed his mathematical activities, which he intended to continue at Krakow, but he died from lung cancer in 1945.

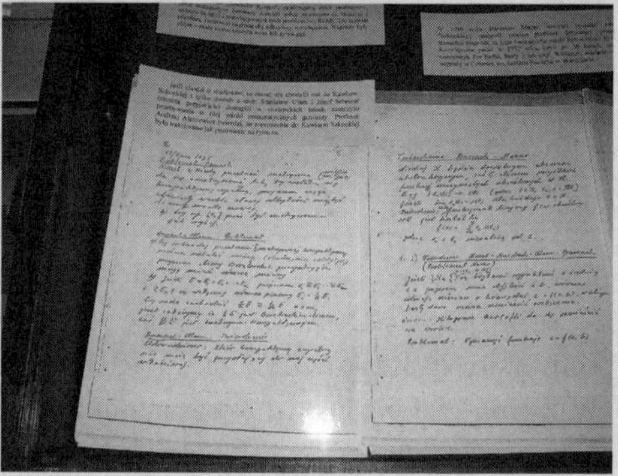

Per Enflo receiving a live goose from Stanisław Mazur in 1972 for solving problem 153 (a separable Banach space without a basis) in the "Scottish Book." The book served as a mathematical diary for the Lwów analysis seminar.

Several mathematical concepts bear Banach's name: the *Banach–Steinhaus*, *Hahn–Banach*, and *Banach fixed point theorems*, the *Banach–Tarski paradox* (on the decomposition of a ball), *Banach spaces*, *Banach algebras* (also defined and investigated by Israel Gelfand under the name of "normed rings"), the *Banach indicatrix*, etc.

(a) Definition. In the case where the set of indices is $\mathbb{Z}_+$, a sequence X is a *Schauder basis* of X, if for every $x \in X$ there exists a unique numerical sequence $(a_n(x))$ such that

$$\lim_k \left\| x - \sum_{n=0}^{k} a_n(x)x_n \right\| = 0.$$

In the case where the set of indices is $\mathbb{Z}$, a sequence X is a *symmetric (respectively, non-symmetric)* Schauder basis of X, if for every $x \in X$ there exists a unique numerical sequence $(a_n(x))$ such that

$$\lim_k \left\| x - \sum_{n=-k}^{k} a_n(x)x_n \right\| = 0 \quad \left(\text{resp. } \lim_{k,l} \left\| x - \sum_{n=-k}^{l} a_n(x)x_n \right\| = 0\right).$$

We admit the following classical theorem.

(b) Theorem (Banach, 1932). *A Schauder basis X of a Banach space X indexed by $\mathbb{Z}_+$ is a uniformly minimal sequence and $a_n(x) = \langle x, x'_n \rangle$ where x'_n is a coordinate functional of X, and hence*

$$x = \sum_{n \geq 0} \langle x, x'_n \rangle x_n$$

is a Fourier series of x convergent for the norm of X.

The same holds for bases indexed by $\mathbb{Z}$, in the sense of non-symmetrical convergence.

(c) Lemma. Let $(X, \mathcal{F})$ be a biorthogonal pair in X. Then:

(1) If X is a basis of X, then X is complete in X and $\mathcal{F}$ is total on X, i.e.
   $(x \in X, \langle x, x'_n \rangle = 0 \forall n) \Rightarrow x = 0$.
(2) X is a basis of X if and only if X is complete in X and $\sup_{k \geq 0} \|P_{0,k}\| < \infty$
   (when indexed by $\mathbb{Z}_+$), $\sup_{k \geq 0} \|P_{-k,k}\| < \infty$ (when indexed by $\mathbb{Z}$ and with
   symmetrical convergence) $\sup_{k,l \geq 0} \|P_{-k,l}\| < \infty$ (when indexed by $\mathbb{Z}$ and
   with non-symmetrical convergence).

Proof Property (2) is immediate by (b) above, the Banach–Steinhaus theorem (see Appendix D), and the fact that $x = \lim_k P_{0,k}x$ for every $x \in \mathrm{Lin}(X)$ (the linear hull of X), with the obvious modifications when indexed by $\mathbb{Z}$.

If, for every n, $\langle x, x_n' \rangle = 0$, then $P_{0,k} x = 0$ for any k, hence the property (1). ∎

4.2 Skew Projections

In analysis in a Banach (or Hilbert) space, an important role is played by the metric geometry of the space in question. For a Hilbert space this is a fact we could of course expect (since this is the case for elementary analysis in $\mathbb{R}^n$ and $\mathbb{C}^n$, $n = 1, 2, 3, \ldots$, and the relations between two or three elements of a Hilbert space are the same as in $\mathbb{R}^n$ and $\mathbb{C}^n$, $n = 2, 3$), and thus everything is ready to define and exploit geometrical concepts such as angles, orthogonality, etc. In a Banach space, this is not so evident because everything must be derived from metrical relations only. In this short section, we introduce the usage of one of these metrical tools: the skew projections. An angle between subspaces can be defined as a function of the skew projections: see § 4.3 below.

Definition 4.2.1 *Let L, M be two subspaces of a vector space X such that $L \cap M = \{0\}$ (there is no hypothesis of any norm or metric on X). Then, the mapping $P = P_{L\|M} : L + M \to X$ defined by*

$$P(x + y) = x \quad (x \in L, y \in M)$$

is called a (skew) projection onto L parallel to M.

4.2.1 Properties of $P_{L\|M}$

Let L, M be two subspaces of a Banach space X such that $L \cap M = \{0\}$.

(a) $P = P_{L\|M}$ is linear, $P \mid L = \mathrm{id}$, $P \mid M = 0$, $P^2 = P$ (these properties justify the name "projection" in Definition 4.2.1).

Proof Evident. ∎

(b) $P_{L\|M}$ is continuous if and only if $P_{\overline{L}\|\overline{M}}$ is continuous (where $\overline{A} = \mathrm{clos}_X(A)$ is the closure of $A \subset X$).

Proof Clear since continuity is equivalent to $\|x\| \le C\|x + y\|$ for every $x \in L$, $y \in M$. ∎

(c) If L and M are closed subspaces, $P_{L\|M}$ is continuous if and only if $L + M$ is closed, or $L + M = \mathrm{clos}_X(L + M)$.

Proof This is a standard consequence of the closed graph theorem. ∎

(d) *Suppose that $X = H$ is a Hilbert space. Then*

$$\|P_{L\|M}\| = \sup_{x \in L} \frac{\|x\|}{\|(I - P_{\overline{M}})x\|} = \sup_{x \in L} \frac{\|x\|}{\|P_{\overline{M}^{\perp}} x\|},$$

where $P_{\overline{M}}$ is an orthogonal projection on $\overline{M}$.

Proof

$$\|P_{L\|M}\| = \sup_{x \in L, y \in M} \frac{\|x\|}{\|x + y\|} = \sup_{x \in L} \frac{\|x\|}{\inf_{y \in M} \|x + y\|},$$

and, clearly, $\inf_{y \in M} \|x + y\| = \|(I - P_{\overline{M}})x\|$. ∎

4.3 The Angle Between the Past and the Future

The title of this section refers to one of the main applications of the technique developed in this chapter: stationary processes. In this specific application, the subspaces L and M in the following definition are taken to be the past $L = \mathrm{span}_H(x_n : n < 0)$ and the future $M = \mathrm{span}_H(x_n : n \geq 0)$ of a processes $(x_n)_{n \in \mathbb{Z}}$ in a Hilbert space H: see § 4.7.

Definition 4.3.1 (angle between two subspaces) *Let H be a Hilbert space and L, M two subspaces of H. The angle (or, the minimal angle) between L and M is a number $A = A(L, M)$ defined by the properties $0 \leq A \leq \pi/2$ and*

$$\cos(A) = \sup_{x \in L, y \in M} \frac{|(x, y)|}{\|x\| \cdot \|y\|}.$$

4.3.1 Properties of the Angle

Let H be a Hilbert space and L, M two subspaces of H.

(a) $A(L, M) = A(\overline{L}, \overline{M})$, and $L \perp M \Leftrightarrow A(L, M) = \pi/2$.

Proof Clear. ∎

(b) $\cos A(L, M) = \|P_{\overline{M}} P_{\overline{L}}\| = \|P_{\overline{L}} P_{\overline{M}}\|$.

Proof If $x \in L$, $y \in M$, then $(x, y) = (P_{\overline{M}} x, y)$ and hence

$$\cos A(L, M) = \sup_{x \in L} \sup_{y \in M} \frac{|(P_{\overline{M}} x, y)|}{\|x\| \cdot \|y\|} = \sup_{x \in L} \frac{\|P_{\overline{M}} x\|}{\|x\|}$$

$$= \sup_{x \in L} \frac{\|P_{\overline{M}} P_{\overline{L}} x\|}{\|x\|} = \sup_{x \in H} \frac{\|P_{\overline{M}} P_{\overline{L}} x\|}{\|x\|} = \|P_{\overline{M}} P_{\overline{L}}\|. \qquad ∎$$

(c) $\sin A(L, M) = \|P_{L\|M}\|^{-1}$, *where $P_{L\|M}$ is the skew projection of* § 4.2.

Proof We have

$$\sin^2 A(L, M) = 1 - \cos^2 A(L, M) = 1 - \sup_{x \in L} \frac{\|P_{\overline{M}}x\|^2}{\|x\|^2} = \inf_{x \in L} \frac{\|(I - P_{\overline{M}})x\|^2}{\|x\|^2}$$

$$= \frac{1}{\sup_{x \in L}(\|x\|^2 / \|(I - P_{\overline{M}})x\|^2)} = \|P_{L\|M}\|^{-2},$$

where the last equation follows from § 4.2.1(d).   ∎

(d) *The skew projection $P_{L\|M}$ is bounded if and only if $\|P_{\overline{M}}P_{\overline{L}}\| < 1$, and if and only if $A(L, M) > 0$.*

Proof Apply (c), then (b).   ∎

4.4 The Case of the Exponentials: A Reduction to P_+

We now turn to the principal subject of this chapter: the exponential bases in the spaces $L^2(\mathbb{T}, \mu)$, where μ is a finite Borel measure on $\mathbb{T}$. It hence concerns the sequence of exponentials $X = \mathcal{E}$,

$$\mathcal{E} = (z^k)_{k \in \mathbb{Z}} = (e^{ikt})_{k \in \mathbb{Z}},$$

which, for any μ, is complete in $L^2(\mathbb{T}, \mu)$: $L^2(\mu) = \mathrm{span}_{L^2(\mu)}(z^n : n \in \mathbb{Z}) = \mathrm{clos}_{L^2(\mu)} \mathcal{P}$.

Lemma 4.4.1 (Kolmogorov, 1941) *Let $\mu = \mu_s + \mu_a = \mu_s + w \cdot m$ be a finite Borel measure on $\mathbb{T}$ with its Radon–Nikodym decomposition.*

(1) *The family of exponentials $\mathcal{E}$ is minimal in $L^2(\mathbb{T}, \mu)$ if and only if it is minimal in $L^2(\mathbb{T}, wm)$, and if and only if*

$$1/w \in L^1(\mathbb{T})$$

     (in particular, the last point holds if $\mathcal{E}$ is a basis in the sense of symmetrical summation).

(2) *The uniform minimality constant is*

$$\delta(\mathcal{E}) = \left(\int_{\mathbb{T}} d\mu \right)^{-1/2} \left(\int_{\mathbb{T}} \frac{1}{w} \, dm \right)^{-1/2},$$

     and the dual sequence is

$$\mathcal{E}' = \left(\frac{(z^n)_a}{w} \right) \subset L^2(\mathbb{T}, \mu_a) = L^2(\mathbb{T}, wm).$$

(3) $\mathcal{E}'$ is complete in $L^2(\mathbb{T}, \mu)$ if and only if $\mu_s = 0$.

(4) If $\mathcal{E}$ is a basis (in the symmetrical sense or not) then $\mu_s = 0$.

Proof (1) Suppose that $\mathcal{E}$ is minimal in $L^2(\mathbb{T}, \mu) = L^2(\mu_s) \oplus L^2(wm)$. Then, for every $n \in \mathbb{Z}_+$, the subspace $E_n := \operatorname{span}_{L^2(\mu)}(z^k : k > n)$ is M_z-invariant, and moreover $E_n = z^{n+1} H^2(\mu) \supset L^2(\mu_s)$ (for the last inclusion see Lemma 1.5.2(2)). Hence, if $x_n' \in L^2(\mu)$ is a dual sequence of $\mathcal{E}$ (it is unique given that $\mathcal{E}$ is complete in $L^2(\mathbb{T}, \mu)$), we have $x_n' \perp E_n$, thus $x_n' \perp L^2(\mu_s)$ for every $n \in \mathbb{Z}$. This implies

$$\delta_{n,k} = (z^k, x_n')_{L^2(\mu)} = (z^k, x_n')_{L^2(wm)} = \int_{\mathbb{T}} z^k \overline{x_n'} w \, dm.$$

Note that $x_n' \in L^2(wm) \subset L^1(wm)$, i.e. $\overline{x_n'} w \in L^1(\mathbb{T})$, and thus the function $\overline{x_n'} w$ has the same Fourier coefficients as $\overline{z}^n$, so that $\overline{x_n'} w = \overline{z}^n$ and

$$x_n' = z^n / w.$$

Now the inclusion $x_n' \in L^2(wm)$ gives $1/w \in L^1(\mathbb{T})$, which proves the necessity of the condition.

Conversely, if $1/w \in L^1(\mathbb{T})$, then clearly the functions $x_n' = (x_n')_a + (x_n')_s$ with $(x_n')_a = z^n/w$ and $(x_n')_s = 0$ are biorthogonal in $L^2(\mathbb{T}, \mu)$ with (z^k), hence $\mathcal{E}$ is minimal in $L^2(\mathbb{T}, \mu)$.

(2) The explicit formula above for x_n' and § 4.1.1(d) show that

$$1/\delta(\mathcal{E}) = \|z^n\|_{L^2(\mu)} \|x_n'\|_{L^2(\mu)} = \|1\|_{L^2(\mu)} \|1/w\|_{L^2(wm)},$$

which is equivalent to the stated formula.

(3) For the completeness of $\mathcal{E}'$, we know that, for every n, $x_n' \in L^2(wm)$ and, moreover, if $f \in L^2(wm)$ and $0 = (f, x_n')_{L^2(wm)}$ for every n, then $f = 0$ as a function of $L^1(\mathbb{T})$ with all of its Fourier coefficients zero. Thus, $\operatorname{span}_{L^2(\mu)}(x_n' : n \in \mathbb{Z}) = L^2(wm)$, and the result follows.

(4) Suppose that $\mathcal{E}$ is a basis. We keep the notation of part (1). If $x \in \bigcap_{n \in \mathbb{Z}} E_n$, then $x \in E_n$, thus $(x, x_n') = 0$ for every $n \in \mathbb{Z}$, and hence $x = \sum_{n \in \mathbb{Z}} (x, x_n') z^n = 0$. As was seen in (1), $L^2(\mu_s) \subset \bigcap_{n \in \mathbb{Z}} E_n$, so that $L^2(\mu_s) = \{0\}$, i.e. $\mu_s = 0$. ∎

Corollary 4.4.2 *Let a function $w \in L^1(\mathbb{T})$ be such that $1/w \in L^1(\mathbb{T})$ and let $(x_n = z^n, x_n' = z^n/w)$, $n \in \mathbb{Z}$ be a biorthogonal pair of exponentials in the weighted space $L^2(\mathbb{T}, wm)$. If $f \in L^2(\mathbb{T}, wm)$, then $f \in L^1(\mathbb{T})$ and its generalized Fourier series*

$$\sum_n (f, x_n')_{L^2(w)} z^n$$

coincides with the classical Fourier series:

$$(f, x'_n)_{L^2(w)} = \widehat{f}(n), \quad \forall n \in \mathbb{Z}.$$

Indeed, this is clear by Lemma 4.4.1.   ∎

We conclude this § 4.4 by connecting the question of bases of exponentials with the angles and skew projections of § 4.2 and § 4.3. We first introduce notation for "analytic" and "anti-analytic" polynomials:

$$\mathcal{P}_+ = \mathcal{P}_a = \mathrm{Lin}(z^k : k \geq 0), \quad \mathcal{P}_- = \mathrm{Lin}(z^k : k < 0),$$

so that $\mathcal{P} = \mathcal{P}_+ + \mathcal{P}_-$. We recall the notation for the partial sum and Riesz projections (see § 4.1.1(e), § 2.8.4(e)): if $f \in \mathcal{P}$, $f = \sum_{n \in \mathbb{Z}} \widehat{f}(n) z^n$, then

$$P_{k,l}f = \sum_{k \leq n \leq l} \widehat{f}(n)z^n, \quad P_+f = \sum_{n \geq 0} \widehat{f}(n)z^n,$$

where $k, l \in \mathbb{Z}$, $k \leq l$. Clearly, in the sense of Definition 4.2.1, we have

$$P_+ = P_{\mathcal{P}_+ \| \mathcal{P}_-}.$$

Similarly, by defining $P_-f = \sum_{n < 0} \widehat{f}(n)z^n$, we obtain

$$P_- = I - P_+ = P_{\mathcal{P}_- \| \mathcal{P}_+}.$$

Lemma 4.4.3 *Let $w \in L^1(\mathbb{T})$, $w \geq 0$. The following assertions are equivalent:*

(1) *$\mathcal{E} = (z^k)_{k \in \mathbb{Z}}$ is a basis of $L^2(wm)$ for the non-symmetric partial sums.*
(2) *$\mathcal{E} = (z^k)_{k \in \mathbb{Z}}$ is a basis of $L^2(wm)$ for the symmetric partial sums.*
(3) *$b(\mathcal{E}) := \sup_{k,l} \|P_{k,l}\| < \infty$ (see § 4.1.1(e) for the definition of the basis constant b).*
(4) *$\sup_{k \geq 0} \|P_{-k,k}\| < \infty$.*
(5) *$A_{L^2(w)}(\mathcal{P}_+, \mathcal{P}_-) > 0$.*
(6) *$A_{L^2(w)}(H^2_+(w), H^2_-(w)) > 0$ where $H^2_\pm(w) = \mathrm{clos}_{L^2(w)}(\mathcal{P}_\pm)$.*
(7) *P_+ is continuous on $L^2(wm)$.*

Moreover,

$$\|P_+\| = (\sin(A_{L^2(w)}(\mathcal{P}_+, \mathcal{P}_-)))^{-1},$$

$$\|P_+\| \leq b(\mathcal{E}) = \sup_{k,l} \|P_{k,l}\| \leq \min\{\|P_+\|^2, 2\|P_+\|\}.$$

Proof Clearly (1) $\Rightarrow$ (2), (2) $\Leftrightarrow$ (4) (by the Banach–Steinhaus theorem, see also § 4.1.2(c)), (1) $\Leftrightarrow$ (3) and (5) $\Leftrightarrow$ (6) $\Leftrightarrow$ (7) (by § 4.3.1(d)).

$(4) \Rightarrow (7)$ since for any $f \in \mathcal{P}$ there exists $k = k(f)$ such that $P_+ f = z^k P_{-k,k} z^{-k} f$, and hence

$$\|P_+ f\| \le \|P_{-k,k} z^{-k} f\| \le \|P_{-k,k}\| \cdot \|f\| \le \left(\sup_{k \ge 0} \|P_{-k,k}\| \right) \|f\|.$$

$(7) \Rightarrow (3)$ since, for any polynomial $f \in \mathcal{P}$, we have

$$P_{k,l} f = z^k P_+ z^{-k} f - z^{l+1} P_+ z^{-l-1} f,$$

and thus $\|P_{k,l} f\| = \|z^k P_+ z^{-k} f\| + \|z^{l+1} P_+ z^{-l-1} f\| \le 2\|P_+\| \cdot \|f\|$, giving

$$b(\mathcal{E}) = \sup_{k,l} \|P_{k,l}\| \le 2\|P_+\|.$$

The equality for $\|P_+\|$ is § 4.3.1(c).

The lower estimate of $b(\mathcal{E})$ is evident, since $P_+ f = \lim_{l \to \infty} P_{0,l} f$ for every $f \in \mathcal{P}$.   ∎

4.5 The Hilbert Operator: The Classical Case of $L^2(\mathbb{T})$

The last prerequisite for the principal result of this chapter (§ 4.6) concerns harmonic conjugation (see § 2.8.4), this time in the Hilbert space $L^2(\mathbb{T})$. Recall that for a real function $u \in L^p(\mathbb{T})$, $1 < p < \infty$, the harmonic conjugation mapping H of § 2.8.4(d) (also called the *Hilbert operator*) is defined by the properties

$$u + iHu \in H^p(\mathbb{T}), \quad \widehat{Hu}(0) = 0.$$

Clearly H is linear (over the field $\mathbb{R}$): see § 2.8.4(d), where it was shown that $\|Hu\|_p \le A_p \|u\|_p$ with a certain constant $A_p < \infty$. H can be "complexified" by defining $H(u + iv) = Hu + iHv$ for every function $u + iv \in L^p(\mathbb{T})$, where u, v are real. Then, H becomes linear over the field $\mathbb{C}$, $H \colon L^p(\mathbb{T}) \to L^p(\mathbb{T})$ and bounded with norm $\|H\| \le 2A_p$.

Lemma 4.5.1 *The mapping H of § 2.8.4(d) admits the representation*

$$Hf = \frac{1}{i}(P_+ f - P_- f) - \frac{1}{i}\widehat{f}(0),$$

and, in the space $L^2(\mathbb{T})$, satisfies the properties

$\|Hf\|_2 \le \|f\|_2$ $(\forall f \in L^2(\mathbb{T}))$ *and* $\|Hf\|_2 = \|f\|_2$ $(\forall f \in L^2(\mathbb{T}), \widehat{f}(0) = 0).$

Proof Given a polynomial $f \in \mathcal{P}$, denote the right side of the formula by g. Then, $f + ig = 2P_+ f - \widehat{f}(0) \in H^p$ and $\widehat{g}(0) = 0$. Hence, $g = Hf$ by the definition of Hf cited before the statement of the theorem. Consequently, $\widehat{Hf}(n) = \pm\widehat{f}(n)/i$ for $\pm n > 0$, and we obtain the properties of the norm $\|Hf\|_2$ by applying Parseval's identity $\|Hf\|_2^2 = \sum_{n\neq 0}|\widehat{f}(n)|^2$. ∎

Remark 4.5.2 The operator H is often called the *Hilbert operator*. It can be represented in the form of a singular integral operator by using the formulas of § 2.8.4. Indeed, we have seen in § 2.8.4(d) that for a real function $f \in L^2(\mathbb{T})$, $Hf = \mathrm{Im}(\Gamma f)$ where Γ is the Herglotz operator,

$$\Gamma f(z) = \int_{\mathbb{T}} \frac{\zeta + z}{\zeta - z} f(\zeta)\, dm(\zeta), \quad z \in \mathbb{D}.$$

Hence, for $z = re^{i\theta}$, $0 < r < 1$, we have

$$Hf(re^{i\theta}) = \int_{\mathbb{T}} \mathrm{Im}\left(\frac{\zeta + z}{\zeta - z}\right) f(\zeta)\, dm(\zeta) = \int_0^{2\pi} Q_r(t - \theta) f(e^{it})\, dt/2\pi,$$

where

$$Q_r(t) = \mathrm{Im}\left(\frac{e^{it} + r}{e^{it} - r}\right) = \frac{2r\sin(t)}{1 - 2r\cos(t) + r^2}.$$

Since $\Gamma f \in H^2(\mathbb{T})$, there exist a.e. on $\mathbb{T}$ the boundary limits of $Hf(re^{i\theta})$ as $r \to 1$, and it can be shown that for the right-hand side of the equality, there also exists a limit which is equal to the "Cauchy principal value of the integral":

$$Hf(re^{i\theta}) \to V.P. \int_0^{2\pi} f(e^{it}) \cotan\left(\frac{t - \theta}{2}\right) \frac{dt}{2\pi}.$$

We will not use this form of the operator H, and refer the reader to the vast theory of singular integrals; see Zygmund (1959) or Duoandikoetxea (2001).

4.6 Exponential Bases in $L^2(\mathbb{T}, \mu)$

It is easy to show a certain number of measures with density $\mu = wm$, $w \in L^1(\mathbb{T})$, such that the exponentials $\mathcal{E} = (z^k)_{k \in \mathbb{Z}}$ form a basis for the space $L^2(\mu)$. For example, any w with $0 < \inf_{\mathbb{T}} w \le \sup_{\mathbb{T}} w < \infty$ is of this type (see Exercise 4.9.2 for more details). It is much more difficult to find such a weight w with a "singularity" (either $\inf_{\mathbb{T}} w = 0$, or $\sup_{\mathbb{T}} w = \infty$). The very first example was given only in 1950 by Ivan Babenko (see Example 4.6.6 below). The complete resolution of the question came only ten years later with the following result of Helson and Szegő.

Theorem 4.6.1 (Helson and Szegő, 1960) *Let $\mu = \mu_s + w \cdot m$ be a finite Borel measure on $\mathbb{T}$ with its Radon–Nikodym decomposition. The following assertions are equivalent.*

(1) *$\mathcal{E} = (z^k)_{k\in\mathbb{Z}}$ is a Schauder basis of the space $L^2(\mu)$ (either symmetrical or non-symmetrical).*

(2) *The Riesz projection P_+ is well-defined and bounded on $L^2(\mu)$.*

(3) *$A_{L^2(\mu)}(\mathcal{P}_+,\mathcal{P}_-) > 0$.*

(4) *$\mu_s = 0$, $w = |h|^2$ where $h \in H^2$ is an outer function such that*

$$\operatorname{dist}_{L^\infty(\mathbb{T})}(\overline{h}/h, H^\infty) < 1.$$

(5) *$\mu_s = 0$ and $w = e^{u+Hv}$ where $u, v \in L^\infty(\mathbb{T})$ are two real functions, $\|v\|_\infty < \pi/2$.*

Proof   (1) $\Leftrightarrow$ (2) $\Leftrightarrow$ (3) by Lemma 4.4.3.

(3) $\Leftrightarrow$ (4) If $\mathcal{E}$ is a basis, by § 4.1.2(b) it is minimal, and by Lemma 4.4.1 $\mu_s = 0$ and $1/w \in L^1(\mathbb{T})$, in particular $\log(w) \in L^1(\mathbb{T})$. Thus there exists an outer function $h \in H^2$ such that $w = |h|^2$ (see Corollary 2.6.2).

Now, we simply calculate $\cos(A_{L^2(\mu)}(\mathcal{P}_+,\mathcal{P}_-))$. For any $f \in \mathcal{P}_+$, $g \in \mathcal{P}_-$ we have

$$(f,g)_{L^2(\mu)} = \int_{\mathbb{T}} f\overline{g}w\,dm = \int_{\mathbb{T}} f h\overline{g}h\frac{\overline{h}h}{h^2}\,dm = \int_{\mathbb{T}} f h\overline{g}h\frac{\overline{h}}{h}\,dm := \int_{\mathbb{T}} FG\frac{\overline{h}}{h}\,dm.$$

Here, we have $F = fh \in H^2$, $G = \overline{g}h \in H_0^2 = zH^2$ (as $\overline{g} \in \overline{\mathcal{P}}_- = z\mathcal{P}_+$). Moreover, since the function h is outer, the set

$$L := \{F = fh \colon f \in \mathcal{P}_+\}$$

is dense in H^2 and the set

$$M := \{G = \overline{g}h \colon g \in \mathcal{P}_-\}$$

is dense in H_0^2. Letting B^2 denote the unit ball of H^2, $B^2 = \{\varphi \in H^2 \colon \|\varphi\|_2 < 1\}$, we obtain that $L \cap B^2$ is dense in B^2, and $M \cap B^2$ dense in $B_0^2 := zB^2$. It is also easy to see that the set of products $(L \cap B^2) \cdot (M \cap B^2)$ is dense in the unit ball B_0^1 of the space $H_0^1 = zH^1$ (see § 2.8.2). In view of the isometries

$$\|f\|_{L^2(\mu)} = \|F\|_{L^2(m)} \text{ and } \|g\|_{L^2(\mu)} = \|G\|_{L^2(m)},$$

all this leads to

$$\cos A_{L^2(\mu)}(\mathcal{P}_+,\mathcal{P}_-)$$
$$= \sup\{|(f,g)_{L^2(\mu)}| \colon f \in \mathcal{P}_+,\ g \in \mathcal{P}_-, \|f\|_{L^2(\mu)} < 1, \|f\|_{L^2(\mu)} < 1\}$$

$$= \sup\left\{ \left| \int_{\mathbb{T}} FG\frac{\overline{h}}{h}\, dm \right| : F \in L \cap B^2, G \in M \cap B_0^2 \right\}$$

$$= \sup\left\{ \left| \int_{\mathbb{T}} u\frac{\overline{h}}{h}\, dm \right| : u \in B_0^1 \right\} = \|\Phi \mid H_0^1\|,$$

where the linear functional Φ is defined on the space $L^1(\mathbb{T})$ by the formula

$$\Phi(u) = \int_{\mathbb{T}} u\frac{\overline{h}}{h}\, dm \quad (u \in L^1(\mathbb{T})).$$

We now use the duality $(L^1)^* = L^\infty$ with respect to the bilinear form $\langle \varphi, \psi \rangle = \int_{\mathbb{T}} \varphi\psi\, dm$ (see Appendix D) and the fact that

$$(H_0^1)^\perp := \{\psi \in L^\infty(\mathbb{T}): \langle \varphi, \psi \rangle = 0 \; \forall \varphi \in H_0^1\} = H^\infty$$

(verify!). By the Hahn–Banach theorem (see Appendix D),

$$\|\Phi \mid H_0^1\| = \mathrm{dist}_{L^\infty(\mathbb{T})}\left(\frac{\overline{h}}{h}, (H_0^1)^\perp\right) = \mathrm{dist}_{L^\infty(\mathbb{T})}\left(\frac{\overline{h}}{h}, H^\infty\right),$$

hence

$$\cos(A_{L^2(\mu)}(\mathcal{P}_+, \mathcal{P}_-)) = \mathrm{dist}_{L^\infty(\mathbb{T})}\left(\frac{\overline{h}}{h}, H^\infty\right).$$

Given that $(3) \Leftrightarrow (1)\&(2)$, it follows that $(3) \Leftrightarrow (4)$.

$(4) \Rightarrow (5)$ It follows from (4) that there exists $g \in H^\infty$ such that $\|\overline{h}/h - g\|_\infty < 1$, and hence with a certain $\epsilon > 0$, $|\overline{h}/h - g| < 1 - \epsilon$ and consequently, $\left||h|^2 - gh^2\right| < (1 - \epsilon)|h|^2$ a.e. on $\mathbb{T}$. Let $\zeta \in \mathbb{T}$ and $a = |h(\zeta)|^2$, then $|a - gh^2| < (1 - \epsilon)a$ (outside of a negligible set). If we denote $\alpha = \arcsin(1 - \epsilon)$ $(0 < \alpha < \pi/2)$ and $A = \{z \in \mathbb{C}: |\arg(z)| \le \alpha\}$, then the preceding inequality implies that

$$h^2(\zeta)g(\zeta) \in A \quad (a.e. \; \zeta \in \mathbb{T}).$$

By Corollary 2.2.5, we can be sure that $h^2 g(\mathbb{D}) \subset A$, and hence there exists a holomorphic function $f = \log(h^2 g)$. Set $v = -\mathrm{Im}(f)$. Then

$$|v| = |\arg(h^2 g)| \le \arcsin(1 - \epsilon) < \frac{\pi}{2},$$

$$Hv = \log|h^2 g| + c, \text{ where } c \in \mathbb{R}.$$

It follows that $\log(h^2 g) = Hv - iv - c$, $h^2 g = e^{Hv - iv - c}$. Moreover, we have $|\overline{h}/h - g| < 1 - \epsilon$ and, consequently, $|1 - |g|| < (1 - \epsilon)$ a.e. on $\mathbb{T}$, which implies $\epsilon \le |g| \le 1 + \epsilon$. Hence $|h|^2 = e^{Hv - c}/|g| = e^{Hv + u}$ where $u = -\log|g| - c$, $u \in L^\infty$ and $\|v\|_\infty \le \arcsin(1 - \epsilon) < \pi/2$, and the proof is complete.

$(5) \Rightarrow (4)$ Let u, v be two functions satisfying the conditions of (5). Then, by § 2.8.4(j), $w \in L^1(\mathbb{T})$, as well as $\log(w) \in L^1(\mathbb{T})$, and hence there exists an outer function $h \in H^2$ such that $|h|^2 = w$. Since $\log|h|^2 = u + Hv$,

$$\log(h^2) = u + Hv + iH(u + Hv) = u + Hv + i(Hu - v + c) \text{ where } c \in \mathbb{R}.$$

Setting $g = e^{-(u+iHu)-ic}$, we obtain $g \in H^\infty$ (since $|g| = e^{-u}$ is bounded) and, on $\mathbb{T}$,

$$\frac{h}{\overline{h}}g = \exp(i(Hu - v + c) - u - iHu - ic) = \exp(-u - iv) \text{ where } \|v\|_\infty < \frac{\pi}{2},$$

hence

$$e^{-\|u\|_\infty} \leq \left|\frac{h}{\overline{h}}g\right| \leq e^{\|u\|_\infty}, \quad \left|\arg\!\left(\frac{h}{\overline{h}}g\right)\right| = |v| \leq \|v\|_\infty < \frac{\pi}{2}.$$

It follows that the values of $(h/\overline{h})g$ are in a domain Ω defined by

$$\Omega = \{z \in \mathbb{C}: e^{-\|u\|_\infty} \leq |z| \leq e^{\|u\|_\infty}, \quad |\arg(z)| \leq \|v\|_\infty\}.$$

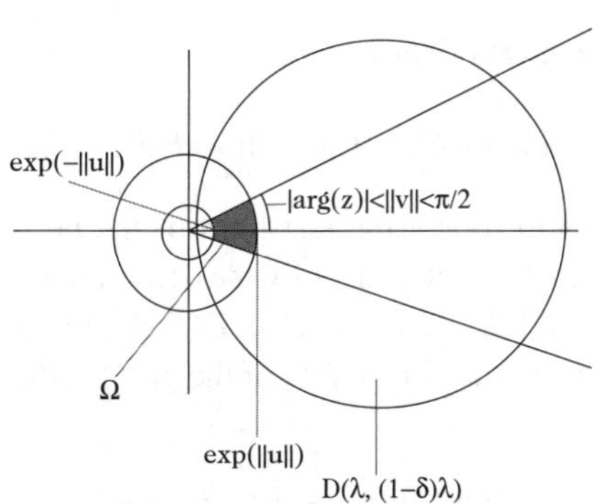

The shaded domain Ω contains the values of the ratio $g h/\overline{h}$ and is contained in the large disk $\mathbb{D}(\lambda, (1 - \delta)\lambda)$ which, in turn, lies in the right half-plane.

Clearly Ω is contained in a disk of large radius lying in the half-plane $\operatorname{Re}(z) > 0$: there exists $\lambda > 0$ and δ, $0 < \delta < 1$, such that $\Omega \subset D(\lambda, (1 - \delta)\lambda)$, and hence

$$\frac{1}{\lambda}\Omega \subset D(1, 1 - \delta).$$

This implies $\lambda^{-1}(h/\overline{h})g \in D(1, 1-\delta)$ a.e. on $\mathbb{T}$, and then $|\lambda^{-1}(h/\overline{h})g - 1| < 1 - \delta$, $|\lambda^{-1}g - \overline{h}/h| < 1 - \delta$ a.e. on $\mathbb{T}$. As $g \in H^\infty$,

$$\operatorname{dist}_{L^\infty(\mathbb{T})}(\overline{h}/h, H^\infty) \leq 1 - \delta < 1. \qquad \blacksquare$$

Definition 4.6.2 *A function $w \geq 0$ is said to be a Helson–Szegő weight, written $w \in (HS)$, if the equivalent conditions (1)–(5) of Theorem 4.6.1 are satisfied.*

Corollary 4.6.3 *With the notation of Theorem 4.6.1, if $w \in (HS)$ then*

$$\cos(A_{L^2(\mu)}(\mathcal{P}_+, \mathcal{P}_-)) = \mathrm{dist}_{L^\infty(\mathbb{T})}\left(\frac{\overline{h}}{h}, H^\infty\right),$$

$$\|P_+\| = \left(1 - \mathrm{dist}^2_{L^\infty(\mathbb{T})}\left(\frac{\overline{h}}{h}, H^\infty\right)\right)^{-1/2}.$$

Indeed, the first equality is established in the proof, and the second is a consequence of the first and of Lemma 4.4.3.   ∎

Corollary 4.6.4 (continuity of $\|P_+\|$ as a function of $w = e^{u+Hv}$) *Let $w = e^{u+Hv}$, and let $\|P_+\|_w$ be the norm of P_+ and $b_w(\mathcal{E})$ be the the basis constant of $\mathcal{E}$ in the space $L^2(wm)$. Then,*

$$\lim_{\substack{\|u\|_\infty \to 0 \\ \|v\|_\infty \to 0}} b_w(\mathcal{E}) = 1, \qquad \lim_{\substack{\|u\|_\infty \to 0 \\ \|v\|_\infty \to 0}} \|P_+\|_w = 1.$$

In particular, if $w \in (HS)$, then

$$\lim_{\epsilon \to 0} b_{w^\epsilon}(\mathcal{E}) = 1, \qquad \lim_{\epsilon \to 0} \|P_+\|_{w^\epsilon} = 1.$$

Indeed, it suffices to follow the last lines of the proof of the implication $(5) \Rightarrow (4)$ of Theorem 4.6.1. It is clear by the definition of Ω that $\forall \epsilon > 0$, $\exists \eta > 0$ such that $(\|u\|_\infty < \eta, \|v\|_\infty < \eta) \Rightarrow (\Omega \subset D(1, \epsilon))$. In the proof, ϵ is denoted $1 - \delta$, and hence the very last inequality of the proof is $\mathrm{dist}_{L^\infty(\mathbb{T})}(\overline{h}/h, H^\infty) \leq \epsilon$. It follows that

$$\lim_{\substack{\|u\|_\infty \to 0 \\ \|v\|_\infty \to 0}} \mathrm{dist}_{L^\infty(\mathbb{T})}(\overline{h}/h, H^\infty) = 0,$$

thus by the second formula of Corollary 4.6.3,

$$\lim_{\substack{\|u\|_\infty \to 0 \\ \|v\|_\infty \to 0}} \|P_+\|_w = 1.$$

The continuity of $b_w(\mathcal{E})$ follows from the last bounds of Lemma 4.4.3.   ∎

Remark 4.6.5 It is easy to see that for a non-null projection $P \colon H \to H$ on a Hilbert space $(P^2 = P)$ we always have $\|P\| \geq 1$, and $\|P\| = 1$ if and only if P is an orthogonal projection on a closed subspace of H. Thus, for a sequence $\mathcal{E}$, $b(\mathcal{E}) = 1$ if and only if $\mathcal{E}$ is an orthogonal basis. Hence

Corollary 4.6.4 expresses an intuitively evident fact: the closer w is to 1, the more $\mathcal{E}$ resembles an orthogonal basis. Nevertheless, it is instructive to note that in the last corollary, if w is a weight with a singularity (i.e. if at least one of the equations $\inf_{\mathbb{T}} w = 0$, $\sup_{\mathbb{T}} w = \infty$ holds), so is w^ϵ ($\epsilon > 0$).

Example 4.6.6 (a power-like weight: Babenko, 1950)   Let $w_\alpha(\zeta) = |1 - \zeta|^\alpha$, $\alpha \in \mathbb{R}$. Then, $w_\alpha^{\pm 1} \in L^1(\mathbb{T})$ if and only if $|\alpha| < 1$. Setting

$$u := \log(w_\alpha(\zeta)) = \alpha \log |1 - \zeta| = \alpha \operatorname{Re}(\log(1 - \zeta)),$$

and selecting the logarithm holomorphic in $\mathbb{C} \setminus (-\infty, 0]$, for $\zeta = e^{it}$, $|t| < \pi$, we obtain

$$Hu(e^{it}) = \alpha \cdot \operatorname{Im}(\log(1 - e^{it})) = \alpha \cdot \arg(1 - e^{it}) = \alpha \cdot \arg(e^{it/2}(e^{-it/2} - e^{it/2}))$$

$$= \begin{cases} \alpha(t/2 - \pi/2) & \text{if } 0 < t < \pi, \\ \alpha(\pi/2 + t/2) & \text{if } -\pi < t < 0. \end{cases}$$

It follows that $\|Hu\|_\infty = |\alpha|\pi/2$.

For the weight $w(e^{it}) = |t|^\alpha$ the property results from the fact that given two weights w and W, the identity mapping $f \longmapsto f$ is an isomorphism $L^2(wm) \to L^2(Wm)$ if and only if the *weights are equivalent*, i.e.

$$cw \le W \le Cw,$$

where $c > 0, C > 0$ are constants. This implies that $\mathcal{E} = (z^k)_{k \in \mathbb{Z}}$ is simultaneously a basis or not in $L^2(wm)$ and $L^2(Wm)$. By setting $w = |1 - e^{it}|^\alpha$, $W(e^{it}) = |t|^\alpha$ ($|t| < \pi$), we obtain the following conclusion.

Conclusion $\mathcal{E} = (z^k)_{k \in \mathbb{Z}}$ is a Schauder basis of the space $L^2(w_\alpha m)$, as well as of $L^2(|t|^\alpha m)$, if and only if $|\alpha| < 1$.

4.7 Prediction and Hankel Operators

We have already applied the techniques of Hardy spaces to stationary processes. The technique of this chapter was also in large part motivated by the needs of the theory of processes and of signal processing. Once established, it resolved a number of problems of prediction. In reality, the appropriate technique to treat the degree of mixing and ergodic properties of a process, such as regularity and singularity, is the use of the *Hankel operators*: see § 4.7.2 below, or sources such as Nikolski (1986).

4.7.1 Strongly Regular Processes

Let $(x_n)_{n\in\mathbb{Z}}$ be a stationary sequence in a Hilbert space H. Its past at time n is

$$H_n^- = \mathrm{span}_H(x_k : k < n)$$

(see Definition 1.6.1), with its "past" being

$$H_- = H_0^- = \mathrm{span}_H(x_n : n < 0).$$

Similarly, H_n is its future at time n and and H_+ the "future" of the process, where

$$H_n = \mathrm{span}_H(x_k : k \geq n), \quad H_+ = H_0.$$

We present below a number of properties that distinguish the set of regular processes (see Definition 1.6.1 for the definition and Corollary 2.7.2 for a spectral description) using the techniques developed in this chapter.

Recall that given a subspace $E \subset H$, P_E denotes the orthogonal projection onto E.

(1) *Let $x \in H_+$ (a state in the future of the process). Then the optimal prediction with respect to the past H_- is the vector $P_{H_-}x$ satisfying the following property (and is well-defined by it)*

$$P_{H_-}x \in H_-, \quad \mathrm{dist}_H(x, H_-) = \|x - P_{H_-}x\|.$$

Indeed, this is practically the definition of the orthogonal projection.

(2) Definition. A stationary process $(x_n)_{n\in\mathbb{Z}}$ is said to be *strongly regular* if the optimal predictions are comparable with the state vectors themselves:

$$\|x - P_{H_-}x\| \geq c\|x\| \quad (\forall x \in H_+),$$

where $c > 0$ is a constant.

The following theorem shows, among other things, that a strongly regular process is regular, and provides a criterion of strong regularity as a function of the spectral measure of the process.

(3) Theorem (Helson and Szegő, 1960) *Let $(x_n)_{n\in\mathbb{Z}}$ be a stationary process, and μ its spectral measure. The following assertions are equivalent.*

    (i) $(x_n)_{n\in\mathbb{Z}}$ *is strongly regular.*
    (ii) $\|P_{H_-}P_{H_+}\| < 1$.
    (iii) $A_H(H_-, H_+) > 0$.
    (iv) $\mu = wm$ $(\mu_s = 0)$ *and* $w \in (HS)$ *(see Definition 4.6.2).*

Proof We have $\|x\|^2 = \|P_{H_-}x\|^2 + \|x - P_{H_-}x\|^2$ $(\forall x \in H_+)$ and hence $\|x - P_{H_-}x\| \geq c\|x\|$ if and only if $(1 - c^2)\|x\|^2 \geq \|P_{H_-}x\|^2$ $(\forall x \in H_+)$, which is equivalent to $1 - c^2 \geq \|P_{H_-}P_{H_+}\|^2$. Thus, (i) $\Leftrightarrow$ (ii).

The equivalences of (ii), (iii), and (iv) were established in Theorem 4.6.1.

$\blacksquare$

4.7.2 Angular Operators and Hankel Operators

As is clear from the proofs of Theorems § 4.7.1(3) and 4.6.1, everything boils down to the norm of the operator

$$\mathcal{H}x = P_{H_-}x, \quad x \in H_+, \quad \mathcal{H}: H_+ \to H_-,$$

for which we have $\cos(A_H(H_+, H_-)) = \|\mathcal{H}\|$. The operator $\mathcal{H}$ is called the *angular operator between the future and the past.*

The majority of the ergodic properties of a stationary process (*mixing properties*, as well as the different forms of regularity) can be expressed as a function of the properties of $\mathcal{H}$ (more sophisticated than the norm, such as the singular numbers, etc.), and then translated into the language of spectral measures. Some references are given in § 4.9.

In principle, the angular operator $\mathcal{H}$ can clearly be defined for an arbitrary pair of subspaces H_-, H_+, but there is a price to pay: it will no longer be a *Hankel operator*, as is the case for the past and future of a process. The following theorem (4.7.1) is the "easy part" of a theorem of Nehari (1957), who established rich and profound links between the Hardy spaces and a vast theory of integral operators (such as those of Hankel and of Toeplitz), which is the subject of numerous monographs; see for example Nikolski (2002) and Peller (2003). See also the comments in § 4.9. In fact, we will only introduce here a restricted notion of a Hankel operator, by pre-supposing the existence of an *operator symbol*. The theorem itself will simply be a reinterpretation of a portion of Theorem 4.6.1.

For our limited goal, we use the following definition of a Hankel operator. *Let $\varphi \in L^\infty(\mathbb{T})$ and*

$$H_\varphi f = P_-(\varphi f), \quad f \in H^2,$$

where P_- is the orthogonal projection in $L^2(\mathbb{T})$ onto the subspace $H^2_- = L^2(\mathbb{T}) \ominus H^2$. H_φ is called the Hankel operator of symbol φ. It is easy to see that a symbol is not unique, and moreover, the following theorem holds.

Hermann Hankel (1839–1873) was a German mathematician. He studied under Möbius, and then Riemann, Weierstrass, and Kronecker. He defended his thesis at Leipzig in 1861, received his habilitation in 1862, and was named professor at the University of Tübingen in 1869. He worked on Grassmann's linear algebra, on integration theory (preparing the road for measure theory) and on the transformations of functions of complex variables. His name is associated with the *Hankel transformation* (on the functions in a Euclidean space depending only on $\|x\|$), the *Hankel function* (i.e. Bessel functions of the third kind), and – especially – the *Hankel matrices (and operators)*. The latter are defined as being the matrices $A = \{a(i, j): 1 \leq i, j \leq n\}$ whose elements are constant on the diagonals perpendicular to the principal diagonal, i.e. $a(i, j) = \varphi(i + j)$ where φ is a function of a single variable. Today, these matrices and their continuous analogs are ubiquitous in several analytic disciplines and in their applications (harmonic analysis, the theory of holomorphic interpolation, optimal control, random processes, signal processing, etc.). Stricken with meningitis, Hankel died at 34 of a cerebral hemorrhage while traveling with his wife.

Theorem 4.7.1 (Nehari, 1957: the norm of a Hankel operator) *Let $\varphi \in L^\infty(\mathbb{T})$ and let H_φ be a Hankel operator with symbol φ. Then,*

$$\|H_\varphi\| = \inf\{\|\psi\|_\infty : H_\psi = H_\varphi\} = \mathrm{dist}_{L^\infty(\mathbb{T})}(\varphi, H^\infty).$$

Proof The equality with the "dist" follows from the simple observation that, for every $f \in H^2, g \in H^2_-$, we have

$$(H_\varphi f, g)_{L^2(\mathbb{T})} = (P_-(\varphi f), g) = (\varphi f, g) = \int_{\mathbb{T}} \varphi f \overline{g} \, dm,$$

and hence

$$\|H_\varphi\| = \sup\{|(H_\varphi f, g)| : f \in H^2, \|f\|_2 \le 1; g \in H^2_-, \|g\|_2 \le 1\}$$

$$= \sup\left\{\left|\int_{\mathbb{T}} \varphi f \overline{g}\, dm\right| : f \in H^2, \|f\|_2 \le 1; g \in H^2_-, \|g\|_2 \le 1\right\}.$$

The rest of the reasoning follows exactly the proof of the implication $(3) \Rightarrow (4)$ in Theorem 4.6.1. ∎

4.8 $b(X)$ Versus $ub(X)$

Here, we require a little more Banach-related terminology, i.e. relative to the more general framework of Banach spaces.

Let $X = (x_\alpha)$ $(\alpha \in A)$ be a family of elements of a Banach space X indexed by a set of indices A (formally arbitrary, but in the examples we usually have $A = \mathbb{Z}, \mathbb{Z}_+, \mathbb{Z}^n$, etc.). Unless otherwise stated, X is assumed to be *separable* and A at most countable.

A family $X = (x_\alpha)$ is called an *unconditional basis of X* if for any $x \in X$ there exists a unique complex family (c_α) such that

$$x = \lim_\sigma \sum_{\alpha \in \sigma} c_\alpha x_\alpha,$$

where σ runs over the set of finite subsets of A ordered by inclusion. More precisely, this disordered convergence means that

$$\forall \epsilon > 0, \ \exists \sigma_\epsilon \subset A \ \text{(finite) such that} \ \forall \sigma \ \text{(finite)}, \sigma \supset \sigma_\epsilon$$

$$\text{we have} \ \left\|x - \sum_{\alpha \in \sigma} c_\alpha x_\alpha\right\| < \epsilon. \quad (4.1)$$

Clearly an unconditional basis is a Schauder basis with respect to any numbering of $A = (\alpha_j)_{j \ge 1}$ (and the result of the summation does not depend on the choice of the numbering, $x = \lim_{n \to \infty} \sum_{j=1}^{n} c_{\alpha_j} x_{\alpha_j}$); in particular, an unconditional basis $X = (x_\alpha)$ is minimal. If $X' = (x'_\alpha)$ is a dual family of functionals,

$$\langle x_\alpha, x'_\beta \rangle = \delta_{\alpha\beta},$$

then, by Banach's theorem § 4.1.2(b), the projections P_σ of the partial sums,

$$P_\sigma x = \sum_{\alpha \in \sigma} \langle x, x'_\alpha \rangle x_\alpha, \quad x \in X,$$

are continuous, and

$$ub(X) := \sup_{\sigma} \|P_\sigma\| < \infty.$$

The number $ub(X)$ is called the *unconditional basis constant*. When choosing a particular numbering $A = (\alpha_j)_{j\geq 1}$, we have also a corresponding basis constant $b(X)$ (see § 4.1.1(e) for the definition). Clearly,

$$b(X) \leq ub(X).$$

Also, an analog of the lemma in § 4.1.2(c) holds, and thus conversely, a minimal family X with $ub(X) < \infty$ is an unconditional basis.

In this section we consider *finite bases* $X_n = (x_j)_{j=1}^n$ (hence, simply, the free finite sequences) for large n and we examine the question of the possible upper bounding of $ub(X)$ in terms of $b(X)$. By a change of notation if necessary, we can suppose that

$$X = \mathrm{span}(X_n).$$

The question is important for the application of bases to approximation theory, and it is not banal even in Hilbert spaces. Note, in passing, that the orthogonal bases are characterized by $b(X) = 1$ and/or $ub(X) = 1$. Moreover, the links between $b(X)$ and $ub(X)$ in an *arbitrary* Banach space, without detailing its geometric nature, can be quite simple.

Theorem 4.8.1

(1) *Let $X_n = (x_j)_{j=1}^n$ be a basis in a Banach space X. Then*
$$ub(X_n) \leq n\, m(X_n) \leq n\, b(X_n).$$
(2) *The upper bound of (1) is sharp up to a numerical constant: there exists a free sequence $X_n = (x_j)_{j=1}^n$ in a Banach space such that*
$$ub(X_n) \geq b(X_n)n/2.$$

Proof (1) is immediate since by definition $b(X_n) \geq m(X_n) = \max \|P_{k,k}\|$, where (recall)

$$P_{k,l}x = \sum_{j=k}^{l} \langle x, x_j' \rangle x_j, \quad x \in X_n := \mathrm{span}(x_j : 1 \leq j \leq n),$$

and $b(X_n) = \max_{1\leq k\leq l\leq n} \|P_{k,l}\|$ (the norm of $P_{k,l}$ and P_σ as always represent the norm of the restrictions $P_{k,l} \mid X_n$ and $P_\sigma \mid X_n$). It follows that for any $\sigma \subset \{1, 2, \ldots, n\}$, $\|P_\sigma\| = \|\sum_{j\in\sigma} P_{k,k}\| \leq m(X_n)\,\mathrm{card}(\sigma) \leq n\, m(X_n)$.

(2) Let $X = \mathbb{C}^n$ be equipped with the norm

$$\|y\| = \sum_{j=0}^{n-1} |y_j - y_{j+1}|,$$

where $y = (y_j)_{j=1}^n \in \mathbb{C}^n$ and $y_0 = 0$. Let $X_n = (x_j)_{j=1}^n$ be the natural $0 - 1$ basis of $\mathbb{C}^n$, $x_j = (\delta_{ij})_{i=1}^n$. For every k ($1 \leq k \leq n$) we have $y_k = \sum_{j=1}^k (y_j - y_{j-1})$ and hence

$$\|P_{k,l}y\| = \|(0,\ldots,y_k,\ldots,y_l,0,\ldots,0)\| = |y_k| + \sum_{j=k}^{l-1} |y_j - y_{j+1}| + |y_l|$$

$$\leq \sum_{j=1}^k |y_j - y_{j-1}| + \sum_{j=k}^{l-1} |y_j - y_{j+1}| + \sum_{j=1}^l |y_j - y_{j-1}| \leq 2\|y\|,$$

so that $\|P_{k,l}\| \leq 2$ and $b(X_n) \leq 2$; moreover, if $\mathbb{1} = (1,1,\ldots,1)$ then $\|\mathbb{1}\| = 1$, but $\|P_{odd}\mathbb{1}\| = n$, hence $ub(X_n) \geq n$.

∎

For a finite-dimensional Hilbert space the situation is much more interesting: in an unexpected manner the upper bound of Theorem 4.8.1(1) can be considerably improved; indeed $ub(X_n)$ always increases with *sublinear growth* as $n \to \infty$, as is shown by the following theorem. Below, in Lemma 4.8.5, we shall see that this new upper bound is sharp in a sense.

Theorem 4.8.2 (McCarthy and Schwartz, 1965) *Let* $X_n = (x_j)_{j=1}^n$ *be a basis in a Hilbert space H. Then*

$$ub(X_n) \leq 2m(X_n)n^{1-(0,32)/(b(X_n)^2)}.$$

Jacob Schwartz (1930–2009) was an American mathematician and computer scientist. He conducted research in operator theory, von Neumann algebras, parallel computing (where he was one of the pioneers), the creation of programming languages, the theory of programming, and other domains of pure and applied mathematics. In pure mathematics, he is known as the co-author (with Nelson Dunford) of a fundamental text, *Linear Operators* I–III (1958–1971), of roughly 2 500 pages – a turning point in the development of twentieth-century functional analysis in terms of its rigor, its breadth, and its universality. Among his other important achievements, we find the above-stated classical result on bases in Hilbert spaces (see Theorem 4.8.2). In computer science, Schwartz was the creator of the programming language SETL and of the NYU (New York University) Ultracomputer, and he was the founder (and director for 15 years) of NYU's Department of Computer Science.

After submitting his thesis at Yale University (1952, supervised by Dunford), Schwartz obtained a position as professor at the Courant Institute of Mathematical Sciences, NYU (1958–2000), and was elected to the US National Academy of Sciences (1976) and National Academy of Engineering (2000). He published 18 monographs on a variety of subjects, in pure mathematics (such as von Neumann algebras), computer science, engineering, mathematical biology, etc., not to mention around 100 research articles of the highest quality.

His passing did not go unnoticed by the public at large: he merited an obituary in the *New York Times* (published March 3, 2009). Soon after, at a celebration of his career, his sister Judith (Dunford) said: "The intelligence [of J.S. was] so great that it seemed to weigh his head down, his omnivorous interests, his kindness and generosity – a stunning generosity that not only did not look for thanks but was puzzled and even annoyed when they came."

We begin with a lemma. To simplify the notation, we write

$$P_k = P_{k,k} = (\cdot, x_k')x_k.$$

Lemma 4.8.3 *Let $X_n = (x_j)_{j=1}^n$ be a basis in a Hilbert space H, with* $\dim H = n$. *Then:*

(1) $P_k^* = (\cdot, x_k)x_k'$, $P_k^* P_k = (\cdot, x_k')\|x_k\|^2 x_k'$, $P_k P_k^* = (\cdot, x_k)\|x_k'\|^2 x_k$.

(2) *The operator*

$$B = \sum_{k=1}^n P_k^* P_k$$

is positive Hermitian $((Bx, x) > 0,\ \forall x \neq 0)$, *hence invertible,*

$$B^{-1} = \sum_{k=1}^n \frac{P_k P_k^*}{\|P_k\|^2},$$

and for every k, $BP_k = P_k^* B$.

(3) *There exists a positive square root of B,*

$$A = B^{1/2} = \left(\sum_{k=1}^n P_k^* P_k \right)^{1/2},$$

and for every k, j: $\overline{P}_k\overline{P}_j = \delta_{kj}\overline{P}_k$, $\overline{P}_k^* = \overline{P}_k$ where $\overline{P}_k = AP_kA^{-1}$ (and hence (Ax_k) is an orthogonal basis of H).

(4) $ub(X_n) \leq \|A^{-1}\| \cdot \|A\|$.

Proof (1) Evident computation.

(2) For any $x \in H$, we have $(Bx, x) = \sum_{k=1}^{n} \|P_kx\|^2 \geq 0$, and if $(Bx, x) = 0$, then $\|P_kx\| = 0$ for every k, hence $x = 0$ (X_n is a basis). Using (1) gives

$$\left(\sum_{k=1}^{n} \frac{P_kP_k^*}{\|P_k\|^2}\right)Bx = \sum_{k=1}^{n} \|P_k\|^{-2}(Bx, x_k)\|x_k'\|^2 x_k$$

$$= \sum_{j,k=1}^{n} \|P_k\|^{-2}(x, x_j')\|x_j\|^2(x_j', x_k)\|x_k'\|^2 x_k$$

$$= \sum_{k=1}^{n} \|P_k\|^{-2}(x, x_k')\|x_k\|^2\|x_k'\|^2 x_k = \sum_{k=1}^{n}(x, x_k')x_k = x,$$

and the result follows (B is also left-invertible, since it is self-adjoint). For the last identity, for every j, we have

$$BP_j = \sum_{k=1}^{n} P_k^*P_kP_j = P_j^*P_j = \sum_{k=1}^{n} P_j^*P_k^*P_k = P_j^*B.$$

(3) As B is positive definite, it admits an orthogonal basis of eigenvectors: $Be_k = \lambda_k e_k$, $\lambda_k > 0$. Clearly $Ae_k = \sqrt{\lambda_k}e_k$ defines a square root of B, and $A^* = A$. Thus

$$A^{-1}\overline{P}_k^*A = A^{-2}P_k^*A^2 = B^{-1}P_k^*B = B^{-1}BP_k = P_k, \text{ hence } \overline{P}_k^* = \overline{P}_k;$$

$$\overline{P}_k\overline{P}_j = AP_kA^{-1}AP_jA^{-1} = \delta_{kj}AP_kA^{-1} = \delta_{kj}\overline{P}_k.$$

(4) By (3), for any complex numbers c_k, we have

$$\left\|\sum_{k=1}^{n} c_kP_kx\right\|^2 = \left\|A^{-1}\sum_{k=1}^{n} c_k\overline{P}_kAx\right\|^2 \leq \|A^{-1}\|^2\left\|\sum_{k=1}^{n} c_k\overline{P}_kAx\right\|^2$$

$$= \|A^{-1}\|^2 \sum_{k=1}^{n} |c_k|^2\|\overline{P}_kAx\|^2 \leq \|A^{-1}\|^2(\max_k |c_k|^2)\sum_{k=1}^{n} \|\overline{P}_kAx\|^2$$

$$= \|A^{-1}\|^2(\max_k |c_k|^2)\|Ax\|^2 \leq \|A^{-1}\|^2(\max_k |c_k|^2)\|A\|^2\|x\|^2.$$

By choosing $c_k = 1$ if $k \in \sigma$ and $c_k = 0$ otherwise, we obtain $\|P_\sigma\| \leq \|A^{-1}\| \cdot \|A\|$ for every $\sigma \subset \{1, 2, \ldots, n\}$, and the result follows. ∎

The following fact is a crucial observation because it allows control of the upper bound of $B = B(X_n)$ as a function of the lower bound of $B = B(X_n')$.

Corollary 4.8.4 *Let $X_n = (x_j)_{j=1}^n$ be a basis such that $\sum_{k=1}^n P_k P_k^* \geq \alpha I$, where $\alpha > 0$. Then*

$$B = \sum_{k=1}^n P_k^* P_k \leq \frac{1}{\delta(X_n)^2 \alpha} I = \frac{m(X_n)^2}{\alpha} I,$$

where $\delta(X_n)$ is the uniform minimality constant (for the equality $\delta(X_n)^{-1} = \max_k \|P_k\|$, see § 4.1.1(d)).

Indeed, $\delta = \delta(X_n) = \delta(X_n')$ and by Lemma 4.8.3(2)

$$\alpha \delta(X_n)^2 I \leq \sum_{k=1}^n \frac{P_k P_k^*}{\|P_k\|^2} = B^{-1}. \qquad \blacksquare$$

Lemma 4.8.5 *Let $X_n = (x_j)_{j=1}^n$ be a basis and $n = 2^m$. Then*

$$B = \sum_{k=1}^n P_k^* P_k \geq \frac{1}{n}(1 + (2b(X_n))^{-2})^m I := \alpha I,$$

and hence $B^{-1} \leq (1/\alpha)I$.

Proof   First note that if P is a bounded projection on a Hilbert space H and $P' = I - P$, then $\|P\| = \|P'\|$ (Exercise 4.9.3(a)) and $(P - P')^{-1} = P - P'$, hence $\|(P - P')^{-1}\| \leq 2\|P\|$ and, for any $x \in H$,

$$\|(P - P')x\| \geq (2\|P\|)^{-1}\|x\|.$$

The parallelogram identity

$$2(\|Px\|^2 + \|P'x\|^2) = \|(P + P')x\|^2 + \|(P - P')x\|^2 = \|x\|^2 + \|(P - P')x\|^2$$

implies that for any $x \in H$ we have

$$\|Px\|^2 + \|P'x\|^2 \geq \frac{1}{2}(1 + (2\|P\|)^{-2})\|x\|^2.$$

We apply this lower estimate successively for $P = P_{1,n/2}$, $P = P_{1,n/4}$, $P = P_{n/2+1,3n/4}$, etc. and use $\|P_{k,l}\| \leq b(X_n)$ to obtain

$$\left\|\sum_1^{2^{m-1}} P_k x\right\|^2 + \left\|\sum_{2^{m-1}}^{2^m} P_k x\right\|^2 \geq \frac{1}{2}(1 + (2b(X_n))^{-2})\|x\|^2,$$

and then

$$\left\|\sum_{k=1}^{2^{m-2}} P_k x\right\|^2 + \left\|\sum_{k=2^{m-2}+1}^{2^{m-1}} P_k x\right\|^2 + \left\|\sum_{k=2^{m-1}+1}^{2^{m-1}+2^{m-2}} P_k x\right\|^2 + \left\|\sum_{k=2^{m-1}+2^{m-2}+1}^{2^m} P_k x\right\|^2$$

$$\geq \frac{1}{2^2}(1 + (2b(X_n))^{-2})^2 \|x\|^2,$$

etc., by repeating m times,

$$\|P_1 x\|^2 + \|P_2 x\|^2 + \cdots + \|P_{2^m} x\|^2 \geq \frac{1}{2^m}(1 + (2b(X_n))^{-2})^m \|x\|^2,$$

or

$$\left(\sum_{k=1}^{n} P_k^* P_k x, x\right) \geq \frac{1}{2^m}(1 + (2b(X_n))^{-2})^m \|x\|^2. \qquad \blacksquare$$

4.8.1 Proof of Theorem 4.8.2

Let l be such that $n \leq 2^l < 2n$ and let X_{2^l} be a family $X_{2^l} = X_n \cup X$ obtained by adding to X_n an orthonormal basis X in a space orthogonal to H. Then, clearly $m(X_{2^l}) = m(X_n)$, $b(X_{2^l}) = b(X_n)$, $ub(X_{2^l}) = ub(X_n)$. By Lemma 4.8.5, $B(X_{2^l}) \geq \frac{1}{2^l}(1 + (2b(X_{2^l}))^{-2})^l I := \alpha I$ and

$$B(X_{2^l})^{-1} \leq (1/\alpha)I.$$

Since $b(X_{2^l}) = b(X'_{2^l})$ and $m(X_{2^l}) = m(X'_{2^l})$, we also have $B(X'_{2^l}) \geq \alpha I$ and hence, by Corollary 4.8.4:

$$B(X_{2^l}) \leq (m(X_{2^l})^2/\alpha)I.$$

Now, by Lemma 4.8.3(4), we have

$$ub(X_{2^l}) \leq \|B(X_{2^l})\|^{1/2} \|B(X_{2^l})^{-1}\|^{1/2}$$

$$\leq m(X_{2^l})/\alpha = m(X_{2^l}) \, 2^l (1 + (2b(X_{2^l}))^{-2})^{-l},$$

and then

$$ub(X_n) \leq m(X_n) \, 2^l (1 + (2b(X_n))^{-2})^l \leq 2m(X_n)n(1 + (2b(X_n))^{-2})^{-\ln(n)/\ln 2}$$

$$\leq 2m(X_n) \, n^{1-(0,32)/((b(X_n))^2)},$$

since $b(X_n) \geq 1$ and for $0 < t \leq 1/4$, $\ln(1 + t) \geq 4t \cdot \ln(5/4)$ (the function $t \mapsto t^{-1} \ln(1 + t)$ is decreasing), and finally $\ln(5/4)/\ln 2 > 0,32$. $\qquad \blacksquare$

4.8.2 Gram Matrices

When studying the approximation properties of sequences $X_n = (x_j)_{j=1}^n$ in a Hilbert space H, we often use the corresponding *Gram matrix*

$$G = ((x_j, x_k)_H)_{1 \leq j,k \leq n}.$$

Jørgen Gram (1850–1916), a Danish mathematician, is known for his research in probability and statistics, numerical analysis, and number theory (Euler zeta function). He obtained his Master's degree in 1873 (University of Copenhagen) and submitted his doctoral thesis in 1879. Research in mathematics never became his profession; however, as an amateur mathematician he became a member of the Danish Academy of Sciences (1888), received the Gold Medal of the Academy in 1885 (for his results on the zeta function), and was an editor of *Tidsskrift for Mathematik* (1883–1889). Gram's professional career was spent in insurance companies: he started as an assistant and climbed the ladder to the post of company president; in 1884 he founded his own company, Skjold, and in 1910 became President of the Danish Insurance Council. We know relatively little about him other than his mathematical and professional activities, except that Gram was very active in research in forestry, long before his German colleagues developed mathematical models for the exploitation of forests.

Gram is principally known for the *Gram–Schmidt orthogonalization procedure* (Gram (1883) and Schmidt (1907), who refers back to Gram), the *Gram determinants*, and the *Gram matrix* (see § 4.8.2). (Later, it was found that the orthogonalization procedure was already known by Laplace in 1816 and by Cauchy in 1836.) He was killed in 1916, run over by a cyclist as he was walking to a meeting at the Danish Academy of Sciences.

The result of the Gram–Schmidt orthogonalization procedure can be expressed using the Gram determinants D_j ($D_0 = 1$):

$$u_j = \frac{1}{D_{j-1}} \begin{vmatrix} \langle v_1, v_1 \rangle & \langle v_2, v_1 \rangle & \cdots & \langle v_j, v_1 \rangle \\ \langle v_1, v_2 \rangle & \langle v_2, v_2 \rangle & \cdots & \langle v_j, v_2 \rangle \\ \vdots & \vdots & \ddots & \vdots \\ \langle v_1, v_{j-1} \rangle & \langle v_2, v_{j-1} \rangle & \cdots & \langle v_j, v_{j-1} \rangle \\ v_1 & v_2 & \cdots & v_j \end{vmatrix},$$

$$D_j = \begin{vmatrix} \langle v_1, v_1 \rangle & \langle v_2, v_1 \rangle & \cdots & \langle v_j, v_1 \rangle \\ \langle v_1, v_2 \rangle & \langle v_2, v_2 \rangle & \cdots & \langle v_j, v_2 \rangle \\ \vdots & \vdots & \ddots & \vdots \\ \langle v_1, v_j \rangle & \langle v_2, v_j \rangle & \cdots & \langle v_j, v_j \rangle \end{vmatrix}.$$

Here are some properties of the Gram matrices and their link with the techniques used in the proof of Theorem 4.8.2 (operators B and A). Up to a change of notation if necessary, we can always suppose that $H = \operatorname{span}(X_n)$.

(1) For $a, b \in \mathbb{C}^n$, we have $(Ga, b)_{\mathbb{C}^n} = (\sum_j a_j x_j, \sum_k b_k x_k)_H$, hence G is positive, and X_n is a basis (free in H) if and only if G is positive definite.

Conversely, it is easy to see that any positive definite matrix is a Gram matrix of a free sequence (hence of a basis). It suffices to take $x_j = G^{1/2} e_j$ where (e_j) is the canonical basis of $\mathbb{C}^n$.

(2) Given two sequences $X_n = (x_j)_1^n$, $\mathcal{Y}_n = (y_j)_j^n$, a linear operator $U \colon H \to H$ defined by $U x_j = y_j$ $(\forall j)$ is unitary (i.e. $(Ux, Uy) = (x, y)$, $\forall x, y \in H$) if and only if $G(X_n) = G(\mathcal{Y}_n)$.

(3) By defining a linear mapping $T \colon \mathbb{C}^n \to H$ by

$$T e_j = x_j, \quad 1 \le j \le n,$$

where (e_j) is the canonical basis of $\mathbb{C}^n$, we obtain $(Ta, Tb)_H = (Ga, b)_{\mathbb{C}^n}$, hence

$$T^*T = G.$$

The operator T^{-1} is said to be the "orthogonalizer" (for obvious reasons). The equations $\delta_{jk} = (x_j, x'_k) = (T e_j, x'_k) = (e_j, T^* x'_k)$ show that, for every k,

$$T^* x'_k = e_k,$$

hence T^* is an orthogonalizer of the dual sequence X'_n. Moreover, for every k, we have

$$T T^* x'_k = T e_k = x_k,$$

then

$$(T T^* x'_k, x'_j) = (T^* x'_k, T^* x'_j) = (e_k, e_j) = \delta_{jk},$$

thus the sequence $(T T^*)^{1/2} X'_n$ is an orthonormal basis of H.

Conclusion $(T T^*)^{1/2}$ is also an orthogonalizer of X'_n.

(4) By supposing that X_n is a normalized sequence, $\|x_j\| = 1$ $(\forall j)$ (which does not change $b(X_n)$, or $ub(X_n)$, or the other geometrical properties of X_n), by the definition in Lemma 4.8.3(1), we have $B x_j = x'_j$ $(\forall j)$, and hence

$$B^{-1} = T T^*.$$

Next, as the operators $T T^*$ and T^*T are unitarily equivalent (Appendix E), so are G and B^{-1}, and hence

$$ub(X_n) \le \|G\|^{1/2} \|G^{-1}\|^{1/2}.$$

(5) To conclude, note that the operator $B^{1/2}$ is also an orthogonalizer (already mentioned in Lemma 4.8.3(3)):

$$(B^{1/2}x_j, B^{1/2}x_k) = (Bx_j, x_k) = (x'_j, x_k)\|x_j\|^2 = \delta_{jk}\|x_j\|^2.$$

Remark 4.8.6 (sharpness of the inequality of Theorem 4.8.2)   It is important to note that Theorem 4.8.2 implies a fairly unusual upper estimate: given an (infinite) basis $X = (x_j)$ in a Hilbert space, we always have a sublinear estimation on the growth of $ub(X_n)$: $ub(X_n) \le Cn^{1-\epsilon}$ where $\epsilon > 0$ depends only on the "quality of X as a Schauder basis" ($\epsilon \ge$ constant $/b(X)^2$). Below we will see that any $\epsilon > 0$ is realizable: for every $\epsilon > 0$ there exists a weight $w \in (HS)$ such that for the exponential basis $\mathcal{E} = (z^k)_{k\ge 0}$ in the space $L^2(\mathbb{T}, w)$ we have

$$ub(\mathcal{E}_n) \ge cn^{1-\epsilon}$$

where $\epsilon > 0$ is of the order of $const/b(\mathcal{E})$. This result of Spijker et al. (2003) will be obtained as the consequence of a series of exercises (§ 4.9.4) concerning the Helson–Szegő Theorem 4.6.1.

4.9 Exercises

We systematically use the notation of this chapter, in particular the Radon–Nikodym decomposition $\mu = \mu_s + wm$ ($w \in L^1(\mathbb{T})$) of a positive measure on $\mathbb{T}$, $\mathcal{E} = (z^k)_{k\in\mathbb{Z}}$ for the family of exponentials, etc.

4.9.1 Criterion of Linear Dependence of Exponentials

Show that $\mathcal{E} = (z^k)_{k\in\mathbb{Z}}$ is linearly dependent (not free) in $L^p(\mathbb{T}, \mu)$ if and only if the closed support supp(μ) *is finite* ($\Leftrightarrow \mu$ *is a finite sum of Dirac measures:* $\mu = \sum_{j=1}^N c_j\delta_{\lambda_j}, |\lambda_j| = 1$).

SOLUTION: $\mathcal{E}$ is not free in $L^p(\mathbb{T}, \mu)$ if and only if there exists a linear combination of the z^k, and hence a polynomial $f = \sum_{k=m}^n a_k z^k$, $a_n \ne 0$, such that $\|f\|_p^p = \int_{\mathbb{T}} |f|^p \, d\mu = 0$. The last equation is equivalent to saying that $|f|^p\mu = 0$, meaning supp(μ) $\subset Z(f)$ where $Z(f)$ is the set of zeros of f on $\mathbb{T}$; this set is finite. ∎

4.9.2 Multipliers Versus Bases

Let $X = (x_\alpha)$ ($\alpha \in A$) be a family of elements of a Banach space X indexed by a set of indices A (formally arbitrary but in the examples we usually have

$A = \mathbb{Z}, \mathbb{Z}_+, \mathbb{Z}^n$, etc.). Unless otherwise specified, X is assumed to be *separable* and A is at most countable.

A family of complex numbers (λ_α) is said to be a *multiplier of X* if the mapping T defined by

$$T x_\alpha = \lambda_\alpha x_\alpha, \quad \alpha \in A,$$

can be extended to a bounded linear operator $\mathrm{span}_X(X) \to \mathrm{span}_X(X)$. We denote this extension by the same letter T and also write $\lambda_\alpha = \lambda_\alpha(T)$; hence x_α are the eigenvectors of T corresponding to the eigenvalues λ_α.

We systematically identify T and $(\lambda_\alpha(T))$ and assume (up to a change of notation) that

$$\mathrm{span}_X(X) = X.$$

The *set of the multipliers of X* is denoted $\mathrm{Mult}(X)$:

$$\mathrm{Mult}(X) = \{(\lambda_\alpha) : \exists T \text{ bounded and linear on } X \text{ such that } T x_\alpha = \lambda_\alpha x_\alpha, \forall \alpha\}.$$

Equipped with the operator norm

$$\|(\lambda_\alpha)\|_{\mathrm{Mult}} := \|T\|,$$

$\mathrm{Mult}(X)$ becomes a normed space.

In what follows, we use the definitions and notation found at the beginning of § 4.8: P_σ, $b(X)$, $ub(X)$, etc.

(a) *Show that $\mathrm{Mult}(X)$ is a Banach algebra (isometrically isomorphic to a sub-algebra of $L(X)$, the algebra of bounded operators on X).*

SOLUTION: Clear by the definitions. ∎

(b) *Show that $\mathrm{Mult}(X) \subset l^\infty(A)$. Equality holds if and only if (x_α) is an unconditional basis of X.*

SOLUTION: As $\lambda_\alpha(T)$ is an eigenvalue of T, we have

$$|\lambda_\alpha(T)| \le \|T\|,$$

and hence $(\lambda_\alpha) \in l^\infty(A)$. In the case of equality, $\mathrm{Mult}(X) = l^\infty(A)$, to each $\lambda = (\lambda_\alpha) \in l^\infty(A)$ corresponds an operator $T_\lambda \in \mathrm{Mult}(X) \subset L(X)$ and the mapping j defined by $j(\lambda) = T_\lambda$, $j : l^\infty(A) \to L(X)$ is linear and closed, since $T \longmapsto (\lambda_\alpha(T))$ is bounded. Hence by the closed graph theorem (see Appendix E) there exists $C > 0$ such that $\|T_\lambda\| \le C\|\lambda\|_{l^\infty(A)}$. In particular, for every finite subset $\sigma \subset A$, we have $T_{\chi_\sigma} = P_\sigma$ (where χ_σ is the characteristic function of σ), $P_\sigma(\sum_{\alpha \in A} c_\alpha x_\alpha) = \sum_{\alpha \in \sigma} c_\alpha x_\alpha$, and $\|P_\sigma\| \le C$; this implies that X is minimal (set $\sigma = \{\alpha\}$, $\alpha \in A$), and thus, by an analog of § 4.1.2(c) already mentioned in § 4.8, X is an unconditional basis. ∎

(c) *Let $\mu = \mu_s + wm$ be a measure on $\mathbb{T}$ and $\mathcal{E} = (z^k)_{k\in\mathbb{Z}}$ the exponentials in the space $L^2(\mu)$. Show that $\mathcal{E}$ is an unconditional basis if and only if $\mu_s = 0$ and $w \approx 1$, i.e. $w^{\pm 1} \in L^\infty(\mathbb{T})$.*

Hint Use (b).

SOLUTION: The sufficiency is clear, since if $w^{\pm 1} \in L^\infty(\mathbb{T})$, then $L^2(\mu) = L^2(\mathbb{T})$ (with equivalence of norms). For the necessity, we use (b): every $\lambda \in l^\infty(\mathbb{Z})$ is a multiplier T_λ and $\|T_\lambda\| \leq C\|\lambda\|_{l^\infty(\mathbb{Z})}$. Let $\zeta \in \mathbb{T}$ and $\lambda = (\bar{\zeta}^n)_{n\in\mathbb{Z}}$, then the corresponding multiplier T_λ is a rotation: for any polynomial $p \in \mathcal{P}$, $p = \sum_{n\in\mathbb{Z}} c_n z^n$, we have $(T_\lambda p)(z) = \sum_{n\in\mathbb{Z}} c_n(\bar{\zeta}z)^n = p(\bar{\zeta}z)$, and moreover $\|T_\lambda p\|_{L^2(\mu)} \leq C\|p\|_{L^2(\mu)}$. Since $\mathcal{P}$ is dense in $L^2(\mu)$, we obtain

$$\int_{\mathbb{T}} |f(\bar{\zeta}z)|^2 d\mu \leq C^2 \int_{\mathbb{T}} |f(z)|^2 d\mu, \quad \forall f \in L^2(\mu),$$

which implies (with $f = \chi_E$, $E \subset \mathbb{T}$ measurable) $\mu(\zeta E) \leq C\mu(E)$ for all $\zeta \in \mathbb{T}$, and then (replacing E with $\bar{\zeta}E$) $\mu(E) \leq C\mu(\bar{\zeta}E)$. By integrating over $dm(\zeta)$, we obtain $C^{-1}\mu(E) \leq (\mu * m)(E) \leq C\mu(E)$ for any measurable E; however, $\mu * m = cm$ where c is a constant, and the result follows. ∎

(d) *Let μ be a measure on $\mathbb{T}$ and let $\mathcal{E} = (z^k)_{k\in\mathbb{Z}}$ be the exponentials in the space $L^2(\mu)$; use the abbreviated notation*

$$\mathrm{Mult}(\mathcal{E}, L^2(\mu)) = \mathrm{Mult}(\mu)$$

and $\|T_\lambda\|_\mu$ for the norm of the multiplier T_λ in the space $L^2(\mu)$. Show that for two arbitrary measures μ and ν we have

$$\mathrm{Mult}(\mu) \subset \mathrm{Mult}(\mu * \nu) \text{ and } \|T_\lambda\|_{\mu*\nu} \leq \|T_\lambda\|_\mu \quad (\forall \lambda \in \mathrm{Mult}(\mu)).$$

SOLUTION: Let $f \in \mathcal{P}$, $f_\zeta(z) = f(\bar{\zeta}z)$, $T = (t_k) \in \mathrm{Mult}(\mu)$, then $Tf_\zeta = (Tf)_\zeta$ and hence

$$\|Tf\|^2_{\mu*\nu}$$
$$= \int_{\mathbb{T}} |Tf(z)|^2 d(\mu * \nu) = \int_{\mathbb{T}} \int_{\mathbb{T}} |Tf(\zeta z)|^2 d\mu(z) d\nu(\zeta)$$
$$= \int_{\mathbb{T}} d\nu(\zeta) \int_{\mathbb{T}} |(Tf)_{\bar\zeta}(z)|^2 d\mu(z) = \int_{\mathbb{T}} d\nu(\zeta) \int_{\mathbb{T}} |(Tf_{\bar\zeta})(z)|^2 d\mu(z)$$
$$= \int_{\mathbb{T}} \|Tf_{\bar\zeta}\|^2_\mu d\nu(\zeta) \leq \|T\|^2_\mu \int_{\mathbb{T}} \|f_{\bar\zeta}\|^2_\mu d\nu(\zeta)$$
$$= \|T\|^2_\mu \int_{\mathbb{T}} d\nu(\zeta) \int_{\mathbb{T}} |f(\zeta z)|^2 d\mu(z)$$
$$= \|T\|^2_\mu \int_{\mathbb{T}} \int_{\mathbb{T}} |f(\zeta z)|^2 d\mu(z) d\nu(\zeta) = \|T\|^2_\mu \int_{\mathbb{T}} |f(z)|^2 d(\mu * \nu)(z)$$
$$= \|T\|^2_\mu \|f\|^2_{\mu*\nu}.$$

Thus $T \in \mathrm{Mult}(w * \nu)$ and $\|T_\lambda\|_{\mu*\nu} \leq \|T_\lambda\|_\mu$. ∎

(e) *Deduce that if $\mathcal{E} = (z^k)_{k\in\mathbb{Z}}$ is a Schauder basis in the space $L^2(\mu)$ with a basis constant $b(\mathcal{E}_\mu)$ (see § 4.1.1(e) for the definition) and v is a positive measure on $\mathbb{T}$, then $\mathcal{E} = (z^k)_{k\in\mathbb{Z}}$ is a basis in $L^2(\mu * v)$ with*

$$b(\mathcal{E}_{\mu*v}) \leq b(\mathcal{E}_\mu).$$

SOLUTION: Clear by (d) and Lemma 4.4.3, because a partial sum projection $P_{k,l}$ is a multiplier, hence $\|P_{k,l}\|_{\mu*v} \leq \|P_{k,l}\|_\mu$. ∎

(f) *Deduce that $w \in (HS) \Rightarrow w * v \in (HS)$, for any positive measure v.*

SOLUTION: Clear by (e) and Theorem 4.6.1. ∎

4.9.3 Projections on a Hilbert Space

Let H be a Hilbert space and let P be a bounded projection on H ($P^2 = P$).

(a) *Show that $\|P\| = \|1 - P\|$. Show with an example that on a Banach space we can have $\|P\| \neq \|1 - P\|$.*

SOLUTION: Denote $L = PH$ and $M = (I - P)H$ so that $P = P_{L\|M}$, $I - P = P_{M\|L}$. Then, by § 4.3.1(c), $\|P_{L\|M}\|^{-2} = \sin^2(A(L, M)) = 1 - \cos^2(A(L, M))$. Clearly $\cos(A(L, M)) = \cos(A(M, L))$, hence $\|P_{L\|M}\| = \|P_{M\|L}\|$. For the example in a Banach space, let $X = C[0, 1]$ be the space of continuous functions on $[0, 1]$ equipped with the uniform norm $\|f\|_\infty = \max_{[0,1]} |f|$, and $Pf = \int_0^1 f(x)\,dx$. Then $P^2 = P$, $\|P\| = 1$ but $\|I - P\| = 1 + \|P\| = 2$: choosing a function f_n continuous and piecewise linear such that $f_n(0) = 1$ and $f_n(x) = -1$ for $1/n \leq x \leq 1$ we obtain $(I - P)f_n = f_n + (1 - 1/n)$, hence $\|(I - P)f_n\|_\infty \geq f_n(0) + 1 - 1/n = 2 - 1/n$ and $\|f_n\|_\infty = 1$. ∎

(b) *Let $\mathcal{H} = P_{H_-\|(PH)}$, where $H_- = (I - P)H$. Show that $\|\mathcal{H}\|^2 + 1/\|P\|^2 = 1$.*

SOLUTION: Indeed, by § 4.7.2 we have $\|\mathcal{H}\| = \cos(A(PH, (I-P)H))$ and by § 4.3.1(c) $1/\|P\| = \sin(A(PH, (I - P)H))$. ∎

4.9.4 The Sharpness of the McCarthy–Schwartz Inequality

We propose to show, via a concrete computation linked to Theorem 4.6.1, that the upper estimate of $ub(X_n)$ in terms of $b(X_n)$ given by Theorem 4.8.2 is sharp. We have already explained the sense of this sharpness in Remark 4.8.6.

(a) *Let $0 < \alpha < 1$ and*

$$w_\alpha(t) = \left|\frac{1 - t}{1 + t}\right|^\alpha.$$

Show that $w_\alpha = |h|^2$ with an outer function h, $h \in H^2$, such that

$$\mathrm{dist}_{L^\infty(\mathbb{T})}(\bar{h}/h, H^\infty) \le \sin(\alpha\pi/2),$$

and deduce that

$$\|P_+\| \le \frac{1}{\sin(\frac{\pi}{2}(1-\alpha))}.$$

Hint Follow the proof of $(5) \Rightarrow (4)$ of Theorem 4.6.1.

SOLUTION: First observe that

$$\log(w_\alpha) = \alpha \cdot \log\left|\frac{1-t}{1+t}\right| = Hv \quad \text{where } v = \alpha \cdot \arg\frac{1-t}{1+t}.$$

Since $t \longmapsto (1-t)/(1+t)$ is a mapping of the disk $\mathbb{D}$ to the half-plane $\mathbb{C}^+ = \{\mathrm{Re}(z) > 0\}$, we have $\|v\|_\infty = \alpha\pi/2 < \pi/2$ and, following the proof of $(5) \Rightarrow (4)$ of Theorem 4.6.1 (with the same notation) we obtain $w_\alpha = |h|^2$, $\bar{h}/h = e^{-iv+ic}$. Hence the domain Ω (the crucial object for our estimations) is

$$\Omega = \{z \in \mathbb{C}: |z| = 1, |\arg(z)| \le \|v\|_\infty\}.$$

Define the disk $D(\lambda, R)$, $R = (1-\delta)\lambda$, $\lambda > 0$ such that $\Omega \subset D(\lambda, (1-\delta)\lambda)$ (as described in the proof of Theorem 4.6.1) and $R/\lambda = 1 - \delta$ attains its minimum: by the proof, this implies that $\mathrm{dist}_{L^\infty(\mathbb{T})}(\bar{h}/h, H^\infty) \le 1 - \delta$. By Pythagoras, $R^2 = (\lambda - \cos(\alpha\pi/2))^2 + \sin^2(\alpha\pi/2)$ and

$$(1-\delta)^2 = R^2/\lambda^2 = (1 - \lambda^{-1}\cos(\alpha\pi/2))^2 + \lambda^{-2}\sin^2(\alpha\pi/2),$$

where $\lambda > \cos(\alpha\pi/2)$. It is easy to see that $\min(R^2/\lambda^2)$ is attained for $\lambda = 1/\cos(\alpha\pi/2)$ and gives the value $(1-\delta)^2 = \sin^2(\alpha\pi/2)$, so $\mathrm{dist}_{L^\infty(\mathbb{T})}(\bar{h}/h, H^\infty) \le \sin(\alpha\pi/2)$. To bound $\|P_+\|$ use Corollary 4.6.3,

$$\|P_+\| = \left(1 - \mathrm{dist}^2_{L^\infty(\mathbb{T})}\left(\frac{\bar{h}}{h}, H^\infty\right)\right)^{-1/2} \le (1 - \sin^2(\alpha\pi/2))^{-1/2}$$

$$= \frac{1}{\sin(\frac{\pi}{2}(1-\alpha))}.$$

                                                      ■

(b) *Let $\mathcal{E}_n = (z^k)_0^{n-1}$, $n > 1$, the basis of exponentials in the space*

$$\mathcal{P}_{\alpha,n} = \mathrm{span}(z^k : 0 \le k < n)$$

of the polynomials of degree $< n$ equipped with the $L^2(\mathbb{T}, w_\alpha m)$ norm. Deduce from (a) that

$$b(\mathcal{E}_n) \le \frac{2}{\sin(\frac{\pi}{2}(1-\alpha))}.$$

SOLUTION: By Lemma 4.4.3, $b(\mathcal{E}_n) \leq 2\|P_+\|$, and the result follows. ∎

(c) (Spijker, Tracogna, and Welfert, 2003). *Re-using the notation of (b), let $R_n : \mathcal{P}_{a,n} \to \mathcal{P}_{a,n}$ be the rotation through the angle π, $(R_n p)(z) = p(-z)$, $z \in \mathbb{T}$. Show that, for every n,*

$$ub(\mathcal{E}_n) \geq \frac{1}{2}\|R_n\| \geq \frac{n^\alpha}{11} = \frac{1}{11}n^{1-\epsilon}, \ \text{where } 0 < \epsilon \leq 2/b(\mathcal{E}_n),$$

with $c > 0$, $C > 0$ constants.

Hint Select $p = \sum_{0 \leq k < n} z^k$ for a test function.

SOLUTION: Let $P = P_\sigma$ where $\sigma \subset \{0, 1, \ldots, n-1\}$ is the set of even numbers between 0 and n. Then $R_n = P - (I - P)$ and

$$\|R_n\| \leq \|P\| + \|I - P\| \overset{\text{Ex. 4.9.3(a)}}{=} 2\|P\| \leq 2ub(\mathcal{E}_n),$$

hence the first inequality of the statement. To bound $\|R_n\|$ below, select p as in the *Hint* above, $p = \sum_{0 \leq k < n} z^k = (1 - z^n)/(1 - z)$, and proceed in two steps.

(1) An upper bound for $\|p\| = \|p\|_{L^2(w_\alpha m)}$:

$$\|p\|^2 = 2 \int_0^\pi |1 - e^{int}|^2 |1 - e^{it}|^{-2+\alpha}|1 + e^{it}|^{-\alpha} \frac{dt}{2\pi}$$

$$= 2\left(\int_0^{a/n} + \int_{a/n}^{\pi/2} + \int_{\pi/2}^{\pi} \right),$$

where $a = 1/\sqrt{1-\alpha}$ and $a/n \leq \pi/2$ (if $a/n > \pi/2$ replace $\int_0^{a/n} + \int_{a/n}^{\pi/2}$ with $\int_0^{\pi/2}$). For the first integral, we have

$$|1 - e^{int}|^2 = 4\sin^2(nt/2) \leq n^2 t^2,$$

$$|1 - e^{it}|^{-2+\alpha} = (2\sin(t/2))^{-2+\alpha} \leq (2t/\pi)^{-2+\alpha},$$

$$|1 + e^{it}|^{-\alpha} \leq 2^{-\alpha/2},$$

thus

$$2\int_0^{a/n} \leq (1/\pi)2^{-\alpha/2}\pi^{2-\alpha}2^{-2+\alpha}n^2(1+\alpha)^{-1}[t^{1+\alpha}]_0^{a/n}$$

$$\leq \frac{\pi a^{1+\alpha}}{4}n^{1-\alpha}.$$

For the second, $|1 - e^{int}|^2 \leq 4$ (use the same estimates as above for the other factors), thus

$$2\int_{a/n}^{\pi/2} \leq 4(1/\pi)\pi^{2-\alpha}2^{-2+\alpha/2}(-1+\alpha)^{-1}[t^{-1+\alpha}]_{a/n}^{\pi/2}$$

$$\leq \frac{\pi}{(1-\alpha)a^{1-\alpha}}n^{1-\alpha}.$$

For the third,

$$|1 - e^{it}|^{-2+\alpha} \le 2^{(-2+\alpha)/2},$$

$$|1 + e^{it}|^{-\alpha} = 4^{-\alpha/2} \sin^{-\alpha}((\pi - t)/2) \le 4^{-\alpha/2}((\pi - t)/\pi)^{-\alpha},$$

and hence

$$2 \int_{\pi/2}^{\pi} \le 4(1/\pi)2^{(-2+\alpha)/2}4^{-\alpha/2}(1 - \alpha)^{-1}\pi[(1 - t/\pi)^{1-\alpha}]_{\pi/2}^{\pi}$$

$$\le \frac{\sqrt{2}}{(1 - \alpha)}.$$

Finally, since $a^{1+\alpha} = 1/((1 - \alpha)a^{1-\alpha}) = (1 - \alpha)^{-(1+\alpha)/2}$, we obtain

$$\|p\|^2 \le \frac{\pi a^{1+\alpha}}{4}n^{1-\alpha} + \frac{\pi}{(1 - \alpha)a^{1-\alpha}}n^{1-\alpha} + \frac{\sqrt{2}}{(1 - \alpha)}$$

$$= \frac{5\pi}{4}(1 - \alpha)^{-(1+\alpha)/2}n^{1-\alpha} + \frac{\sqrt{2}}{1 - \alpha}$$

$$< \frac{4}{(1 - \alpha)^{(1+\alpha)/2}}n^{1-\alpha} + \frac{\sqrt{2}}{1 - \alpha}.$$

(2) A lower bound for $\|R_n p\| = \|R_n p\|_{L^2(w_\alpha m)}$:

$$\|R_n p\|^2 = 2 \int_0^{\pi} |1 - (-1)^n e^{int}|^2 |1 + e^{it}|^{-2}|1 - e^{it}|^\alpha |1 + e^{it}|^{-\alpha} \frac{dt}{2\pi}$$

$$\ge 2 \int_{\pi/2}^{\pi} \ge 2^{1+\alpha/2} \int_{\pi/2}^{\pi} |1 - (-1)^n e^{int}|^2 |1 + e^{it}|^{-2-\alpha} \frac{dt}{2\pi}$$

$$\overset{s=\pi-t}{=} \frac{2^{\alpha/2}}{\pi} \int_0^{\pi/2} |1 - e^{-ins}|^2 |1 - e^{-is}|^{-2-\alpha} ds$$

$$= \frac{2^{\alpha/2}}{\pi} \int_0^{\pi/2} 4\sin^2(ns/2)(4\sin^2(s/2))^{-1-\alpha/2} ds$$

$$\ge \frac{2^{\alpha/2}}{\pi} \int_0^{\pi/n} 4(ns/\pi)^2(4(s/2)^2)^{-1-\alpha/2} ds$$

$$= \frac{2^{\alpha/2}4n^2}{\pi^3} \int_0^{\pi/n} s^{-\alpha} ds = \frac{2^{\alpha/2}4n^2}{\pi^3(1 - \alpha)}(\pi/n)^{1-\alpha}$$

$$= \frac{2^{2+\alpha/2}n^{1+\alpha}}{\pi^{2+\alpha}(1 - \alpha)}.$$

To complete the lower estimate for $ub(\mathcal{E}_n)$, use the inequalities already established:

$$ub(\mathcal{E}_n)^2 \ge \left(\frac{1}{2}\|R_n\|\right)^2 \ge \frac{1}{4}\|R_n p\|^2/\|p\|^2$$

$$\ge \frac{2^{\alpha/2}n^{1+\alpha}}{\pi^{2+\alpha}(1 - \alpha)} \bigg/ \left(\frac{4}{(1 - \alpha)^{(1+\alpha)/2}}n^{1-\alpha} + \frac{\sqrt{2}}{(1 - \alpha)}\right)$$

$$= n^{1+\alpha}\left(\frac{2^{1/2}}{\pi}\right)^\alpha \bigg/ \pi^2(4n^{1-\alpha}(1 - \alpha)^{(1-\alpha)/2} + \sqrt{2}).$$

Since $x^x \leq 1$ for $0 \leq x \leq 1$, we have

$$ub(\mathcal{E}_n)^2 \geq \frac{2^{1/2}n^{1+\alpha}}{\pi^3(4 + \sqrt{2})n^{1-\alpha}} = \frac{2^{1/2}n^{2\alpha}}{\pi^3(4 + \sqrt{2})},$$

hence

$$ub(\mathcal{E}_n) \geq n^{\alpha}\left(\frac{2^{1/2}}{\pi^3(4 + \sqrt{2})}\right)^{1/2} \geq \frac{n^{\alpha}}{11}.$$

It only remains to note that $\alpha = 1 - (1 - \alpha) := 1 - \epsilon$ and hence, according to (b) above, we have $\epsilon = 1 - \alpha \leq \sin(\frac{\pi}{2}(1 - \alpha)) \leq 2/b(\mathcal{E}_n)$.

∎

Remark There is no known example that provides a complete converse to the McCarthy–Schwartz inequality of Theorem 4.8.2 (of the type $ub(\mathcal{X}_n) \geq b(\mathcal{X}_n)n^{1-\epsilon}$ with $\epsilon \leq$ constant $/b(\mathcal{X}_n)^2$).

4.10 Notes and Remarks

The Fourier series (and integrals) form the basis of harmonic analysis: ever since their appearance at the beginning of the nineteenth century in the works of Fourier, they have remained extremely powerful techniques in the service of applications. The reason is of course found in a profound idea (which dates back to Euler, almost a century earlier): an operator L (differential, or any other) is analyzed by reducing it to a diagonal form, i.e. by using its series development (or sums) in terms of its eigenvectors, $Lx_j = \lambda_j x_j$ ($j \in J$). For a self-adjoint operator L, we are led to orthogonal series, as was the case for Laplace and Fourier with the a-periodic *Laplace operator* $Lx = -x''$, $x(0) = x(a)$ ($a > 0$) where $x_j(t) = e^{2\pi ijt/a}$, $j \in \mathbb{Z}$. For non-self-adjoint operators, the eigenvectors are no longer orthogonal and the *generalized Fourier series* appear. We could again handle the situation with the help of orthogonal series of eigenvectors of a "neighboring" self-adjoint operator, but this becomes ever more complicated, and we end up turning to other developments and transformations of Fourier type (wavelets, etc.). In this chapter we have only developed the very beginning of the theory, but invite the interested reader to continue with more advanced texts such as those of Kenig (1994), Duoandikoetxea (2001), Stein (1993), Meyer (1992), and Kahane and Lemarié-Rieusset (1998).

For § 4.1–§ 4.3 we refer the reader to the great classics of functional analysis such as Banach (1932), Riesz and Sz.-Nagy (1955), and Lindenstrauss and Tzafriri (1977). The angle between two subspaces of a Hilbert space was defined by Friedrichs (1937), when developing a geometric approach to the problems of perturbation of operators.

Section 4.4 is found, for the most part, in Kolmogorov (1941), but we have taken into account important later developments: Helson and Szegő (1960), Helson and Sarason (1967), Ibragimov and Rozanov (1970), and Peller and Khruschev (1982).

For comments on the *Hilbert operator H* (§ 4.5) see Notes and Remarks 2.9. We can also add that Hilbert himself defined the operation P_+ (which is now known as the *Hilbert transform*, or the *Riesz projection*) in his famous course on integral equations (1904–1908 (Hilbert, 1912)) where he made progress on the *Riemann problem* dealing with the basic equation in singular integrals (Cauchy kernel). Today, the techniques linked to the Riemann–Hilbert problem are crucial tools in a dozen mathematical disciplines: from complex analysis, integral operators, and integrable models in mathematical physics through to probabilistic combinatorics, not to mention orthogonal polynomials, random matrices, and signal theory. It is not surprising that throughout the twentieth century this field attracted such enormous attention; in pure analysis, it culminated in the theory of the *Calderón–Zygmund integral operators*.

The treatment in § 4.6 is close to that of Nikolski (1986), but the main contents comes from Helson and Szegő (1960). It is interesting to note that while Theorem 4.6.1 provides a complete characterization of whether the subspaces $H^2(w)$ and $H^2_-(w)$ are at a strictly positive angle to each other, the natural question "When is the projection $P_+(x_- + x_+) = x_+$ well-defined on the vector sum $H^2(w) + H^2_-(w)$?", i.e. when is

$$H^2(w) \cap H^2_-(w) = \{0\},$$

remains open: see Sarason (1994).

We must also point out that since the appearance, thanks to Burkholder, Hunt, Muckenhoupt and Wheeden (see Hunt et al., 1973), of an approach to weighted analysis totally different from that of Helson and Szegő, the two complement each other for different applications and generalizations. The Helson–Szegő (HS) approach is based on the operation of harmonic conjugation which is not "local" and whose existence requires a group structure. By contrast, the Hunt–Muckenhoupt–Wheeden (HMW) approach is predominantly "local," realizable on a more or less arbitrary metric space, and is based on the technique of *scaling* – an operation close to many "dyadic" techniques, e.g. wavelets. More precisely, given a weight $w \geq 0$ on $\mathbb{T}$, the condition (HMW), known as the *Muckenhoupt condition* (A_2), which is equivalent to the condition

$w \in (HS)$ of Theorem 4.6.1(5), consists of

$$A_2(w) := \sup_I \left(\frac{1}{m(I)} \int_I w \, dm \right) \left(\frac{1}{m(I)} \int_I \frac{1}{w} \, dm \right) < \infty,$$

where I runs over the set of sub-intervals (arcs) of $\mathbb{T}$.

To mention only a few selected points of "competition" between these two approaches, we note the following.

(i) (HS) works better at providing simple examples of weights $w \in (HS) = (A_2)$ (beginning with the Babenko Example 4.6.6), and it is more easily adaptable to harmonic analysis in several variables (on the circle $\mathbb{T}^n$, for example) and for the *two-weight problem*, consisting of characterizing the pairs $w_1, w_2 \geq 0$ such that

$$\int_{\mathbb{T}} |Hf|^2 w_1 \, dm \leq C^2 \int_{\mathbb{T}} |f|^2 w_2 \, dm, \quad \forall f \in \mathcal{P}$$

(see Cotlar and Sadosky, 1979).

(ii) By contrast, (HMW) is indispensable in the analysis of singular integrals on manifolds without a group structure, and is better adapted for applications to multi-dimensional random processes, etc.

It is quite easy to see that $(HS) \Rightarrow (A_2)$, but there is no direct proof for the converse $(A_2) \Rightarrow (HS)$.

During the twentieth century, the applications to stationary processes (§ 4.7) and signal processing were the principal motors of development of Fourier analysis, both classical and generalized. For the period of the "youth of the theory" (which is all that is presented in this book), refer to Kolmogorov (1941) and Wiener and Masani (1957, 1958), and for subsequent developments, to Rozanov (1963) and Ibragimov and Rozanov (1970); for an innovative survey article, see Peller and Khruschev (1982). In particular, the very tight links between processes and Hankel operators were discovered in the last-cited article. The Hankel operators (see § 4.7.2 and Theorem 4.7.1, but we have not given the general definition here), thanks to the efforts of Nehari, Krein, Adamyan, Arov, Peller, and others, have been transformed into a powerful and indispensable tool for the study of a certain number of analytic phenomena such as random processes, signal processing and H^∞ optimal control, interpolation theory, theory of best approximations, etc. The references for § 4.7.2 are Nehari (1957), Peller (2003), Power (1982), and Nikolski (2002). It is interesting to note that several properties of the operators $H_\varphi = P_{-\varphi} \mid H^2$ (in principle, quite special objects closely linked to harmonic analysis on $\mathbb{T}$) are shared with abstract expressions of the type $H = P_{E^\perp} T \mid E$ where E is a subspace of a Hilbert space on which operates a bounded operator T (for H_φ this is the operator of multiplication by φ); see Devinatz and Shinbrot (1969) or the monographs mentioned above.

The question of the "quantitative qualities" of the bases in a vector space (and in particular, in a Hilbert space), tackled in § 4.8, is extremely important for any application to numerical analysis and matrix analysis, but also for purely theoretical problems in high-dimensional geometry. We have limited ourselves to the beautiful result of McCarthy and Schwartz (1965) (Theorem 4.8.2), and to its converse by Spijker, Tracogna, and Welfert (2003) (§ 4.9.4(c)). It is interesting to note that this result remains somewhat mysterious (despite the extreme transparency of the original proof), because it does not provide an explanation of the origin of this improvement in the order of $ub(X_n)/b(X_n)$ in a Hilbert space compared to a general Banach space. It is also curious to know that the authors of this analytic gem from 1965 were convinced that their result was quite approximate and that the true rate of growth of $ub(X_n)$ would be logarithmic; they put forward the conjecture $ub(X_n) = O((\log n)^{c(b(X_n))} b(X_n))$ as $n \to \infty$. It is also very instructive that the result of § 4.9.4(c) (the converse of Theorem 4.8.2) was found 35 years later by specialists in applied numerical analysis (Spijker, Tracogna, and Welfert, 2003). (The calculations of Spijker et al., 2003 are different from those presented in § 4.9.4(c).) The result of § 4.9.4(c) is based on the inequality § 4.9.4(a) which is, in fact, the equality $\|P_+\| = 1/\sin(\frac{\pi}{2}(1 - \alpha))$ (Hollenbeck and Verbitsky, 2000); these authors found exact values for norms of many other operators of harmonic analysis).

Another aspect of the question of the "quality of a basis" is to regard it in the form of a *summation basis*. More precisely, let $X = (x_k)$, $X' = (x_k')$ be a complete and total biorthogonal pair, and let $V = (v_{\alpha k})$ be a "matrix" of complex numbers such that

$$\sum_k |v_{\alpha k}| \cdot \|x_k\| \cdot \|x_k'\| < \infty \quad (\forall \alpha) \quad \text{and} \quad \lim_\alpha v_{\alpha k} = 1 \quad (\forall k).$$

Then, V is said to define a *summation method* and the V-sum of the series $\sum_k \langle x, x_k' \rangle x_k$ is taken as the limit (if it exists)

$$(V) \sum_k \langle x, x_k' \rangle x_k = \lim_\alpha \sum_k v_{\alpha k} \langle x, x_k' \rangle x_k.$$

We only consider the "hereditary" methods, i.e. where the convergence of a series $\sum_k \langle x, x_k' \rangle x_k$ implies $(V) \sum_k \langle x, x_k' \rangle x_k = \sum_k \langle x, x_k' \rangle x_k$. For example, the method of arithmetic means (Cesàro and Fejér) corresponds to $v_{nk} = A_{nk}$ where $A_{nk} = \max(0, 1 - |k|/(n + 1))$, and that of Abel and Poisson to $v_{rk} = P_{rk}$,

$P_{rk} = r^{|k|}$, $0 < r < 1$ (so that, for example, $(P) \sum_{k\geq0}(-1)^k(k+1) = 1/4$, but $(A) \sum_{k\geq0}(-1)^k(k+1)$ does not exist, whereas if the (A)-sum exists then the (P)-sum also, and they are equal). Rosenblum (1962) proved that $\mathcal{E} = (z^k)_{k\in\mathbb{Z}}$ in $L^2(wm)$ is a basis for the Abel–Poisson method (P) if and only if $w \in (HS)$ (hence it is already a Schauder basis). It is not known how to characterize the weights w on $\mathbb{T}$ for which there exists a summation method for the Fourier series in $L^2(wm)$ (we suppose, of course, that $w^{\pm1} \in L^1(\mathbb{T})$). The same is true for the following property of *spectral synthesis* (the weakest property guaranteeing a "reconstruction" of every function $f \in L^2(wm)$ from its Fourier series $\sum_{k\in\mathbb{Z}} \widehat{f}(k)z^k$):

$$f \in \mathrm{span}_{L^2(wm)}(\widehat{f}(k)z^k : k \in \mathbb{Z}) \quad (\forall f \in L^2(wm)).$$

In the literature, this property is also known as *hereditary completeness*, or *strong M-basis*. For recent news on the hereditary (non)completeness of nonharmonic exponentials and systems of reproducing kernels see Baranov, Belov, Borichev (2013) and Baranov, Yakubovich (2016)

The Gram matrix techniques of § 4.8.2 are classical in this group of ideas and ubiquitous in all applied matrix analysis, as well as in approximation theory. Indeed, it is easy to see that any positive matrix $((Ax, x) \geq 0, \forall x)$ is a Gram matrix of a sequence of vectors. For Gram matrices, see Golub and Van Loan (1996), Akhiezer (1965), and Gantmacher (1966).

The multipliers, and especially the Fourier multipliers (i.e. for the family of exponentials $\mathcal{E}$) count among the indispensable subjects of harmonic analysis, as they can be identified with the eigenvalues of the convolution operators (invariant with respect to translations). We refer to Zygmund (1959) for an introductory presentation.

5

Harmonic Analysis and Stationary Filtering

Topics. Filters (finite-power, stable, causal), harmonic signals, time and frequency domains, transfer function, Wiener's fundamental theorems, synthesis of filters, band-pass filters, Rudin–Carleson theorem, Helson sets, inverse problems, five differences between C and W, a brief overview of sampling.

The mathematical theory of stationary filtering was founded by Wiener in the 1930s, but also by Kolmogorov, Masani, and others, and, on the engineering side, by Kotelnikov and (independently, but 15 years later) by Shannon; see Notes and Remarks 5.7 at the end of this chapter (including the biographies). Throughout the twentieth century, filtering theory amply nourished all of harmonic analysis by proposing fundamental problems to be solved. In particular, this is the case for the theory of Hardy spaces, so that separating the theoretical applications of filtering from the theory of Hardy spaces itself has today become delicate. This is why the contents of this chapter can be considered more a "filtering interpretation" of the theory already developed, rather than a new subject. More precisely, in this chapter we only consider discrete-time signals. For continuous-time signals, there are a few references given in § 5.7. Finally, we warn the reader that in the different mathematical presentations of signal processing, the terminology linked with the physical nature of the signals (such as the energy, power, etc.) can vary. Here, we follow the language of the founders of the theory (see the references in § 5.7).

5.1 The Language of Linear Filters

By definition, *a signal x* is a complex-valued function of a variable called the *time*. We will study signals of discrete time $\mathbb{Z}$, i.e. of complex sequences

150

$n \longmapsto x_n$, $n \in \mathbb{Z}$, hence

$$x = (x_n)_{n \in \mathbb{Z}}.$$

The principal operation on signals is *filtering* – which consists of passing a signal x through an apparatus Φ (a "box") which transforms it into another signal y:

$$\Phi : x \longmapsto \Phi x = y.$$

The *energy of a signal* is defined as $\sum_{n \in \mathbb{Z}} |x_n|^2$. We will only consider filters Φ that are *stationary* and *of finite power*; these are described by the following axioms.

Definition 5.1.1 *A finite-power stationary filter Φ is a mapping of numerical sequences $(x_k)_{k \in \mathbb{Z}}$ satisfying the axioms (A_1)–(A_4):*

(A_1) *Φ is of finite power, i.e. it transforms a signal with finite energy into another signal with finite energy, hence Φ is a mapping of the space $l^2(\mathbb{Z})$ into itself,*

(A_2) *Φ is linear,*

(A_3) *Φ is stationary (invariant by translation on $\mathbb{Z}$), i.e. for every $n \in \mathbb{Z}$*

$$\Phi \tau_n = \tau_n \Phi,$$

where $\tau_n((x_k)_{k \in \mathbb{Z}}) = (x_{k-n})_{k \in \mathbb{Z}}$ is a translation of step n on $\mathbb{Z}$,

(A_4) *Φ is correctly observable, i.e. the observation of a coordinate $x \longmapsto (\Phi x)_0$ is continuous on $l^2(\mathbb{Z})$.*

Another important class of filters is that of "stable" filters. A filter Φ is said to be stable stationary if satisfies (A_2)–(A_4) and the following axiom (A_1') in place of (A_1). In (A_1) we replace $l^2(\mathbb{Z})$ by $l^\infty(\mathbb{Z})$:

(A_1') *Φ is well-defined on the bounded signals $x = (x_k)_{k \in \mathbb{Z}}$, $x \in l^\infty(\mathbb{Z})$, and transforms them into bounded signals, hence Φ is a mapping of the space $l^\infty(\mathbb{Z})$ into itself.*

Remark 5.1.2 (diagonalization) The eigenvectors of the group of translations $(\tau_n)_{n \in \mathbb{Z}}$, i.e. the sequences $x = (x_k)_{k \in \mathbb{Z}}$ satisfying

$$\tau_n x = \lambda_n x, \quad n \in \mathbb{Z},$$

where $\lambda_n \in \mathbb{C}$, play a particularly important role in the analysis of filters. It is easy to find them:

$$\tau_1 x = \lambda x \Leftrightarrow x = a \cdot x_\lambda, \text{ where } x_\lambda := (\lambda^{-k})_{k \in \mathbb{Z}} \text{ (here } a \in \mathbb{C}, \lambda \in \mathbb{C}^* = \mathbb{C} \setminus \{0\}).$$

The signal $x_\lambda := (\lambda^{-k})_{k \in \mathbb{Z}}$ is called the *input harmonic signal of frequency* λ (or of frequency $\arg(\lambda)$).

Clearly the harmonic signals (or, more briefly, the *harmonics*) in a way diagonalize a stationary filter: $\lambda(\Phi x_\lambda) = \Phi \tau_1 x_\lambda = \tau_1(\Phi x_\lambda)$ and hence

$$\Phi x_\lambda = a(\lambda) x_\lambda,$$

for any x_λ in the domain of definition of Φ; the numerical function $\lambda \longmapsto a(\lambda)$ is an important characteristic of the filter (see below). This is a key idea for the development of a theory of stationary filtering, but in this form, it remains somewhat heuristic. A weak point in the above reasoning is that we do not know the x_λ for which Φ is well-defined: it could be that this set is simply empty (in particular, there is no x_λ with finite energy, but there are cases where it is possible to directly follow the above reasoning – for example, for the stable filters).

To work around this difficulty of "infinite energies," we use the discrete Fourier transform.

5.1.1 The Fourier Transform and the Frequency Domain

By the axioms (A_1)–(A_2), a stationary filter Φ is a linear mapping

$$\Phi \colon l^2(\mathbb{Z}) \to l^2(\mathbb{Z})$$

satisfying $\Phi \tau_n = \tau_n \Phi$ for every $n \in \mathbb{Z}$. The Fourier transform $\mathcal{F}$ is defined in the space $L^1(\mathbb{T})$ by

$$\mathcal{F} f = (\widehat{f}(n))_{n \in \mathbb{Z}},$$

and the inverse Fourier transform by

$$\mathcal{F}^{-1}((x_n)_{n \in \mathbb{Z}}) = \sum_{n \in \mathbb{Z}} x_n \zeta^n, \quad \zeta \in \mathbb{T}.$$

The series is formal, however – in more precise contexts – we can find a manner in which it converges. For example, we know that $\mathcal{F}$ is a unitary mapping between the spaces L^2:

$$\mathcal{F} \colon L^2(\mathbb{T}) \to l^2(\mathbb{Z}), \quad \mathcal{F} L^2(\mathbb{T}) = l^2(\mathbb{Z}), \quad \|\mathcal{F} f\|_{l^2(\mathbb{Z})} = \|f\|_{L^2(\mathbb{T})} \quad (\forall f \in L^2(\mathbb{T})),$$

(theorems of Fourier–Plancherel, or of Riesz–Fischer: see Appendix A).

In signal processing, the space $l^2(\mathbb{Z})$ is known as the *time domain*, and the space $L^2(\mathbb{T})$ as the *frequency domain* (or *spectral domain*). Signals with finite energy $x \in l^2(\mathbb{Z})$ are also called *pulse signals*. The following diagram defines a representation of a filter in the frequency domain:

$$
\begin{array}{ccc}
\Phi\colon l^2(\mathbb{Z}) & \longrightarrow & l^2(\mathbb{Z}) \\
\uparrow{\scriptstyle \mathcal{F}} & & \downarrow{\scriptstyle \mathcal{F}^{-1}} \\
\widehat{\Phi}\colon L^2(\mathbb{T}) & \longrightarrow & L^2(\mathbb{T})
\end{array}
$$

i.e.

$$
\widehat{\Phi} = \mathcal{F}^{-1}\Phi\mathcal{F}\colon L^2(\mathbb{T}) \to L^2(\mathbb{T}).
$$

The stationarity, i.e. the fact that Φ commutes with τ_n, is transformed into

$$
\widehat{\Phi}M_z^n = M_z^n\widehat{\Phi},
$$

since, for every $n \in \mathbb{Z}$, $\mathcal{F}^{-1}\tau_n\mathcal{F} = M_z^n$, where M_z is the operator of multiplication by z.

Now, we are able to find the temporal and frequency characterizations of the stationary filters.

5.2 Characterization of Stationary Filters

The following characterizations reduce the analysis of filtering to the techniques of convolution operations.

Theorem 5.2.1 (Wiener, 1933)

(1) *A finite-power stationary filter Φ is a bounded linear operator on $l^2(\mathbb{Z})$, and its frequency representation $\widehat{\Phi}$ is bounded on $L^2(\mathbb{T})$.*

(2) *Let Φ be a finite-power stationary filter and $\varphi = \mathcal{F}^{-1}S$, where $S = \Phi e_0$, $e_0 = (\delta_{0j})_{j\in\mathbb{Z}}$. Then $\varphi \in L^\infty(\mathbb{T})$, $\|\Phi\| = \|\varphi\|_\infty$ and*

$$
\Phi x = x * S := \left(\sum_{k\in\mathbb{Z}} x_k S_{n-k}\right)_{n\in\mathbb{Z}} \quad (\forall x \in l^2(\mathbb{Z})), \quad \widehat{\Phi}f = \varphi f \ (\forall f \in L^2(\mathbb{T})).
$$

(3) *Conversely, for any $\varphi \in L^\infty(\mathbb{T})$, the mapping $\Phi x = x * S$, where $S = \mathcal{F}\varphi$, is a finite-power stationary filter.*
   The correspondence $\Phi \leftrightarrow \varphi$ is bijective.

Proof (1) By the closed graph theorem (see Appendix E), it suffices to show that the mapping $\Phi\colon l^2(\mathbb{Z}) \to l^2(\mathbb{Z})$ is closed, i.e.

$$
(u_k \in l^2(\mathbb{Z}), \quad \lim_k \|u_k\|_2 = 0, \quad \lim_k \Phi u_k = v) \quad \Rightarrow \quad v = 0.
$$

However, the functional $x \longmapsto (\Phi x)_0$ is continuous, hence so is

$$x \longmapsto (\Phi(\tau_n x))_0 = (\tau_n \Phi x)_0 = (\Phi x)_{-n} \quad (\forall n \in \mathbb{Z}),$$

hence $0 = \lim_k (\Phi u_k)_n = (v)_n$ for every $n \in \mathbb{Z}$. Thus $v = 0$, and Φ is continuous.

(2) By the stationarity commutation relation, for any $f \in L^2(\mathbb{T})$, $\widehat{\Phi}(z^n f) = z^n \widehat{\Phi} f$, hence $\widehat{\Phi}(pf) = p\widehat{\Phi} f$ for any polynomial $p \in \mathcal{P}$. By taking $f = 1 = \mathcal{F}^{-1} e_0$, we obtain $\widehat{\Phi}(p) = p\widehat{\Phi} 1 = p\varphi$ for every polynomial $p \in \mathcal{P}$, where $\widehat{\Phi} 1 = \varphi \in L^2(\mathbb{T})$. We show that, for every $f \in L^2(\mathbb{T})$, $\widehat{\Phi}(f) = f\varphi$. As $\varphi \in L^2(\mathbb{T})$, it suffices to show that the Fourier coefficients coincide: for every $n \in \mathbb{Z}$,

$$(\widehat{\Phi(f)})(n) = (\widehat{f\varphi})(n).$$

Indeed, this is the case for the set of polynomials $f \in \mathcal{P}$ which is dense in $L^2(\mathbb{T})$. Moreover, both sides are continuous in $L^2(\mathbb{T})$: the left by (A_4) and the right by the Cauchy–Schwarz inequality (since $\varphi \in L^2(\mathbb{T})$), hence $\widehat{\Phi}(f) = f\varphi$, for every $f \in L^2(\mathbb{T})$. Hence, φ is a multiplier of $L^2(\mathbb{T})$ and by Exercise 1.8.3(a) (or see Appendices A and D), $\varphi \in L^\infty(\mathbb{T})$ and $\|\Phi\| = \|\widehat{\Phi}\| = \|\varphi\|_\infty$.

The formula $(\Phi x)_n = \sum_{k \in \mathbb{Z}} x_k S_{n-k}$, where $S = \mathcal{F}\varphi = \mathcal{F}\widehat{\Phi} 1 = \Phi e_0$, is immediate since it holds for $x = e_j = \tau_j e_0 = (\delta_{kj})_{k \in \mathbb{Z}}$:

$$(\Phi \tau_j e_0)_n = (\tau_j \Phi e_0)_n = (\tau_j S)_n = S_{n-j}.$$

(3) Clearly, the mapping $\widehat{\Phi} f = f\varphi$ is bounded $L^2(\mathbb{T}) \to L^2(\mathbb{T})$, and for every $n \in \mathbb{Z}$, $M_z^n \widehat{\Phi} = \widehat{\Phi} M_z^n$, hence $\Phi = \mathcal{F}\widehat{\Phi}\mathcal{F}^{-1}$ is a stationary filter. ∎

There is an analog of Theorem 5.2.1 for stable filters.

Theorem 5.2.2 (Wiener, 1933)

(1) *A stable stationary filter Φ is a bounded linear operator on $l^\infty(\mathbb{Z})$.*
(2) *Let Φ be a stable stationary filter. Then $S = \Phi e_0 \in l^1(\mathbb{Z})$, $e_0 = (\delta_{0j})_{j \in \mathbb{Z}}$, $\|\Phi\| = \|S\|_{l^1(\mathbb{Z})}$ and*

$$\Phi x = x * S := \left(\sum_{k \in \mathbb{Z}} x_k S_{n-k} \right)_{n \in \mathbb{Z}} \quad (\forall x \in l^\infty(\mathbb{Z})).$$

(3) *Conversely, for every $S \in l^1(\mathbb{Z})$, the mapping $\Phi x = x * S$ is a stable stationary filter.*

The correspondence $\Phi \leftrightarrow S$ is bijective.

Proof The proof is similar to that of Theorem 5.2.1, with the difference that $\mathcal{F}^{-1}l^\infty(\mathbb{Z})$ is no longer a space of functions on $\mathbb{T}$, but of distributions; we thus try to avoid it.

(1) We use the same proof as in Theorem 5.2.1 (replacing $l^2(\mathbb{Z})$ by $l^\infty(\mathbb{Z})$).
(2) By setting $S = \Phi e_0$ and using (A_3), we obtain $\Phi e_n = \tau_n S = S * e_n$, hence by linearity, $\Phi c = S * c$ for any signal with finite support (i.e. a linear combination $c = \sum_k c_k e_k$, where

$$\operatorname{supp}(c) = \{k : c_k \neq 0\}$$

is a finite set). Moreover, by (1), $\|\Phi\| < \infty$, hence $\|S * c\|_\infty \leq \|\Phi\| \cdot \|c\|_\infty$. The matrix of the mapping

$$x \longmapsto x * S := \left(\sum_{k \in \mathbb{Z}} x_k S_{n-k} \right)_{n \in \mathbb{Z}}$$

is $A = (a_{nk})$ where $a_{nk} = S_{n-k}$ (a Toeplitz matrix on $\mathbb{Z}$, also called a Laurent matrix), thus by Lemma 5.2.4 below we obtain $S \in l^1(\mathbb{Z})$ and $\|\Phi\| = \|S\|_{l^1(\mathbb{Z})}$.
(3) Φ is bounded in $l^\infty(\mathbb{Z})$ by Lemma 5.2.4; Φ is stationary because $\tau_n \Phi x = e_n * S * x = S * e_n * x = \Phi \tau_n x$, according to well-known properties of convolutions (see Appendix A). $\blacksquare$

Remark 5.2.3 The program of *diagonalization* of a filter, described in § 5.1.1, can also be given for *stable filters*. Indeed, there is a family of eigenvectors $x_\lambda = (\lambda^{-k})_{k \in \mathbb{Z}}$ of the group of translations (τ_n) that belong to the space $l^\infty(\mathbb{Z})$:

$$x_\lambda \in l^\infty(\mathbb{Z}) \Leftrightarrow |\lambda| = 1 \quad \Leftrightarrow \quad \lambda \in \mathbb{T}.$$

We start with the convolution operator S corresponding to a stable filter Φ by Theorem 5.2.1 (the *impulse response* of Φ, according to the terminology of § 5.3.1) and its Fourier transform

$$\varphi(\zeta) = (\mathcal{F}^{-1} S)(\zeta) = \sum_{k \in \mathbb{Z}} S_k \zeta^k, \quad \zeta \in \mathbb{T},$$

the (*transfer function* of Φ, by § 5.3.1). Then, for any $\lambda \in \mathbb{T}$, we have

$$\Phi x_\lambda = \left(\sum_{k \in \mathbb{Z}} \lambda^{-k} S_{n-k} \right)_{n \in \mathbb{Z}} = \varphi(\lambda)(\lambda^{-n})_{n \in \mathbb{Z}} = \varphi(\lambda) x_\lambda.$$

Hence, the value $\varphi(\lambda)$ is an "amplifying factor" of the harmonic

$$x_\lambda = (\lambda^{-k})_{k \in \mathbb{Z}} = (e^{-ik\theta})_{k \in \mathbb{Z}}$$

by the filter Φ; here $\lambda = e^{i\theta}$: θ (or λ itself) is called the *frequency* of x_λ.

The family of harmonics x_λ is *weak-$*$ complete* in the space of the bounded signals $l^\infty(\mathbb{Z})$ (with respect to the weak-$*$ topology, $\sigma(l^\infty, l^1)$: see Appendix D):

$$\mathrm{span}_{\sigma(l^\infty, l^1)}(x_\lambda : \lambda \in \mathbb{T}) = l^\infty(\mathbb{Z}).$$

Indeed, if $a \in l^1(\mathbb{Z})$ and $0 = \langle a, x_\lambda \rangle = \sum_{k \in \mathbb{Z}} a_k \lambda^{-k} = (\mathcal{F}^{-1}a)(\lambda^{-1})$ for every $\lambda \in \mathbb{T}$, then $a = 0$ (since a_k are the Fourier coefficients of $\mathcal{F}^{-1}a$, an absolutely convergent Fourier series).

Lemma 5.2.4 *Let $A = (a_{jk})_{j,k \in J}$ be a matrix on an at most countable set of indices J, and let $c_{00} = c_{00}(J)$ be the vector space of finitely supported functions on J equipped with the norm $\|x\|_\infty = \sup_j |x_j|$. Suppose*

$$Ax = \left(\sum_{k \in J} a_{jk} x_k \right)_{j \in J} \in l^\infty(J)$$

for every $x \in c_{00}(J)$. Then,

(1) $\|A : c_{00} \to l^\infty(J)\| = \sup_{j \in J} \sum_{k \in J} |a_{jk}|.$
(2) $\|A : l^\infty(J) \to l^\infty(J)\| = \sup_{j \in J} \sum_{k \in J} |a_{jk}|.$

Proof (1) By definition,

$$\|A : c_{00} \to l^\infty(J)\| = \sup\{\|Ax\|_\infty : c \in c_{00}, \|x\|_\infty \leq 1\}$$

$$= \sup_{\|x\|_\infty \leq 1} \sup_j \left| \sum_{k \in J} a_{jk} x_k \right| = \sup_j \sup_{\|x\|_\infty \leq 1} \left| \sum_{k \in J} a_{jk} x_k \right|$$

$$= \sup_{j \in J} \sum_{k \in J} |a_{jk}|.$$

(2) follows from (1) and the fact that $|\sum_{k \in J} a_{jk} x_k| \leq \|x\|_\infty \sum_{k \in J} |a_{jk}|.$ ∎

5.3 What Can Filtering Do?

This section describes a number of problems that arise in the mathematical theory of filters.

5.3.1 A Bit More Terminology for Filters

In addition to the preceding notation of Theorems 5.2.1–5.2.2, the following language is also used.

- $S = \Phi e_0$ is the *impulse response* of a stationary filter Φ.
- The function $\varphi = \mathcal{F}^{-1}\Phi e_0$ is the *transfer function*, or *frequency characteristic*, or *frequency response*, or *impedance*, or *voltage gain*. For

finite-power filters, we have $\varphi \in L^\infty(\mathbb{T})$, and for stable filters,

$$\varphi \in \mathcal{F}^{-1}l^1(\mathbb{Z}) := W,$$

the *Wiener algebra of absolutely convergent Fourier series.*

- The function $e^{i\lambda} \longmapsto |\varphi(e^{i\lambda})|$ is called the *energy spectrum.*
- The number $\|\varphi\|_\infty$ is the *amplitude distortion.*
- A *band-pass filter* is defined by the condition $|\varphi| = \chi_\sigma$, $\sigma \subset \mathbb{T}$ (σ is a band of frequencies that pass without distortion of the amplitudes). The *ideal band-pass filter* is $\varphi = \chi_\sigma$ (there is no distortion of either the amplitude or the phase on σ).
- An *all-pass filter* (or *dephasing filter*, or *phase correction filter*) satisfies $|\varphi| = 1$.
- A signal $x = (x_k)_{k \in \mathbb{Z}}$ is said to be *causal* (or *positive time*) if $x_k = 0$ for $k < 0$.
- A filter Φ is said to be *causal,* or *physically realizable,* if x *causal* $\Rightarrow \Phi x$ *causal.*
- Φ is said to be *stable* stationary if it satisfies (A_2)–(A_4), is well-defined on the bounded signals $x = (x_k)_{k \in \mathbb{Z}}$, $x \in l^\infty(\mathbb{Z})$, and transforms them into bounded signals.
- The *phase lag* at the frequency λ of a filter Φ is by definition $\arg(\varphi(e^{i\lambda}))$.

Similar terminology exists for the *pulse* signals $x \in l^2(\mathbb{Z})$.

- The function $\mathcal{F}^{-1}x \in L^2(\mathbb{T})$ is called the *energy spectrum* of x.
- $|\mathcal{F}^{-1}x(e^{i\lambda})|^2$ is called the *energy density at the frequency* λ (for it to be defined everywhere, it suffices to require $x \in l^1(\mathbb{Z})$).
- $\arg(\mathcal{F}^{-1}x(e^{i\lambda}))$ is a phase of x at the frequency λ.

5.3.2 Some Typical Problems in Filtering

Here is a short list of problems in the mathematical theory of filters.

(1) *Direct problem.* Construct a ("white box") filter having a given frequency response on a particular band of frequencies (and as an option, physically realizable). In particular, describe a filter that detects a useful signal against background noise.

(2) *Inverse problem.* Identify an unknown ("black box") filter $\Phi \colon x \longmapsto y$ from the harmonic analysis of an observable input/output couple x, y. In particular, study the possibility of reconstructing Φ when the spectral densities $|\mathcal{F}^{-1}x|^2$ and $|\mathcal{F}^{-1}y|^2$ are known.

(3) *Problem of causality.* Study questions (1)–(2) for finite-power and/or stable causal filters.

We first examine problem (1).

5.4 Synthesis of Causal Filters

The construction of a filter having certain desired characteristics is known as the "synthesis of a filter." We begin with a frequency description of the causal filters.

Lemma 5.4.1 (Wiener, 1930) *A stationary filter Φ (finite-power, or stable) is causal if and only if its transfer function $\varphi = \mathcal{F}^{-1}\Phi e_0$ is in H^∞,*

$$\varphi \in H^\infty$$

(in the case of a stable filter, $\varphi \in H^\infty \cap \mathcal{F}^{-1}l^1(\mathbb{Z}) = \mathcal{F}^{-1}l^1(\mathbb{Z}_+) := W_a$).

Proof The necessity is evident since e_0 is a causal signal, and thus so is Φe_0, hence $(\Phi e_0)_k = \widehat{\varphi}(k) = 0$ for every $k < 0$.

For the sufficiency, let $x = (x_k)_{k \in \mathbb{Z}}$ be a causal signal (in l^2 or l^∞), $x_k = 0$ for every $k < 0$. Then, for $n < 0$ and with $S = \Phi e_0$ we have $S_{n-k} = 0$ for every $k \geq 0$, and thus

$$(\Phi x)_n = \sum_{k \in \mathbb{Z}} x_k S_{n-k} = \sum_{k \geq 0} x_k S_{n-k} = 0,$$

and the result follows.   ∎

Corollary 5.4.2 *Let $w \in L^\infty(\mathbb{T})$, $w \geq 0$, $w \neq 0$. The following assertions are equivalent.*

(1) *There exists a finite-power causal filter with energy spectrum w.*
(2) $\log(w) \in L^1(\mathbb{T})$.

This is evident by the theorems of Szegő (Corollary 2.6.2) and Smirnov (§ 3.3.1(g)).   ∎

Corollary 5.4.3 *Let $\Phi \neq 0$ be a causal filter and φ its transfer function. Then:*

(1) Φ *is an ideal band-pass filter if and only if $\Phi = \mathrm{id}$,*
(2) Φ *is a band-pass filter if and only if it is all-pass, and if and only if φ is an inner function,*
(3) Φ *is stable and all-pass if and only if φ is a finite Blaschke product.*

Indeed, (1) and (2) follow directly from the boundary uniqueness Theorem 1.4.4. For (3), if Φ is stable and φ is not reduced to a finite Blaschke product, then the spectrum $\sigma(\varphi_{in})$ contains at least one point λ on the boundary of the disk $\mathbb{D}$ (see Definition 3.2.2). By the description of Corollary 3.2.4, $\underline{\lim}_{z\to\lambda} |\varphi(z)| = 0$. However as it is in $\mathcal{F}^{-1}l^1(\mathbb{Z}_+) = W_a$, the function φ is continuous in $\overline{\mathbb{D}}$, whence $\varphi(\lambda) = 0$ and $\lim_{z\to\lambda, z\in\mathbb{T}} \varphi(z) = 0$, which is not compatible with the property "all-pass." Hence, φ is indeed a finite Blaschke product. $\blacksquare$

Remark 5.4.4 (filter with minimal mean lag)   Real-life applications require filters that allow the passage of a band of frequencies with a minimum of distortion (and suppress – or almost suppress – another band). According to the filter formula

$$y = \Phi x, \quad (\mathcal{F}^{-1}y)(t) = \varphi(t)(\mathcal{F}^{-1}x)(t) \quad \text{(for all } t \in \mathbb{T}),$$

we can distinguish an amplitude distortion $|\varphi(t)| - 1$ and a phase distortion $\arg(\varphi(t))$. If we suppose that the question of amplitude is somehow resolved, *so that the energy spectrum of the filter $|\varphi|$ is chosen*, it remains to minimize the phase lag θ,

$$\varphi(t) := |\varphi(t)|e^{i\theta(t)}.$$

Note that a constant lag $\theta = $ constant does not present a problem since it can be compensated by a multiple of the identity filter $e^{i\tau}$ id. The question is to *minimize the variation of the phase lag*. There is a large variety of concrete situations where one criterion of optimization is chosen over another (for example, only a finite set of frequencies might be considered, etc.). However, here we present only one (linked to the techniques of Hardy spaces), namely, minimize the following *weighted variation of the phase*:

$$P(\varphi) = \int_{\mathbb{T}} \int_{\mathbb{T}} |\varphi(u)| \cdot |\varphi(v)| \sin^2 \frac{\theta(u) - \theta(v)}{2} \, dm(u)\, dm(v).$$

With this formulation of the problem, it is possible to obtain a complete solution.

Theorem 5.4.5 *Let w be a causal energy spectrum, i.e. $w \geq 0$, $w \in L^\infty(\mathbb{T})$ and $\log(w) \in L^1(\mathbb{T})$. Among all causal filters Φ with $|\varphi| = w$, the minimum of the weighted phase variation $P(\varphi)$ is attained (exclusively) for outer transfer functions $\varphi = c[w]$, $c \in \mathbb{T}$, and*

$$\min P(\varphi) = \left(\int_{\mathbb{T}} w \, dm \right)^2 - \exp\left(2 \int_{\mathbb{T}} \log(w) \, dm \right).$$

Proof Write

$$\left(\int_{\mathbb{T}} |\varphi|\, dm \right)^2 - \left| \int_{\mathbb{T}} \varphi\, dm \right|^2 = \int_{\mathbb{T}} \int_{\mathbb{T}} \left(|\varphi(u)| \cdot |\varphi(v)| - \varphi(u)\overline{\varphi(v)} \right) dm(u)\, dm(v)$$

(the integral with $\mathrm{Im}(\varphi(u)\overline{\varphi(v)})$ is canceled given the oddness of the sine function)

$$= \int_{\mathbb{T}} \int_{\mathbb{T}} |\varphi(u)| \cdot |\varphi(v)| (1 - \cos(\theta(u) - \theta(v)))\, dm(u)\, dm(v) = 2P(\varphi),$$

hence the minimum of $P(\varphi)$ among the $\varphi \in H^\infty$, $|\varphi| = w$, is attained if and only if

$$\max \left| \int_{\mathbb{T}} \varphi\, dm \right|^2 = \max |\varphi(0)|^2$$

is attained. However, by Jensen's inequality in Lemma 2.3.1, we always have

$$|\varphi(0)|^2 < \exp\left(2 \int_{\mathbb{T}} \log(w)\, dm \right),$$

with the exception of equality only in the case where $\varphi = c[w]$ (see the criterion of Theorem 2.6.7).   ■

5.4.1 Filters of Optimal "Signal to Noise Ratio"

This problem concerns a signal x of a known form that is polluted by a random parasite signal b, assumed to be independent white noise. More precisely, $b(\cdot) = (b_k(\cdot))_{k \in \mathbb{Z}}$ is a sequence of random independent variables on a probability space $(\Omega, d\omega)$, all with expectation 0 and variance 1.

The problem is to construct a finite-power (or stable) stationary filter Φ, providing, at a fixed moment $n \in \mathbb{Z}$, the best ratio of useful output signal $y_n = (\Phi x)_n$ against the quadratic mean of the noise, i.e. giving

$$\max \frac{|y_n|}{B} = \max \frac{|(\Phi x)_n|}{\left(\int_{\Omega} |(\Phi b(\omega))_n|^2\, d\omega \right)^{1/2}}.$$

A filter Φ providing the maximum, if such exists, is said to be *optimal*. The following theorem provides a solution.

Theorem 5.4.6

(1) *Given x, b and n, an optimal finite-power (respectively, stable) filter exists if and only if $\mathcal{F}^{-1}x \in L^\infty(\mathbb{T})$, hence x is a signal of bounded energy density (respectively, $x \in l^1(\mathbb{Z})$).*

(2) *Suppose $\mathcal{F}^{-1}x \in L^\infty(\mathbb{T})$ (respectively, $x \in l^1(\mathbb{Z})$). Then the only filter Φ giving $\max(|y_n|/B)$ is the filter with impulse response $S_k = (\Phi e_0)_k = c x_{n-k}$ $(k \in \mathbb{Z})$ where $c \neq 0$ is a constant (an "adapted filter").*

Proof Given a finite-power filter Φ, we have $(\Phi b(\omega))_n = \sum_{k \in \mathbb{Z}} b_k(\omega) S_{n-k}$, where the functions $\omega \longmapsto b_k(\omega)$, $\omega \in \Omega$, form an orthonormal sequence, and hence

$$B = \left\| \sum_{k \in \mathbb{Z}} b_k(\cdot) S_{n-k} \right\|_{L^2(\Omega)} = \|S\|_{l^2(\mathbb{Z})}.$$

However, $y_n = \sum_{k \in \mathbb{Z}} x_k S_{n-k} = \sum_{k \in \mathbb{Z}} x_{n-j} S_j = (S, \overline{x})_{l^2}$, where $\overline{x} = (\overline{x}_{n-j})_{j \in \mathbb{Z}}$. Consequently,

$$\sup \left\{ \frac{|y_n|}{B} : \Phi \text{ of finite power} \right\} = \sup \left\{ \frac{|(S, \overline{x})_{l^2}|}{\|S\|_{l^2(\mathbb{Z})}} : S \in \mathcal{F}L^\infty(\mathbb{T}) \right\} = \|x\|_{l^2(\mathbb{Z})},$$

since the image $\mathcal{F}L^\infty(\mathbb{T})$ is clearly dense in $l^2(\mathbb{Z})$ (along with others, it contains the sequences with finite support c_{00}).

It is well-known that the Cauchy–Schwarz inequality

$$|(S, \overline{x})_{l^2}| < \|S\|_{l^2(\mathbb{Z})}\|x\|_{l^2(\mathbb{Z})}$$

is strict with only a single exception, $S = c\overline{x}$ where $c \neq 0$ is a constant. Hence the sup is attained with a finite-power filter if and only if $\mathcal{F}^{-1}\overline{x} \in L^\infty(\mathbb{T})$, in accordance with the statement of the theorem.

Clearly the version for stable filters is also resolved. ∎

5.4.2 Frequency Response on a Very Thin Band

Let $\sigma \subset \mathbb{T}$ be a Borel set (a band of frequencies). Is it possible to find a *finite-power or stable causal filter* Φ having a transfer function φ *equal to an arbitrary bounded* (or continuous) function *on σ?*

An evident restriction is given by the uniqueness theorem of Corollary 1.4.4:

$$\text{necessarily, } m(\sigma) = 0$$

(otherwise, by decomposing $\sigma = \sigma_1 \cup \sigma_2$ where σ_j are disjoint and $m(\sigma_j) > 0$, $j = 1, 2$, it would not be possible to find $\varphi \in H^\infty$ satisfying $\varphi|\sigma = \chi_{\sigma_1}$). Another fact – not really a restriction but merely an inconvenience – is that the transfer function of a generic finite-power filter is defined *almost everywhere* on $\mathbb{T}$ and *not everywhere*.

The outcome of this somewhat ambiguous situation consists in regarding

$$\sigma = \overline{\sigma}, \quad m(\sigma) = 0$$

and the causal filters Φ having a *continuous transfer function* φ (if Φ is stable, φ is automatically continuous), i.e.

$$\varphi \in C_a(\mathbb{D}) = H^\infty \cap C(\mathbb{T}).$$

Other than these two conditions, *there are no other constraints* when constructing a *finite-power filter* having a predetermined frequency response on σ.

It turns out that the response for *stable filters* is completely different: to have an arbitrary continuous frequency response on σ, it is necessary that σ be a "very thin" set with a very specific arithmetical structure. This kind of set σ, such that

$$C(\sigma) = W_a \mid \sigma,$$

is called a *Helson set* (Helson introduced them in 1954). Even today, there does not exist any intelligible description of what makes a set a Helson set. We give below a few examples of Helson and non-Helson sets.

We begin with the following theorem, proved independently by Walter Rudin (also known for his university textbooks in mathematics) and Lennart Carleson, indicating that on the sets of measure zero, the spaces $C_a(\mathbb{D})$ and $C(\mathbb{T})$ are indistinguishable.

Walter Rudin (1921–2010) was an American mathematician, one of the primary experts in the harmonic and complex analysis of the years 1950–1990. He published a series of books (university textbooks and research monographs) of unequaled mathematical and pedagogical quality, whose influence on the teaching of mathematical analysis and the formation of the new generations of mathematicians worldwide is incontestable. The most well known are *Real and Complex Analysis* (1966), *Functional Analysis* (1973), *Principles of Mathematical Analysis* (1953) (nicknamed "Baby Rudin") and also *Fourier Analysis on Groups* (1962). Rudin was rewarded with the Steele Prize for Mathematical Exposition in 1993.

He came from a well-known European Jewish family that had lived in Vienna for centuries. His great-grandfather Aron Pollak, thanks to his charitable actions, was named Chevalier by the Emperor Franz Joseph, and was granted the name von Rudin (1869). Shortly after the Anschluss in 1938, the family fled from Vienna, suffering under the anti-Jewish oppression of the Nazi regime. Rudin obtained his doctorate in 1947 at Duke University (Durham, North Carolina), and then joined the University

Theorem 5.4.7 (Rudin, 1956; Carleson, 1956) *Let $\sigma = \overline{\sigma} \subset \mathbb{T}$, $m(\sigma) = 0$. For every function $f \in C(\sigma)$ there exists $\varphi \in C_a(\mathbb{D})$ such that $\varphi \mid \sigma = f$ and $\|\varphi\|_{C_a(\mathbb{D})} = \|f\|_{C(\sigma)}$.*

Proof Let $Rf = f|\sigma$ be the restriction operator, $R\colon C_a(\mathbb{D}) \to C(\sigma)$. By Banach's theorem (see Appendix E) $RC_a(\mathbb{D}) = C(\sigma)$ if and only if there exists $c > 0$ such that

$$\|R^*\mu\|_{(C_a)^*} \geq c\|\mu\|_{(C(\sigma))^*}$$

for all $\mu \in (C(\sigma))^*$. Moreover, Banach's theorem states that, in the case where the condition is satisfied, for every $f \in C(\sigma)$ there exists a solution of the equation $R\varphi = f$, $\varphi \in C_a(\mathbb{D})$ such that $\|\varphi\| \leq c^{-1}\|f\|$.

The dual space $(C(\sigma))^*$ is the space of complex measures $\mathcal{M}(\sigma)$ equipped with the variation norm, $\|\mu\|_{(C(\sigma))^*} = \mathrm{Var}(\mu)$ (Riesz representation theorem, Appendices A and D). The dual of a subspace $C_a \subset C(\mathbb{T})$ is the quotient space of the dual $(C(\mathbb{T}))^*$, $(C_a)^* = (C(\mathbb{T}))^*/(C_a)^\perp$ where the annihilator $(C_a)^\perp$ is defined by

$$(C_a)^\perp = \{\mu \in (C(\mathbb{T}))^* = \mathcal{M}(\mathbb{T})\colon 0 = \langle z^n, \mu\rangle = \widehat{\mu}(-n) \text{ every } n \geq 0\}.$$

By the Riesz brothers' Theorem 1.5.4, $(C_a)^\perp = H_0^1 = \{h \in H^1\colon h(0) = 0\}$, hence

$$(C_a)^* = \mathcal{M}(\mathbb{T})/H_0^1.$$

It is easy to see that the adjoint of a restriction is an embedding operator: $R^*\colon \mathcal{M}(\sigma) \to \mathcal{M}(\mathbb{T})/H_0^1, \mu \longmapsto \mu + H_0^1$ $(\mu \in \mathcal{M}(\sigma))$. Consequently,

$$\|R^*\mu\| = \inf_{h \in H_0^1} \|\mu + h \cdot m\|_{\mathcal{M}},$$

where $\|\cdot\|_{\mathcal{M}} = \mathrm{Var}$. However, as $m(\sigma) = 0$, the measure μ is singular, thus $\|\mu + h \cdot m\|_{\mathcal{M}} = \|\mu\|_{\mathcal{M}} + \|h \cdot m\|_{\mathcal{M}}$, hence $\|R^*\mu\| = \|\mu\|_{\mathcal{M}}$. Therefore, the Banach condition is satisfied with $c = 1$, which concludes the proof. ∎

5.4.3 Helson Sets: Arbitrary Frequency Response on $\sigma \subset \mathbb{T}$

We treat here the case of stable filters and provide an example of a Helson set σ (allowing arbitrary responses on the frequencies of σ) and an example of a non-Helson set. For this purpose, we again use the same theorem of Banach (see the proof of Theorem 5.4.7): if $Rf = f \mid \sigma$ is the restriction operator, $R \colon W_a \to C(\sigma)$, then $RW_a = C(\sigma)$ if and only if

$$\|R^*\mu\|_{(W_a)^*} \geq c\|\mu\|_{(C(\sigma))^*}.$$

Since $W_a = \mathcal{F}^{-1}l^1(\mathbb{Z}_+)$, the dual $(W_a)^* = (\mathcal{F}^{-1}l^1(\mathbb{Z}_+))^*$ is realized as $l^\infty(\mathbb{Z}_+)$ with the duality expressed as

$$\langle f, c \rangle = \sum_{k \geq 0} \widehat{f}(k)c_k, \quad c = (c_k)_{k \geq 0} \in l^\infty(\mathbb{Z}_+).$$

Hence, a set σ is Helson if and only if there exists a constant $c > 0$ such that, for any $\mu \in \mathcal{M}(\sigma) = (C(\sigma))^*$, we have

$$\sup_{k \geq 0} |\widehat{\mu}(k)| \geq c\|\mu\|_{\mathcal{M}(\sigma)}.$$

Another definition: a set $\sigma \subset \mathbb{T}$ is said to be *independent* (on the field $\mathbb{Q}$) if, for every $\lambda_j \in \sigma$, $\lambda_j \neq \lambda_k$, the equation $\lambda_1^{n_1} \lambda_2^{n_2} \ldots \lambda_s^{n_s} = 1$, where $n_j \in \mathbb{Z}$, implies $n_1 = n_2 = \cdots = n_s = 0$.

(1) (Helson, 1954). *A closed and at most countable independent set $\sigma \subset \mathbb{T}$ is a Helson set, i.e. $C(\sigma) = W_a \mid \sigma$.*

Proof We write the set σ as a sequence, $\sigma = \{\lambda_j \colon j = 1, 2, \ldots\}$, and let $\mu \in \mathcal{M}(\sigma)$ and $\epsilon_j \in \mathbb{T}$ be such that

$$\epsilon_j \mu(\{\lambda_j\}) = |\mu(\{\lambda_j\})|.$$

We then make use of the following approximation theorem, known as the *Kronecker "Solenoid Theorem"*.

Theorem 5.4.8 (Kronecker "Solenoid Theorem") *If $\zeta = (\zeta_1, \zeta_2, \ldots, \zeta_n) \in \mathbb{T}^n$ where $\{\zeta_1, \zeta_2, \ldots, \zeta_n\}$ is independent, then the trajectory (the semigroup)*

$$\{\zeta^k = (\zeta_1^k, \zeta_2^k, \ldots, \zeta_n^k) \colon k = 0, 1, 2, \ldots\}$$

is dense in $\mathbb{T}^n$.

We will use this theorem, but for the proof we refer the reader to Kahane and Salem (1963, pp. 21, 175).

We apply Kronecker's theorem to $\zeta = (\lambda_1, \ldots, \lambda_n)$. Then there exists $k = k_n$ such that $|\lambda_j^k - \epsilon_j| < 1/n$ for $1 \leq j \leq n$, which implies $\lim_n \lambda_j^{k_n} = \epsilon_j$ for every j.

By the dominated convergence theorem (Appendix A)

$$\sup_{k\geq 0} |\widehat{\mu}(k)| \geq \varlimsup_{k\to\infty} |\widehat{\mu}(k)| \geq \lim_{n\to\infty}\left|\int_{\mathbb{T}} \lambda^{k_n} d\mu(\lambda)\right|$$

$$= \sum_{j\geq 1} \epsilon_j \mu(\{\lambda_j\}) = \sum_{j\geq 1} |\mu(\{\lambda_j\})| = \|\mu\|_{\mathcal{M}(\sigma)}.$$

By Banach's theorem cited above, we obtain that for every $f \in C(\sigma)$ there exists $\varphi \in W_a$ such that $\varphi \mid \sigma = f$ and $\|\varphi\|_{W_a} = \|f\|_{C(\sigma)}$. ∎

(2) (Helson, 1954). *Let $\sigma \subset \mathbb{T}$ be a closed set containing arbitrarily long arithmetic progressions, i.e. for every $n \geq 1$ there exist $\zeta, \lambda \in \mathbb{T}$ such that $\zeta\lambda^j \in \sigma$ and $\lambda^{j+1} \neq 1$ for $0 \leq j \leq n$. Then σ is not a Helson set $(C(\sigma) \neq W_a \mid \sigma)$.*

Proof We begin by using Exercise 5.6.2(c): there exist polynomials $p_n \in C_a(\mathbb{D})$ such that $\deg(p_n) = n$, $\|p_n\|_\infty = 1$, $\lim_n \|p_n\|_{W_a} = \infty$. Let

$$p_n(z) = \sum_{j=0}^{n} a_j z^j$$

(where of course, $a_j = a_{j,n}$). We then consider the points (which exist by hypothesis) $\zeta\lambda^j \in \sigma$ for $1 \leq j \leq n$ ($\lambda = \lambda_n, \zeta = \zeta_n$). Set

$$\mu_n = \sum_{j=0}^{n} a_j \delta_{\zeta\lambda^j}.$$

Then, $\mu_n \in \mathcal{M}(\sigma)$ and

$$\widehat{\mu}_n(k) = \sum_{j=0}^{n} a_j(\zeta\lambda^j)^k = \sum_{j=0}^{n} a_j\zeta^k(\lambda^k)^j = \zeta^k p_n(\lambda^k).$$

Thus $\sup_{k\in\mathbb{Z}} |\widehat{\mu}_n(k)| \leq 1$; however, $\|\mu_n\|_{\mathcal{M}(\sigma)} = \sum_{j=0}^{n} |a_j|$, hence $\lim_n \|\mu_n\|_{\mathcal{M}(\sigma)} = \infty$. By Banach's theorem cited above, we obtain $C(\sigma) \neq W_a|\sigma$. ∎

(3) Examples. It is easy to find sets, countable or not, satisfying either the hypothesis of (1), or that of (2). In general, it is very easy to choose *an independent sequence $\sigma = (\lambda_j)$ of behavior on $\mathbb{T}$ prescribed in advance* by using the following recursive construction.

$\lambda_1 \in \mathbb{T} \setminus \{1\}$ is arbitrary; if $\sigma_n = \{\lambda_1, \ldots, \lambda_n\}$ is already chosen, we consider the subgroup of $\mathbb{T}$ generated by σ_n, $G_n = \{\lambda_1^{k_1} \ldots \lambda_n^{k_n} : k_j \in \mathbb{Z}\}$ (G_n is a countable set) and select $\lambda_{n+1} \in \mathbb{T} \setminus G_n$ (in particular, this last set is dense everywhere, thus there is much freedom in the choice of λ_{n+1}). The result $\sigma = (\lambda_j)_{j\geq 1}$ is an independent sequence.

Here are some concrete examples.

Example of an independent sequence. Let $\lambda_k = e^{i\theta_k}$ where $\theta_k = \pi^{k+1}$, $k = 1, 2, \ldots$. Then $(\lambda_k)_{k \geq 1}$ is an independent sequence.

Indeed, if $\lambda_1^{n_1} \lambda_2^{n_2} \ldots \lambda_s^{n_s} = 1$, where $n_j \in \mathbb{Z}$, then $\sum_{k=1}^{s} n_k \pi^{k+1} = 2\pi N$ where $N \in \mathbb{Z}$. Then all the n_k are zero, since π is not the root of any polynomial with rational coefficients (π is a transcendental number).

Example of a convergent sequence containing arbitrarily long arithmetic progressions. Let $\lambda_k = e^{i\theta_k}$, where for $2^n \leq k < 2^{n+1}$

$$\theta_k = \frac{1}{n} - \frac{k - 2^n}{2^n n(n+1)}, \qquad n = 1, 2, \ldots.$$

Clearly, $\lim_k \lambda_k = 1$, the sequence θ_k is monotonically decreasing to 0, and for $0 \leq j < 2^n$, we have $\lambda_{2^n + j} = \zeta_n t_n^j$ where $\zeta_n = e^{i/n}$, $t_n = \exp(-i/2^n n(n+1))$.

5.4.4 Causal Recursive Filters

A filter Φ is said to be *recursive* if the input and output signals $\Phi x = y$ satisfy a recurrence equation:

$$\sum_{j=0}^{m} b_j y_{k-j} = \sum_{j=0}^{m} a_j x_{k-j},$$

where $k \in \mathbb{Z}$ and $a_j, b_j \in \mathbb{C}$. By using the same techniques of Fourier transforms, it is easy to treat this special case (which appears quite often in practical applications).

Theorem 5.4.9 *Let $p(z) = \sum_{j=0}^{m} a_j z^j$, $q(z) = \sum_{j=0}^{m} b_j z^j$, and let Φ be a filter satisfying a recursion as above. Let k_p, k_q be the zero divisors of p, q (defined in Remark 2.4.4).*

(1) *Φ is a stationary filter if and only if, for every $\zeta \in \mathbb{T}$,*

$$k_p(\zeta) \geq k_q(\zeta).$$

(2) *If the above condition is satisfied, Φ is stable and with finite power, and its transfer function is $\varphi = p/q$.*

(3) *Φ is causal if and only if $k_p(\zeta) \geq k_q(\zeta)$ for $\zeta \in \overline{\mathbb{D}}$; Φ realizes a minimum for the weighted phase variation $P(\varphi)$ (of Theorem 5.4.5) if and only if $k_p(\zeta) = k_q(\zeta)$ for $\zeta \in \mathbb{D}$ and $k_p(\zeta) \geq k_q(\zeta)$ for $\zeta \in \mathbb{T}$.*

Proof By passing to the frequency domain, the recursion becomes $q \mathcal{F}^{-1} y = p \mathcal{F}^{-1} x$. The rest follows from Theorems 5.2.1, 5.2.2, 5.4.5, and Lemma 5.4.1. The stability is a consequence of the fact that a rational function p/q bounded on $\mathbb{T}$ is in $C^{\infty}(\mathbb{T}) \subset W$. $\blacksquare$

5.5 Inverse Problem: "Can One Hear the Shape of a Drum?"

The title of the famous article of Mark Kac "Can one hear the shape of a drum?" (Kac, 1966) explained the spirit of inverse problems to the general public.

For filtering, the problem is posed in the following manner. We are confronted with an unknown filter Φ (a "black box," for example, the atmosphere of a distant planet, an optical filter, etc.), to be identified with the aid of an input test signal x whose output signal y can be observed (or possibly, only a portion of y, or certain functions of y). The objective is to choose a test signal x so that such an experiment leads to the recognition of the transfer function φ.

To be more precise, here are different situations in which we would like to recognize an unknown filter:

(1) among all the stationary filters,
(2) among the physically realizable (causal) filters,
(3) when knowing the whole output $y = (y_n)_{n \in \mathbb{Z}}$,
(4) when knowing only the "physically observable" part $y_+ = (y_n)_{n \geq 0}$ of the output,
(5) when knowing a moving average of y.

We begin with the situation where only the energy densities $|\mathcal{F}^{-1}x|^2$, $|\mathcal{F}^{-1}y|^2$ of the signals x, y are known.

Theorem 5.5.1 (identification of a finite-power causal filter from its energy spectrum) *Let $x, y \in l^2(\mathbb{Z})$ and let Φ be an unknown finite-power causal filter such that $\Phi x = y$.*

(1) *For Φ to be uniquely determined by the knowledge of the energy spectra $|\mathcal{F}^{-1}x|^2$, $|\mathcal{F}^{-1}y|^2$, it is necessary that $|\mathcal{F}^{-1}x(\zeta)|^2 \neq 0$ a.e. on $\mathbb{T}$.*
(2) *Suppose $|\mathcal{F}^{-1}x(\zeta)|^2 \neq 0$ a.e. on $\mathbb{T}$. For Φ to be uniquely determined (up to a multiplicative constant) by the knowledge of $|\mathcal{F}^{-1}x|^2$, $|\mathcal{F}^{-1}y|^2$ it is necessary and sufficient that Φ be a filter with minimal weighted phase variation (as in Theorem 5.4.5).*

Proof (1) Let φ be a transfer function of Φ. Then, $|\varphi| \cdot |\mathcal{F}^{-1}x| = |\mathcal{F}^{-1}y|$. If $|\mathcal{F}^{-1}x(\zeta)|^2 = 0$ on a set E of positive measure, then $|\varphi|$ remains arbitrary on E (but nonetheless satisfying $\int_E \log|\varphi|\, dm > -\infty$), and hence φ is not well-defined by the equation $|\varphi| \cdot |\mathcal{F}^{-1}x| = |\mathcal{F}^{-1}y|$.

(2) Given Theorem 5.4.5, this part simply affirms that a function φ of H^∞ is uniquely determined by its modulus if and only if it is outer. ∎

Theorem 5.5.2 (identification of a finite-power filter by its response to a test signal) *Let $x \in l^2(\mathbb{Z})$.*

I. *Identification by the complete response. The following assertions are equivalent.*

    (1) $\Phi x = \Psi x \Rightarrow \Phi = \Psi$ *for any finite-power filters* Φ, Ψ.

    (2) $\mathcal{F}^{-1}x \neq 0$ *a.e. on* $\mathbb{T}$.

II. *Identification by the physically observable response. The following assertions are equivalent.*

    (1) $(\Phi x)_n = (\Psi x)_n, \forall n \geq 0 \Rightarrow \Phi = \Psi$ *for any finite-power filters* Φ, Ψ.

    (2) *The energy density does not vanish,* $\mathcal{F}^{-1}x \neq 0$ *a.e. on* $\mathbb{T}$, *but the "signal entropy" is infinite:*

$$\int_{\mathbb{T}} \log |\mathcal{F}^{-1}x| \, dm = -\infty.$$

Proof I. Let φ, ψ be the transfer functions of Φ and Ψ, respectively (arbitrary functions of $L^\infty(\mathbb{T})$). Then the equation is $\varphi \mathcal{F}^{-1}x = \psi \mathcal{F}^{-1}x$, and this implies $\varphi = \psi$ a.e. if and only if $\mathcal{F}^{-1}x \neq 0$ a.e. on $\mathbb{T}$.

II. Here, the equation is $P_+\varphi\mathcal{F}^{-1}x = P_+\psi\mathcal{F}^{-1}x$, hence $P_+((\varphi - \psi)\mathcal{F}^{-1}x) = 0$ where P_+ is the Riesz projection (the orthogonal projection on H^2). This is equivalent to $g := (\varphi - \psi)\mathcal{F}^{-1}x \in H^2_-$. Let $G \in H^2$ be such that $\overline{G} = g$.

$(2) \Rightarrow (1)$ Indeed, (2) implies

$$\int_{\mathbb{T}} \log |G| \, dm = \int_{\mathbb{T}} \log |(\varphi - \psi)\mathcal{F}^{-1}x| \, dm$$

$$\leq \log \|\varphi - \psi\|_\infty + \int_{\mathbb{T}} \log |\mathcal{F}^{-1}x| \, dm = -\infty,$$

hence $G = 0$ (see Corollary 2.3.3), and then $\varphi = \psi$.

$(1) \Rightarrow (2)$ If $\mathcal{F}^{-1}x = 0$ on a set of positive measure, then the identification is impossible, by part I.

Suppose, on the contrary,

$$\int_{\mathbb{T}} \log |\mathcal{F}^{-1}x| \, dm > -\infty,$$

and set $h = \min(1, |\mathcal{F}^{-1}x|)$, $\varphi = \overline{z[h]}/\mathcal{F}^{-1}x$, where $[h]$ is an outer function of absolute value h. Then, $\varphi \in L^\infty(\mathbb{T})$ and $\varphi\mathcal{F}^{-1}x \in H^2_-$. The filter Φ corresponding to φ satisfies $\Phi \neq 0$ and $(\Phi x)_n = 0$ for all $n \geq 0$. This is a contradiction. $\blacksquare$

Theorem 5.5.3 (identification of a finite-power causal filter) *Let $x \in l^2(\mathbb{Z})$.*

I. *Identification by the complete response. The following assertions are equivalent.*

    (1) $\Phi x = \Psi x \Rightarrow \Phi = \Psi$ *for any finite-power causal filters Φ, Ψ.*
    (2) $x \neq 0$.

II. *Identification by the physically observable response. The following assertions are equivalent.*

    (1) $(\Phi x)_n = (\Psi x)_n, \forall n \geq 0 \Rightarrow \Phi = \Psi$ *for any finite-power causal filters Φ,*
    Ψ.
    (2) $\mathcal{F}^{-1} x \notin (H_-^2 : H^\infty)$, *where* $(H_-^2 : H^\infty) := \{g/h : g \in H_-^2, h \in H^\infty\}$.

Proof I. As in Theorem 5.5.2, the equation is $\varphi \mathcal{F}^{-1} x = \psi \mathcal{F}^{-1} x$, but this time $\varphi, \psi \in H^\infty$. Moreover $x \neq 0 \Leftrightarrow (\mathcal{F}^{-1} x \neq 0$ on a set E, $mE > 0)$. Since $\varphi = \psi$ on E, we have $\varphi = \psi$ (see Theorem 1.4.4 or Corollary 2.3.3).

II. As in Theorem 5.5.2, the equation is $P_+((\varphi - \psi)\mathcal{F}^{-1} x) = 0$, i.e. $(\varphi - \psi)\mathcal{F}^{-1} x \in H_-^2$. Hence $\varphi - \psi \neq 0$ implies $\mathcal{F}^{-1} x = h/(\varphi - \psi)$, where $h \in H_-^2$, thus $\mathcal{F}^{-1} x \in (H_-^2 : H^\infty)$.

Conversely, if $\mathcal{F}^{-1} x = h/\varphi$ where $h \in H_-^2$, $\varphi \in H^\infty$ $(\varphi \neq 0)$, then $\varphi \mathcal{F}^{-1} x \in H_-^2$, and hence there exists a causal filter Φ such that $(\Phi x)_n = 0$ for $n \geq 0$, however $\Phi \neq 0$. ∎

5.5.1 Moving Averages of a Signal

Let $x, y \in l^2(\mathbb{Z})$, where x is interpreted as a finite energy signal and y as measurement instrument making observations at times $n \in \mathbb{Z}$. The results of the observation are written $(\tau_n x, y)_{l^2(\mathbb{Z})}$, and the scalar product is called a *moving average of x*. In the following theorem we consider a version of the identification problem with the aid of moving averages corresponding to conditions (1)&(4) in the introduction of § 5.5. For other combinations of conditions, see § 5.6.

Theorem 5.5.4 (identification of a finite-power filter by means of the physically observable moving average) *Let $x, y \in l^2(\mathbb{Z})$. The following assertions are equivalent.*

(1) $(\tau_n \Phi x, y)_{l^2(\mathbb{Z})} = (\tau_n \Psi x, y)_{l^2(\mathbb{Z})}$ $\forall n \geq 0 \Rightarrow \Phi = \Psi$ *for any finite-power filters Φ, Ψ.*

(2) $\mathcal{F}^{-1}x \neq 0$ *and* $\mathcal{F}^{-1}y \neq 0$ *a.e. on* $\mathbb{T}$ *and at least one of the signals has infinite entropy:*

$$\int_{\mathbb{T}} \log |\mathcal{F}^{-1}x|\, dm = -\infty, \ \ \text{or} \ \int_{\mathbb{T}} \log |\mathcal{F}^{-1}y|\, dm = -\infty.$$

Proof By linearity, we can always suppose $\Psi = 0$, hence the equation in question is $(\tau_n \Phi x, y)_{l^2(\mathbb{Z})} = 0$ for every $n \geq 0$. After Fourier transformation, this is equivalent to $(z^n \varphi f, g)_{L^2(\mathbb{T})} = 0$ for $n \geq 0$, where

$$f = \mathcal{F}^{-1}x \in L^2(\mathbb{T}), \quad g = \mathcal{F}^{-1}y \in L^2(\mathbb{T}),$$

hence to

$$\int_{\mathbb{T}} z^n \varphi f \overline{g}\, dm = 0 \quad (n \geq 0).$$

The latter is equivalent to saying that $\varphi f \overline{g} \in H^1_-$. Now, it is easy to show the equivalence of the stated properties.

(2) $\Rightarrow$ (1) First note that for functions $f, g \in L^2(\mathbb{T})$, we have

$$\left(\int_{\mathbb{T}} \log |f\overline{g}|\, dm = -\infty \right) \Leftrightarrow \left(\int_{\mathbb{T}} \log |f|\, dm = -\infty, \ \text{or} \ \int_{\mathbb{T}} \log |g|\, dm = -\infty \right),$$

since $\int_{\mathbb{T}} \log |f|\, dm < +\infty$, $\int_{\mathbb{T}} \log |g|\, dm < +\infty$. Let $\varphi f \overline{g} \in H^1_-$, where $f, g \in L^2(\mathbb{T})$ and $\varphi \in L^\infty(\mathbb{T})$. We use the same inequality as in Theorem 5.5.2 and the hypothesis of (2):

$$\int_{\mathbb{T}} \log |\varphi f \overline{g}|\, dm \leq \log \|\varphi\|_\infty + \int_{\mathbb{T}} \log |f\overline{g}|\, dm = -\infty,$$

hence $\varphi f \overline{g} = 0$ (by Corollary 2.3.3) and thus $\varphi = 0$.

(1) $\Rightarrow$ (2) Suppose the contrary. If $f\overline{g} = 0$ on a set $E \subset \mathbb{T}$, $mE > 0$, then we obtain $\varphi f \overline{g} = 0$ with $\varphi = \chi_E \neq 0$, which is a contradiction.

Suppose $f\overline{g} \neq 0$ a.e., but $\int_{\mathbb{T}} \log |f\overline{g}|\, dm = -\infty$. As in Theorem 5.5.2, set $h = \min(1, |f\overline{g}|)$ and $\varphi = \overline{z[h]}/f\overline{g}$. Then, $\varphi \in L^2(\mathbb{T})$, $\varphi \neq 0$ and $\varphi f \overline{g} \in H^1_-$./ This is a contradiction.   ∎

5.6 Exercises

5.6.1 Identification of Filters: Moving Averages

I. Let $x \in l^2(\mathbb{Z})$. Describe the moving averages $n \longmapsto (\tau_n \Phi x, y)_{l^2(\mathbb{Z})}$,

(a) over all time ($\forall n \in \mathbb{Z}$), or

(b) physically observable ($n \geq 0$),

that can identify an unknown finite-power filter Φ among

(1) all filters,
(2) the causal filters.

II. Let $y \in l^2(\mathbb{Z})$ define a moving average $n \longmapsto (\tau_n \Phi x, y)_{l^2(\mathbb{Z})}$,

(a) over all time ($\forall n \in \mathbb{Z}$), or
(b) physically observable ($n \geq 0$).

Describe the input signals $x \in l^2(\mathbb{Z})$ that can identify with the aid of these averages a finite-power filter among

(1) all filters,
(2) the causal filters.

5.6.2 The Non-equality $C_a(\mathbb{D}) \neq W_a(\mathbb{D})$

Recall that

$$C_a(\mathbb{D}) = H^\infty \cap C(\mathbb{T})$$

is the *disk algebra* and $W_a(\mathbb{D})$ is the *analytic Wiener algebra* of absolutely convergent Taylor series on $\mathbb{T}$:

$$W_a(\mathbb{D}) = \mathcal{F}^{-1} l^1(\mathbb{Z}_+) = \left\{ f = \sum_{n \geq 0} \widehat{f}(n) z^n : \|f\|_{W_a} = \sum_{n \geq 0} |\widehat{f}(n)| < \infty \right\}.$$

Clearly $W_a(\mathbb{D}) \subset C_a(\mathbb{D})$ and $\|f\|_{C_a} \leq \|f\|_{W_a}$ for every function $f \in C_a(\mathbb{D})$. We propose to work on several different proofs of the fact that $C_a(\mathbb{D})$ is much larger than $W_a(\mathbb{D})$. Even though at first sight the subject appears quite special, the question of differentiating C_a and W_a is basic, and at the time of the birth of modern analysis it played an important role. In particular, it is linked to the question of the convergence of the Fourier series of an arbitrary continuous function. The first counter-example, by Paul du Bois-Reymond in 1873, greatly surprised his contemporaries; we will find it as a corollary of our calculations.

(a) *Show that $C_a(\mathbb{D}) \neq W_a(\mathbb{D})$ if and only if*

$$c := \sup \left\{ \frac{\|f\|_{W_a}}{\|f\|_{C_a}} : f \in \mathcal{P}_a, \, p \neq 0 \right\} = \infty.$$

SOLUTION: If $c < \infty$, the norms $\| \cdot \|_{W_a}$ and $\| \cdot \|_{C_a}$ are equivalent, and as W_a is a complete space and dense in C_a, we have $W_a = C_a$. Conversely, if $W_a = C_a$, the embedding $j(f) = f$, $j\colon C_a \to W_a$ is closed (as the inverse of a continuous mapping) and hence bounded (closed graph theorem, Appendix E). ∎

(b) *Let*

$$S_n(e^{ix}) = \sum_{k=1}^{n} \frac{\sin(kx)}{k} = \sum_{k=1}^{n} \frac{e^{ikx} - e^{-ikx}}{2ik}.$$

Show that

$$\|S_n\|_\infty \le \frac{5\pi}{4} + 1 < 5.$$

Hint Use the Abel transformation

$$\sum_{k=1}^{n} a_k b_k = \sum_{k=1}^{n-1} (a_k - a_{k+1}) B_k + a_n B_n \quad \text{where } B_k = \sum_{1}^{k} b_j,$$

and first show that with $b_j = \sin(jx)$ we have

$$B_k = \frac{\sin(kx/2) \cdot \sin((k+1)x/2)}{\sin(x/2)}.$$

SOLUTION: Clearly, it suffices to bound $S_n(x)$ above for $0 < x < \pi$. By setting $b_j = \sin(jx)$ we obtain

$$B_k \sin(x/2) = (1/2)(\cos(x/2) - \cos((2k+1)x/2)) = \sin(kx/2) \cdot \sin((k+1)x/2).$$

Then

$$S_n = \sum_{k=1}^{n-1} \frac{1}{k(k+1)} B_k(x) + \frac{1}{n} B_n,$$

where $|B_n| \le n$ and hence

$$|S_n| - 1 \le \sum_{k\le 1/x} \frac{1}{k(k+1)} \frac{(kx/2)((k+1)x/2)}{(x/\pi)} + \sum_{k>1/x} \frac{1}{k(k+1)} \frac{1}{(x/\pi)}$$

$$= \sum_{k\le 1/x} \frac{\pi x}{4} + \frac{\pi}{x} \sum_{k>1/x} \left(\frac{1}{k} - \frac{1}{k+1} \right) \le \frac{\pi}{4} + \frac{\pi}{x} x = \frac{5\pi}{4}. \quad ∎$$

(c) First proof of $C_a(\mathbb{D}) \ne W_a(\mathbb{D})$. *Deduce from (b) and (a) that there exist polynomials* $p_n = z^n S_n \in C_a$ *such that*

$$\|p_n\|_\infty \le 5 \ and \ \|p_n\|_{W_a} \ge \log(n+1)$$

($n = 1, 2, \dots$), and hence $C_a(\mathbb{D}) \ne W_a(\mathbb{D})$.

SOLUTION: Indeed, $\|p_n\|_{W_a} = \sum_{1}^{n} k^{-1} \ge \int_1^{n+1} x^{-1} dx = \log(n+1)$, and the rest follows from (a) above. ∎

(d) Second proof of $C_a(\mathbb{D}) \neq W_a(\mathbb{D})$ (du Bois-Reymond, 1873). *Deduce from the solution of (c) that there exists a function $f \in C_a$ whose Fourier series diverges at the point 1. Of course, for such an f we have $f \in C_a \setminus W_a$.*

SOLUTION: A functional $Q_n(f) = \sum_{0 \leq k \leq n} \widehat{f}(k)$ (the partial sum of the Fourier series at the point 1) is clearly continuous in $C_a(\mathbb{T})$ (since, for example, $|\widehat{f}(k)| \leq \|f\|_\infty$), and hence by the Banach–Steinhaus theorem (Appendix E),

$$\left(\forall f \in C_a(\mathbb{T}) \ \sup_n |Q_n(f)| < \infty \right) \Leftrightarrow \left(\sup_n \|Q_n\| < \infty \right).$$

However, by (c) (and with its notation), we have

$$|Q_n(p_n)(1)| = \sum_{k=1}^{n} \frac{1}{2k} \geq \frac{1}{2} \log(n+1),$$

and $\|p_n\|_\infty \leq 5$. Hence, by definition of the norm, $\|Q_n\| \geq \frac{1}{10} \log(n+1)$, and by the result cited, $\exists f \in C_a(\mathbb{T})$ such that $\sup_n |Q_n(f)| = \infty$. ∎

(e) Third proof of $C_a(\mathbb{D}) \neq W_a(\mathbb{D})$ (Hardy and Littlewood, 1916).

(1) *Show that the series*

$$f = \sum_{k \geq 2} \frac{e^{ik \ln k}}{k^{\alpha + 1/2}} z^k,$$

where $0 < \alpha < 1$, converges uniformly on $\overline{\mathbb{D}}$ and represents a function of $C_a(\mathbb{D})$ satisfying the Lip(α) condition, $|f(z) - f(z')| \leq C|z - z'|^\alpha$; and hence, for $0 < \alpha \leq 1/2$, we have $f \in C_a(\mathbb{D}) \setminus W_a(\mathbb{D})$.

(2) *Show that the series*

$$f = \sum_{k \in \mathbb{Z} \setminus \{0\}} \frac{z^k \epsilon_k}{k},$$

where $\epsilon_k = \epsilon_{|k|} \searrow 0$, converges uniformly on $\mathbb{T}$ and represents a function of $C(\mathbb{T})$; and hence $f \in C(\mathbb{T}) \setminus W(\mathbb{T})$ if $\sum_{k \geq 1}(\epsilon_k / k) = \infty$.

SOLUTION: For the solution of (1) (which demands considerable effort), we refer the reader to the treatise of Zygmund (1959, Ch. 5, § 4), or to Hardy and Littlewood (1916). For (2), by performing an Abel transformation, we obtain

$$\sum_{\substack{-n \leq k \leq n \\ k \neq 0}} \frac{z^k \epsilon_k}{k} = \sum_{k=1}^{n-1} (\epsilon_k - \epsilon_{k+1}) S_k + \epsilon_n S_n,$$

where $\|\epsilon_n S_n\|_\infty \leq 5\epsilon_n$, whereas the series $\sum_{k \geq 1}(\epsilon_k - \epsilon_{k+1}) S_k$ converges normally,

$$\sum_{k \geq 1} (\epsilon_k - \epsilon_{k+1}) \|S_k\|_\infty \leq 5 \sum_{k \geq 1} (\epsilon_k - \epsilon_{k+1}) = 5\epsilon_1 < \infty.$$

The result follows. ∎

(f) The Riesz projection P_+ on $C(\mathbb{T})$ and $\mathcal{P}_n$. *Show that the Riesz projection $P_+f = \sum_{k\geq 0} \widehat{f}(k)z^k$ is not bounded on the space of polynomials $\mathcal{P}$ equipped with the norm $\|\cdot\|_\infty$, hence is not bounded on $C(\mathbb{T})$. Moreover,*

$$\|P_+ : \mathcal{P}_n \to \mathcal{P}_n\| \geq \frac{1}{10}\log(n+1),$$

where $\mathcal{P}_n$ is the space of trigonometric polynomials of degree $\leq n$ equipped with the norm $\|\cdot\|_\infty$.

Hint This is a slight variation on the theme of (d) above; compare the result and the proof with Exercises 2.8.3(g) and 2.8.4(g).

SOLUTION: Indeed, $S_n \in \mathcal{P}_n$ and

$$(P_+S_n)(e^{ix}) = \sum_{k=1}^{n} \frac{e^{ikx}}{2ik},$$

thus

$$\|P_+S_n\|_\infty \geq |(P_+S_n)(1)| = \sum_{k=1}^{n} \frac{1}{2k} \geq (1/2)\log(n+1),$$

and the result follows. ∎

(g) Fourth proof of $C_a(\mathbb{D}) \neq W_a(\mathbb{D})$. *Deduce from (f) and (a) that $C_a(\mathbb{D}) \neq W_a(\mathbb{D})$.*

SOLUTION: If $C_a(\mathbb{D}) = W_a(\mathbb{D})$, the norms $\|\cdot\|_{W_a}$ and $\|\cdot\|_{C_a}$ would be equivalent, as would be $\|\cdot\|_W$ and $\|\cdot\|_C$ (the translations $f \longmapsto z^n f$ are isometric in C and W). As the Riesz projection P_+ is clearly bounded, and even contracting, on W, this would have to be the same in C, which is not the case by (f). ∎

(h) Fifth proof of $C_a(\mathbb{D}) \neq W_a(\mathbb{D})$: Littlewood's crocodile. *The sketch of a crocodile (Littlewood, 1953, § 16, p. 46) represents a Jordan domain $\mathrm{CRO} \subset \mathbb{C}$ having the form of a crocodile, whose nose ends at $z = 1$, and whose teeth overlap (say, at half their length) and have infinite total length. Let $f : \mathbb{D} \to \mathrm{CRO}$ be a conformal mapping such that $f(1) = 1$. Show that $f \in C_a(\mathbb{D})\backslash W_a(\mathbb{D})$.*

SOLUTION: Indeed, f exists by a classical theorem of Riemann, and by Carathéodory's theorem (Rudin, 1998, Theorem 14.19, p. 336), $f \in C_a(\mathbb{D})$. Moreover, the image $f([0,1)) = \gamma$ is a curve joining $f(0) \in \mathrm{CRO}$ and $1 = f(1)$. Its length $|\gamma|$ is infinite because γ goes around the teeth, hence

$$\infty = |\gamma| = \int_0^1 |f'(r)|\,dr = \int_0^1 \left|\sum_{k\geq 0} \widehat{f}(k)kr^{k-1}\right| dr$$

$$\leq \int_0^1 \sum_{k\geq 0} |\widehat{f}(k)|kr^{k-1}\,dr = \sum_{k\geq 1} |\widehat{f}(k)|. \qquad \blacksquare$$

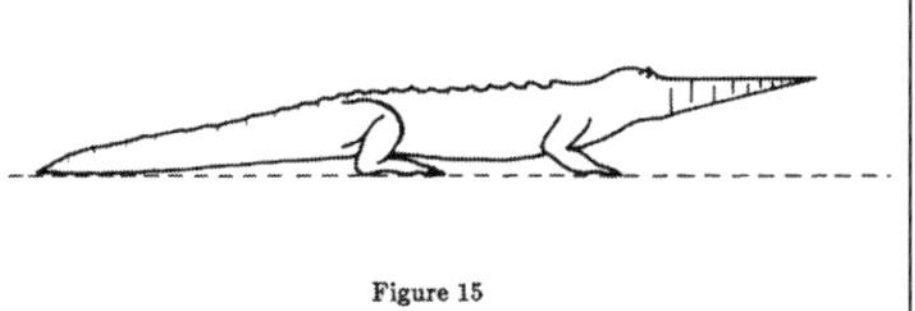

The crocodile (Fig. 15). The teeth overlap and have a total length just infinite. If the domain is represented on a unit circle we have an example of a function $f(z) = \sum c_n z^n$ in which the real part $U(\theta)$ of $f(e^{i\theta})$ is of bounded variation, and the imaginary part $V(\theta)$ is as nearly so as we like. On the other hand

Figure 15

$$\sum |c_n| = \int_0^1 \left(\sum n|c_n|\rho^{n-1}\right) d\rho \geq \int_0^1 |f'(\rho)| d\rho,$$

and the last integral is the length of the image of the radius vector $(0,1)$ of the z-circle. This image, however, is some path winding between the teeth to the nose and has infinite length. Hence $\sum |c_n|$ is divergent.

The Riemann conformal mapping f of the unit disk $\mathbb{D}$ on the interior of Littlewood's crocodile gives an example of a function in the disk algebra whose Fourier series does not converge absolutely.

5.6.3 Helson Sets in the Disk $\mathbb{D}$ (Vinogradov, 1965)

Let $\sigma \subset \mathbb{D}$. Then $W_a(\mathbb{D}) \mid \sigma \subset C(\overline{\sigma}) \mid \sigma$, and

$$W_a(\mathbb{D}) \mid \sigma = C(\overline{\sigma}) \mid \sigma$$

if and only if σ is finite.

SOLUTION: Indeed, if σ is finite, we clearly have $W_a(\mathbb{D})|\sigma = C(\sigma)$ (for example, by Lagrange interpolation). To show the converse, we begin as in Exercise 5.6.2(a) by remarking that $W_a(\mathbb{D})|\sigma = C(\overline{\sigma})|\sigma$ if and only if there exists a constant c that controls the norms of the interpolating functions:

$$\forall f \in C(\overline{\sigma}) \; \exists g \in W_a(\mathbb{D}) \text{ such that } g \mid \sigma = f|\sigma \text{ and } \|g\|_{W_a} \leq c\|f\|_{C(\overline{\sigma})}.$$

If we suppose σ infinite and a Helson set, $(W_a(\mathbb{D}) \mid \sigma = C(\overline{\sigma}) \mid \sigma)$, we can select two disjoint sequences $z_j \in \sigma$ and $w_j \in \sigma$ converging to the same point $\lambda \in \overline{\sigma}$ $(\lambda \neq z_j, w_j)$ and consider the functions $f_n \in C(\overline{\sigma})$ such that $f_n(z_j) = 1$ and $f_n(w_j) = 0$ for $1 \leq j \leq n$ and $f_n(z_j) = f_n(w_j) = 0$ for $j > n$. Then there exists a function $g_n \in W_a(\mathbb{D})$ such that $g_n \mid \sigma = f_n \mid \sigma$ and

$$\sum_{k \geq 0} |\widehat{g_n}(k)| = \|g_n\|_{W_a} \leq c.$$

By the Montel compactness theorem (see Appendix B), we can choose a subsequence (g_{n_i}) converging uniformly on the compact subsets of the disk $\mathbb{D}$ to a holomorphic function g. A passage to the limit in the sums $\sum_{k=1}^N |\widehat{g_{n_i}}(k)|$ leads to $\|g\|_{W_a} \leq c < \infty$ and, for every j, $g(z_j) = 1$ and $g(w_j) = 0$, which is impossible, since g is continuous at the point λ. ∎

5.7 Notes and Remarks

The mathematical theory of filtering began with the works of Wiener (1930, 1933, 1949) and also of Whittaker (1915, 1924) and Kotelnikov (1933), and on the engineering side with the works of Kotelnikov (1956) and Shannon (1948). The survey article by Masani (1966) gives a broad and comprehensible survey of Wiener's impact on the discipline, as well as numerous historical details on the early period of filtering theory in the USA, and in particular the pre-eminent role of the treatise *Extrapolation, Interpolation, and Smoothing of Stationary Time Series* (Wiener, 1949). (In fact this work already existed in 1941–1942 but was classified "Top Secret" until 1949 because of the war; among the students and researchers of the day it was known as the "yellow peril" due to its difficulty and to the yellow color of its cover.) A large part of the theory of random processes, and of optimal control can profitably be interpreted and used for the theory of filtering. In particular, the works of Kolmogorov cited in this book had a strong resonance with the subject. In general, the same remark holds for much of Fourier analysis. Note also that the "theoretical" and "applied" domains of filtering are very different: in the latter the algorithmic and numerical aspects prevail. See Butzer (1983) for a survey of the applied aspects. For more advanced and modern techniques related to wavelets, see Kahane and Lemarié-Rieusset (1998).

The present introduction to the subject is centered on the use of Hardy space techniques, which was important historically. Without mentioning the "continuous version" of the theory presented in this chapter, i.e. covering signals defined on $\mathbb{R}$ (see for example Papoulis (1984)), the principal omission here is *the topic of sampling* ("frames"), closely linked with the techniques developed in this book. Based on a recent survey presentation (Higgins, 1996; Higgins and Stens, 1999), we outline below a few details on the links between sampling and the analysis presented so far in this book.

In fact, the history of the discovery of sampling – a crucial idea today for the transmission of signals (both theoretical and practical) – is quite short, but full of twists and turns worthy of a detective novel. In particular, we still do not know precisely when and by whom the idea originated – even if we separate the mathematical and applied engineering aspects.

The foundation of the *mathematical theory of sampling* (of "frames") is habitually attributed to the British mathematician E. T. Whittaker due to his pioneering paper "On the functions which are represented by the expansions of the interpolation theory" (Whittaker, 1915).

Edmund Taylor (E. T.) Whittaker (1873–1956) was a British mathematician, as well as an astronomer and historian of the sciences. He was president of the London Mathematical Society (1928–1929) and the Edinburgh Mathematical Society (1914), Copley Medalist of the Royal Society (the most prestigious scientific distinction of Great Britain), and a Fellow of the Royal Society (1905) and the Pontifical Academy of Sciences (Vatican, 1935). The name Whittaker was taken from a farm in Lancashire where the family had lived since the year 1236. He was knighted in 1945.

His mathematical results concern relativity theory, representation theory, special functions, partial differential equations and numerical analysis (in particular, interpolation theory). Whittaker's most famous article is without doubt his work on sampling cited in this chapter. He is also known for his university textbook *A Course of Modern Analysis* (1902; from the second edition (1915) co-authored with G. N. Watson) – one of the few mathematics texts in Great Britain to remain in print for 100 years. Another memorable mathematical text is *The Calculus of Observations* (1924). Whittaker's name is linked to several objects in representation theory and special functions, such as the *Whittaker model*, *Whittaker functions*, and *Whittaker integrals*. His research students included Hardy, Bateman, Eddington, Littlewood, Watson, and Hodge.

Beyond mathematics, Whittaker is celebrated as a historian of science and philosophy. His reference text *A History of the Theories of Aether and Electricity* (1910, 1954) contains a chapter entitled "The relativity theory of Poincaré and Lorentz," which gives a history of relativity paying little attention to the results of Einstein (in particular, the formula $E = mc^2$ is attributed to Poincaré). Whittaker was a devout Christian who converted to Catholicism in 1930. He gave several public

However, it was Poussin who, in 1908, discovered the two principal Sampling Theorems A and B (below), not to mention the somewhat less definitive statements of Borel (1898) and Hadamard (1901), which without doubt provided a gateway to the theme of applied analysis (their works contain results equivalent to those of Theorem A). In fact, the roots of the theory are even deeper: we find elements of the sampling theorems in the *Poisson summation formula* (1820)

$$\frac{1}{a} \sum_{n\in\mathbb{Z}} \widehat{f}\left(\frac{2\pi n}{a}\right) = \sum_{k\in\mathbb{Z}} f(k)$$

($a > 0$; $|f(x)| \le c(1 + x^2)^{-1}$ and $f'' \in L^1(\mathbb{R})$) and in Cauchy's trigonometric interpolation formulas (1841). The interested reader can find the exact references in the impressive and informative surveys by Higgins (1985), Butzer, Higgins, and Stens (2000), and Butzer et al. (2011). In the statements below, we follow the terminology of signal processing.

Sampling Theorem A *If a function f does not contain any frequencies beyond $\Lambda/2$ cycles per second (hence if $\widehat{f}(\lambda) = 0$ for $|\lambda| > \Lambda/2$), it is completely determined by its values on a sequence of instances spaced by $1/\Lambda$ seconds:* $(f(k/\Lambda) = 0 \ \forall k \in \mathbb{Z}) \Rightarrow f = 0.$

Sampling Theorem B *A function f whose frequency spectrum is limited to $[-\pi\Lambda, \pi\Lambda]$, i.e. of the form*

$$f(t) = \int_{-\pi\Lambda}^{\pi\Lambda} g(x)e^{ixt}\, dx,$$

is the sum of its sampled "cardinal series"

$$f(t) = \sum_{k\in\mathbb{Z}} f\left(\frac{k}{\Lambda}\right) \operatorname{sinc}(\Lambda t - k),$$

where

$$\operatorname{sinc}(t) = \frac{\sin(\pi t)}{\pi t}$$

is the "sinus cardinal" function, the Fourier transform of $\chi = \chi_{(-\pi,\pi)}$,

$$\widehat{\chi}(t) = (2\pi)^{-1} \int_{\mathbb{R}} \chi(x) e^{-itx}\, dx$$

(= the "spectrum of χ" in the engineering literature).

A "downside" of Whittaker's article is that its level of rigor is not always sufficient: the spectral nature of the *cardinal functions f* of Theorem B (for $\Lambda = 1$) is not made precise, a sufficient condition for the convergence of the same series was not mentioned (he stated that it was sufficient to assume that the function f is entire and bounded on $\mathbb{Z}$, which is not the case), etc. The mathematical evolution of the subject was not "linear": it turns out that some of the fundamental work of the years 1920–1930 was not digested by the community at the time, leading to a slowdown and duplications in research. In fact, Whittaker's results were cultivated in two different ways in two corners of the world, far apart and isolated from each other.

Namely, in Russia, Kotelnikov, a practicing engineer in radio communications, without any knowledge of the results of Whittaker, rediscovered the formula (Kotelnikov, 1933) and proved (without much rigor in his reasoning) the simple convergence of the cardinal series under the Dirichlet condition for the inversion of the Fourier transform (hence, for integrable and piecewise monotone functions, "which is always the case in electrical engineering," wrote Kotelnikov). The works of Kotelnikov were unknown to the rest of the world until the end of the 1950s.

Vladimir A. Kotelnikov (1908–2005) was a Russian (Soviet) mathematician and communications engineer, an inventor and promoter of signal transmission by sampling, and a pioneer of Russian cryptography. His father and grandfather were mathematics professors at the University of Kazan (a region of the Volga); the latter served for a time as an assistant of Nikolai Lobachevsky, one of the pioneers of non-Euclidean geometry. Kotelnikov obtained his university degree at the Moscow Power Engineering Institute (MPEI), and then continued in a postdoctoral position.

In 1932 he prepared a presentation for a conference devoted to improving the communication system of the Red Army; the conference never took place but a collection of the presentations was published (1933). It was a pioneering publication by Kotelnikov on the techniques of sampling in signal engineering which (with the works of Claude Shannon 15 years later) definitively changed the landscape of signal processing. Kotelnikov attempted to have his article published in a widely circulated journal, *Electricity* (in Russian), but it was rejected "because the capacity of the journal is already exceeded and the subject is of limited interest." Today, sampling based on the Whittaker–Ogura–Kotelnikov–Shannon theorem is ubiquitous in the techniques of signal processing.

It was around 1939 that the first communication line using the technique of sampling was launched between Moscow and Khabarovsk (6 100 km). During the Second World War, Kotelnikov worked on problems of encoding/decoding and the encryption of telephonic and radio communications, as scientific director of a team of prisoners of the NKVD (predecessor of the KGB), the infamous Marfino "sharashka" near Moscow. This dark chapter of Stalin's regime was given literary form in the novel *The First Circle* by Aleksandr Solzhenitsyn (1968). We note in passing that the novel contains, among other things, the very first description of "wavelets" (a rather vague description, since literary, but quite recognizable), a decomposition technique based on scaling for stable signals (multi-resolution analysis); for a remarkable presentation of signal processing via wavelet techniques, see Kahane and Lemarié-Rieusset (1998). In 1944 Kotelnikov managed to escape from the confined circle of the prison-laboratories of the NKVD (which was extremely difficult) with the aid of Valeria Golubtsova, director of the MPEI (who by a decisive "coincidence" was the partner of Georgy Malenkov, First Secretary of the Central Committee of the Communist Party). In 1947 he submitted his dissertation, and in 1956 published it (in a more elaborate form) as the monograph Теория потенциальной помехоустойчивости (1956) (published in English as *The Theory of Optimum Noise Immunity*, 1959). During the years 1950–1970 Kotelnikov guided with great success a certain number of major projects such as the observation and cartography of the planets (in particular, he was the chief editor of *Atlas of the Surface of Venus*, 1985) and the communications and control part of the Russian space exploration program. He was elected member of the Academy of Sciences of the USSR (1953), and later became its Vice President (1970–1988). He was also the chief editor of a certain number of journals, and awarded many Soviet and Russian honors, as well as the Eduard Rhein Prize (Germany, 1999) and the IEEE's Alexander Graham Bell Medal (2000).

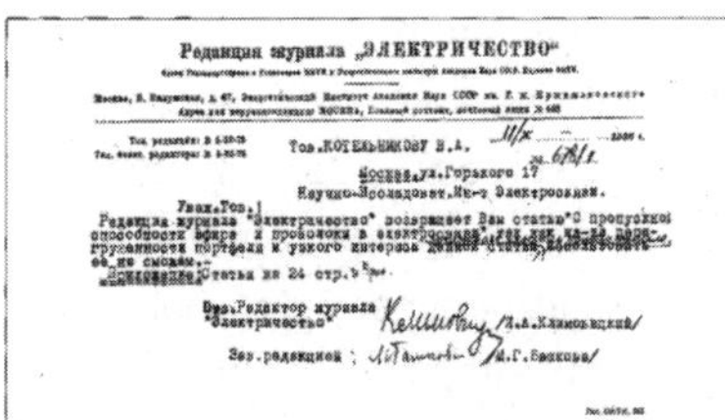

Letter from the journal *Electricity* (in Russian) rejecting Kotelnikov's pioneering paper "because the capacity of the journal is already exceeded and the subject is of limited interest." The techniques of sampling in signal engineering would soon become ubiquitous throughout the world.

The Marfino "sharashka" was a Soviet Gulag research laboratory near Moscow (25 Botanicheskaya Street, Marfino) where, during and after the Second World War, important work was done in encoding/decoding and the encryption of telephonic and radio communications on behalf of the secret services NKVD/KGB.

The principle of sampling (hence the contents of Theorem A above) was also known (at latest in 1928) by Harry Nyquist, a Norwegian engineer who worked at Bell Telephone Laboratories in the USA. However, on the mathematical side, it was Kinnosuke Ogura, a renowned Japanese professional mathematician, who was the most advanced, as early as 1920 (Ogura, 1920): he had corrected the inaccuracies of Whittaker by giving a counter-example to his conditions and by replacing them (but without a proper proof) with a sufficient condition (too strong to be necessary) for the convergence of the cardinal series (f must be an entire function of type less than π bounded on $\mathbb{R}$).

Kinnosuke Ogura.
(From Butzer et al. (2011), reprinted by permission of Springer Nature.)

Kinnosuke Ogura (1885–1962) was a Japanese mathematician with a very large spectrum of interests: from interpolation theory (including his publications on sampling) and other questions in analysis, up to relativity and differential geometry. He authored around 70 articles in pure and applied mathematics (including 42 between 1911 and 1923, including some of major importance) and published 35 books (!) on subjects varying from infinite series, applied analysis and relativity, through to social problems, education, and history.

Ogura studied in Tokyo between 1902 and 1916 (first in chemistry, then in mathematics) and obtained his doctorate in 1916 on a subject of mathematical physics. He spent two years in France, in 1920–1921, motivated in particular by the opportunity to study and collaborate with Borel and subsequently Langevin and Hadamard. His common interests with Borel ranged from the cardinal series to the applications of geometry to relativity. He was invited to lecture at the International Congress in Strasbourg (1920). In 1922, together with physicist colleagues, he organized Einstein's six-week visit to Japan.

Ogura was interested in a broad range of topics in mathematics, but also in philosophy, statistics, and social problems (he was influenced by Marxism and by Leo Tolstoy). During his career as a research professor, he taught at the Siomi Research Institute (Osaka) and at the universities of Hiroshima and Osaka, but his work was often interrupted because of health problems. After his retirement in 1943, he presided over the Society of History of Sciences and Mathematics. With all of these activities, Ogura was one of the key figures of Japanese mathematics between the two

World Wars. A distressing fact in Ogura's biography is that during the Sino-Japanese War of 1937–1945 (which caused the death of more than 20 million people in Asian countries) he took the side of the military power, in particular, engaging in 1940 with the Imperial Rule Assistance Association (a militarist civilian organization) and publishing several articles which mobilized Japanese scientists for the "Greater East Asia War" (an official slogan). When he published a revised version of these articles in 1948, Ogura replaced the words "for the Greater East Asia War" with "for the Democratic Revolution" (see "The Mathematician K. Ogura and the 'Greater East Asia War'," by Tetu Makino (2003)).

The moment of truth for the mathematical aspect of sampling came with the *Paley–Wiener theorem* (Paley and Wiener, 1934: $\mathcal{F}L^2(-a, a)$ is a space of entire functions of exponential type $\leq a$ and square integrable on $\mathbb{R}$; see also § 6.3 below) and, for the cardinal series, with the article of Hardy (1941) where the Whittaker–Ogura–Kotelnikov theorem (above all, a dream ...) took on its definitive form:

$\{\operatorname{sinc}(\cdot - k)\colon k \in \mathbb{Z}\}$ *is an orthonormal basis of the space* $\mathcal{F}L^2(-\pi, \pi)$, *and hence, for every* $f \in \mathcal{F}L^2(-a, a)$,

$$f = \sum_{k \in \mathbb{Z}} f(k)\operatorname{sinc}(\cdot - k), \qquad \int_{\mathbb{R}} |f|^2\, dt = \sum_{k \in \mathbb{Z}} |f(k)|^2,$$

and moreover, $\mathcal{F}L^2(-\pi, \pi) \mid \mathbb{Z} = l^2(\mathbb{Z})$ *("free" interpolation on* $\mathbb{Z}$*).*

Furthermore, in the same article Hardy assigned a name to this space by calling it the *Paley–Wiener space* (a name which it still holds):

$$PW_{\pi\Lambda} = \{\widehat{g}\colon g \in L^2(-\pi\Lambda, \pi\Lambda)\}$$

(a closed subspace of $L^2(\mathbb{R})$). He also showed that $\operatorname{sinc}(t - x)$ is a reproducing kernel for PW_π,

$$f(t) = \int_{\mathbb{R}} f(x)\operatorname{sinc}(t - x)\, dx, \quad \forall f \in PW_\pi.$$

We can also mention that a cardinal series of Theorem B above is simply a special case of the formula of *Lagrangian interpolation*,

$$f(z) = \sum_k \frac{f(z_k)p(z)}{p'(z_k)(z - z_k)},$$

where p is an entire holomorphic function and (z_k) its simple zeros, and which can be justified by the residue theorem of complex analysis.

In *engineering*, the principal practical consequence of Sampling Theorems A and B is that for the treatment (transmission, transformation, etc.) of continuous signals $f(t)$, $t \in \mathbb{R}$ "not containing rapid oscillations," we can restrict ourselves to a discrete sequence $f(k/\Lambda)$ without any loss of information contained in the signal (this is the essence of the digitalization of electrical, optical, and any other type of signals). This discovery was a veritable revolution in signal processing and telecommunications engineering. As a consequence, the evolution of the theory and practice of sampling advanced in leaps and bounds in Russia during the years 1935–1945, especially because during the Second World War signal processing was linked with the encryption and coding of communications (see the biography of Kotelnikov on page 179). The true engineering pioneer of sampling, at the center of all these changes, was Vladimir Kotelnikov; in particular, he implemented the ideas of his seminal work, "On the transmission capacity of 'aether' and wire in electrocommunications" (Kotelnikov, 1933), by installing and putting into operation the secure high-resolution telephone line between Moscow and Khabarovsk.

On the other side of the Atlantic, in the USA, the principle of sampling was discovered by Claude Shannon, an engineer at Bell Telephone Laboratories.

Claude E. Shannon (1916–2001) was an American electrical engineer, cryptographer, and mathematician, and founder of information theory. He was a distant relative of Thomas Edison, both being descended from John Ogden, one of the founding fathers of the American colonies and ancestor to many famous individuals. One of the greatest inventions of the twentieth century is attributed to Shannon: all information and all communications can be coded in the simple language of 0s and 1s. This idea was presented in his Master's thesis at MIT, "A symbolic analysis of relay and switching circuits" (1937), described by Howard Gardner (Harvard) as "possibly the most important, and also the most famous, master's thesis of the century." (To be fair, note that this discovery had also been made in 1935 by a Russian logician and engineer, Victor Shestakov: it figures in his thesis submitted in 1938 at the University of Moscow and published in 1941.)

During the Second World War, Shannon worked at Bell Laboratories, investigating cryptography and the automatic guidance of anti-aircraft fire. Once declassified, his works gave rise to his classical post-war publications "A mathematical theory of communication" (1948, where he introduced the word "bit" for the minimal unit of information, as well as his rediscovery of the Whittaker–Ogura–Kotelnikov theorem), and "Communication theory of secrecy systems" (1949). These publications of Shannon revolutionized the theory and practice of telecommunications (by introducing, in particular, error-correcting codes), cryptography, applied probability and statistics, and then the theory of replication of DNA, etc. The media buzz around this "numerical revolution" was so important that Shannon felt obliged to remark that "Information theory has perhaps ballooned to an importance beyond its actual accomplishments."

Later, Shannon worked on a program to play chess (and published "Programming a computer for playing chess" (Shannon, 1950); the first match was played by the Los Alamos MANIAC machine in 1956), and conducted the very first experiments in artificial intelligence (a "mouse" running a labyrinth with elements of self-learning). He retired early, in 1966, but continued as a consultant for Bell Laboratories. At the end of his life he was stricken with Alzheimer's disease.

Numerous witnesses attest to his acute mind, his sense of humor, and his originality. For example, he enjoyed tearing down the corridors of MIT on a unicycle, scaring the living daylights out of his colleagues, or surprising them with his "Ultimate Machine" – a device whose sole function is to switch itself off.

The list of honors received by Claude Shannon has about 30 entries, including the US National Medal of Science (1966) and the Kyoto Prize (1985). Several concepts in computer science bear Shannon's name: *Shannon's theory of information*, the capacity of a transmission channel, the *Shannon entropy*, etc.

His findings were published around 1948–1949 (probably obtained around 1940, but classified "Top Secret" until 1949) and contained results similar to those of Kotelnikov, again with a presentation from an "engineering perspective," without discussion of the mathematical conditions of validity, or precise details of the classes of signals or the types of convergence of the developments. As a result, in the USA and in Europe, signal processing was known as "Shannon's theory," and it was only very slowly that the more complete history of the subject began to be recognized, with the appearance of the names of Kotelnikov around 1959 (and his article, Kotelnikov (1933), in 2001!) and of Ogura, as late as 1992. For more historical and mathematical details we refer the reader to the surveys of Butzer et al. (2000, 2011), as well as to the book by Kahane and Lemarié-Rieusset (1998).

The results attributed to Wiener, Theorems 5.2.1– 5.2.2 and Lemma 5.4.1, can be found, at times without an explicit statement, in Wiener (1933, 1949). In particular, a significant event of the epoch was the characterization of the energy spectrum of a causal filter by the integrability of the logarithm (Corollary 5.4.2). For Theorem 5.4.7 see Rudin (1956) and Carleson (1956). A large number of generalizations are known: see Garnett (1981), Gamelin (1969), and Havin and Jöricke (1994). For the Helson sets and related problems, see Kahane and Salem (1963), Rudin (1962), and for the presentation of § 5.5, Nikolski (2002). The contents of Exercise 5.6.2 are classical: see Zygmund (1959).

6

The Riemann Hypothesis, Dilations, and H^2 in the Hilbert Multi-disk

Topics. Euler ζ function, integral representations of ζ, the Riemann hypothesis, the H^p spaces in the half-plane, the Paley–Wiener theorem, invariant subspaces generated by the ζ function, distance function and zeros of ζ, Beurling's problem on the completeness of the dilations, the space H^2 in the Hilbert multi-disk, dilations of the polynomials.

In 1737 Leonhard Euler, at the time professor at Saint Petersburg (and member of the Russian Academy of Sciences), wrote an article entitled *Variae observationes circa series infinitas* (published in 1744) where he defined a function that is now called the *Euler zeta function* (or, more frequently, the *Riemann zeta function*: see § 6.8 for comments),

$$\zeta(s) = \sum_{n \geq 1} \frac{1}{n^s}.$$

In his 1737 article, Euler was interested in $\zeta(s)$ uniquely for the values $s \in \mathbb{N}$: $s = 1$ to give a new proof of the infinitude of prime numbers, and $s = 2, 4, \ldots$ to resolve the "Basel problem," consisting precisely of the calculation of the sums $\zeta(2)$, $\zeta(4)$, etc. In 1749, he extended the definition to the real values. Later, in a result published in 1761, Euler presented a crucial tool for the study of ζ – the *Riemann functional equation* of Theorem 6.1.5, which in all justice should be called the Euler–Riemann equation (see the article by Gelfond (1958) on this subject).

Leonhard Euler (1707–1783) was a Swiss, Russian, and German mathematician, a mathematical genius, instigator of several modern disciplines, founder of topology and graph theory, author of approximately 900 original articles (many – but not all – were published in Switzerland in 73 volumes *in quarto*), including very important monographs in mechanics, analysis, naval science, celestial mechanics, integral calculus, algebra, etc., not to mention works for the general public such as *Letters to a German Princess* (three volumes!) explaining the principles of physics, philosophy, and mathematics in a simplified form. Around 30 mathematical objects bear Euler's name: the base *e* (Euler number) of the natural logarithms, the *Euler angles*, the *Euler* Γ *function*, the *Euler constant*, the *Euler–Lagrange equation*, etc. Introducing the function $\zeta(s)$ (see § 6.1), Euler established the product formula of § 6.1.2 and (*de facto*) the functional equation of Theorem 6.1.5. His results in number theory, as well as the famous correspondence between Euler and Christian Goldbach (between Berlin and Saint Petersburg/Moscow), defined the research direction for additive number theory for centuries.

Leonhard Euler was born in 1707 in Basel (Bâle). Under the influence of a family friend, Johann Bernoulli, he studied science at the University of Basel, and submitted a Master's thesis in 1723 on the philosophies of Descartes and Newton. In 1726, under Bernoulli's supervision, he defended his doctoral thesis on the propagation of sound. After losing the competition (!) for a position at the University of Basel, he accepted an offer as professor of Physiology (!) at the Saint Petersburg Academy of Science, which had just opened the previous year. He remained in Russia for 14 years (1727–1741), and returned later (1766–1783) at the personal invitation of the Russian Empress Catherine the Great, after spending the years 1741–1766 in Berlin as head of the Prussian Academy. Euler married Katharina Gsell and had 13 children. He wrote (in fact dictated, because he became blind) 800 pages *in quarto* per year, in addition to participating in 10 to 15 annual conferences. In Russia, his annual salary grew from 200 roubles at the start of his career to 3 000 roubles by the end (a horse – the equivalent of a car today – cost around 5 roubles); this allowed him to support a household of up to 20 people.

Euler had an equable character: when in 1767 he lost the use of his second eye, he confided "Henceforth nothing will be able to distract me from mathematics." The Marquis de Condorcet quoted Euler's reply to a question of the German Queen Mother during a reception at the Berlin court after his return from Russia: "Why will you not speak to me?" "Madame," replied Euler, "because I have come from a country where one can be hanged for what one says." Another anecdote recounts his confrontation with Diderot, who had been invited by Catherine the Great to the court of Saint Petersburg – a confrontation that Euler is said to have begun by announcing "$e^{i\pi} = -1$, hence God exists: reply!"[1]

In his book *A Concise History of Mathematics*, Dirk Struik describes the contemporary response to the publication of Euler's mathematical theory of music (*Tentamen novae theoriae musicae*, 1739): "it was too musical for mathematicians and too mathematical for musicians."

Euler's extraordinary importance was largely recognized, from the time of Laplace, who stated *Lisez Euler, lisez Euler, c'est notre maître à tous* ("Read Euler, read Euler, he is the master of us all"), up to a recent text on the Internet, "Top 10 Greatest Mathematicians" (M. Sexton): "If Gauss is the Prince [of Mathematics], Euler is the King."

A crater on the moon and an asteroid are named after Euler, as well as several scientific prizes and research institutes; many stamps and coins bear his effigy. Euler is buried in the Alexander Nevsky Cemetery in Saint Petersburg.

Euler had already linked the function ζ to the principal question of arithmetic: "How many prime numbers are found in nature and how are they distributed?" Over time, this link has only been strengthened. In 1859, Bernhard

---

[1] The anecdote was recounted by Dieudonné Thiébault in his *Souvenirs de vingt ans de séjour à Berlin* (vol. 3, p. 142 of the 1804 edition). He does not cite Euler by name and gives for the formula $(a + b^n)/z = x$. He points out he cannot ensure the veracity of this story and he simply transcribed what he heard. If this story has an element of truth, we could suppose that Euler invoked a real formula, illustrating the "divine beauty" of mathematics: this is why it is often replaced (as here) with $e^{i\pi} = -1$.

Riemann directly connected the question of the distribution of prime numbers with the properties of ζ. In particular, he showed that $\zeta(s) = 1/(s-1) + F(s)$, where F is an entire holomorphic function in $\mathbb{C}$ (see Theorem 6.1.5 below), and presented the hypothesis:

All zeros of ζ in the half-plane $\mathrm{Re}(s) > 0$ are located on the line $\mathrm{Re}(s) = 1/2$.

Since then, this proposition has been known as the *Riemann hypothesis* (RH); it remains unresolved today.

In this chapter we present an approach to the RH using the invariant subspaces of the space H^2 discovered by the Swedish mathematician Bertil Nyman, in 1949. This approach is based on an integral representation of the function ζ and the Fourier transform.

In what follows, we let $\rho(x)$ denote the *fractional part* of $x \in \mathbb{R}$,

$$\rho(x) = x - [x] \quad ([x] \text{ the integer part of } x).$$

The function ρ is 1-periodic, $\rho(x) = \rho(x+1)$; for $x > 0$ we set

$$\varphi(x) = \rho(1/x).$$

6.1 The Euler ζ Function and the Riemann Hypothesis (RH)

We outline here a few elementary properties of the Euler ζ function.

Definition 6.1.1 *Let $s \in \mathbb{C}$, $\mathrm{Re}(s) > 1$. Set*

$$\zeta(s) = \sum_{n \geq 1} \frac{1}{n^s};$$

this is called the Euler ζ function.

Clearly the series converges absolutely (and uniformly on every half-plane $\mathrm{Re}(s) \geq 1 + \epsilon$, $\epsilon > 0$), and represents a holomorphic function in $\mathrm{Re}(s) > 1$.

6.1.1 Prime Number Decomposition (Euclid, c. 300 BCE; Gauss, 1801)

Let $(p_s)_{s \geq 1}$ be the sequence of consecutive prime numbers

$$p_1 = 2, \quad p_2 = 3, \quad p_3 = 5, \quad \ldots$$

Then, every natural number $n \in \mathbb{N}$ possesses a unique representation of the form

$$n = p_1^{\alpha_1} p_2^{\alpha_2} \cdots p_k^{\alpha_k},$$

where $\alpha_j \in \mathbb{Z}_+ = \mathbb{N} \cup \{0\}$ (and $k = k(n)$).

6.1.2 The Euler Infinite Product

For every $s \in \mathbb{C}$, $\mathrm{Re}(s) > 1$,

$$\zeta(s) = \prod_{k \geq 1} \left(1 - \frac{1}{p_k^s}\right)^{-1},$$

and the product converges absolutely.

Proof Indeed, the product converges absolutely since $p_k \geq k$ and hence $\sum_{k \geq 1} 1/|p_k^s| < \infty$. Let $D(k)$ be the set of integers having all their prime divisors among $p_1, \ldots, p_k$, i.e.

$$D(k) = \{p_1^{\alpha_1} p_2^{\alpha_2} \cdots p_k^{\alpha_k} : \alpha_j \in \mathbb{Z}_+, 1 \leq j \leq k\}.$$

Given the absolute convergence of Definition 6.1.1, we have

$$\zeta(s) = \lim_k \sum_{n \in D(k)} \frac{1}{n^s},$$

$$\sum_{n \in D(k)} \frac{1}{n^s} = \sum_{\alpha_2 \geq 0 \ldots \alpha_k \geq 0} \sum_{\alpha_1 \geq 0} \frac{1}{p_1^{s\alpha_1} p_2^{s\alpha_2} \cdots p_k^{s\alpha_k}}$$

$$= \sum_{\alpha_2 \geq 0 \ldots \alpha_k \geq 0} \left(1 - \frac{1}{p_1^s}\right)^{-1} \frac{1}{p_2^{s\alpha_2} \cdots p_k^{s\alpha_k}} = \prod_{j=1}^{k} \left(1 - \frac{1}{p_j^s}\right)^{-1},$$

and the result follows. ∎

Corollary 6.1.2 $\zeta(s) \neq 0$ *for every $s \in \mathbb{C}$, $\mathrm{Re}(s) > 1$.*

This is clear by § 6.1.2 and the definition of a convergent product. ∎

Lemma 6.1.3 *For every $t \geq 1$, $\mathrm{Re}(s) > 0$, we have*

$$\frac{1}{t(s-1)} - \frac{\zeta(s)}{t^s s} = \int_0^1 \varphi(tx) x^{s-1} \, dx.$$

Proof With $u = 1/tx$ we have

$$\int_0^1 \varphi(tx) x^s \frac{dx}{x} = t^{-s} \int_{1/t}^{\infty} \rho(u) u^{-s} \frac{du}{u} = t^{-s} \left(\int_{1/t}^1 + \int_1^{\infty}\right)$$

$(\rho(u) = u$ for $0 \le u < 1)$

$$= t^{-s}\left(\int_{1/t}^{1} \frac{du}{u^s} + \sum_{n \ge 1} \int_{n}^{n+1} \rho(u)\frac{du}{u^{s+1}} \right)$$

$$= t^{-s}\left(\frac{1}{1-s}\left(1 - \frac{1}{t^{1-s}}\right) + \sum_{n \ge 1} \int_{n}^{n+1} (u-n)\frac{du}{u^{s+1}} \right)$$

(integration by parts)

$$= t^{-s}\left(\frac{1}{1-s}\left(1 - \frac{1}{t^{1-s}}\right) + \sum_{n \ge 1} \frac{1}{s}\left(\int_{n}^{n+1} \frac{du}{u^s} - \frac{1}{(n+1)^s}\right) \right)$$

$$= t^{-s}\left(\frac{1}{1-s}\left(1 - \frac{1}{t^{1-s}}\right) + \frac{1}{s}\int_{1}^{\infty} \frac{du}{u^s} - \frac{1}{s}(\zeta(s) - 1) \right)$$

$$= t^{-s}\left(\frac{1}{1-s}\left(1 - \frac{1}{t^{1-s}}\right) + \frac{1}{s(s-1)} - \frac{1}{s}(\zeta(s) - 1) \right)$$

$$= t^{-s}\left(-\frac{1}{(1-s)t^{1-s}} - \frac{\zeta(s)}{s} \right). \qquad \blacksquare$$

Corollary 6.1.4 *The function ζ can be extended to a meromorphic function in the half-plane $\{\mathrm{Re}(s) > -1\}$ having a single pole at the point $s = 1$ (of residue 1) and the integral representations*

$$\frac{\zeta(s)}{s} = \frac{1}{s-1} - \int_{1}^{\infty} \rho(y)y^{-s-1}\, dy, \text{ for } \mathrm{Re}(s) > 0,$$

$$\frac{\zeta(s)}{s} = -\int_{0}^{\infty} \left(\rho(y) - \frac{1}{2}\right)y^{-s-1}\, dy, \text{ for } -1 < \mathrm{Re}(s) < 0,$$

where the last integral is considered as improper: $\int_{0}^{\infty} = \lim_{a \to \infty} \int_{0}^{a}$.

Indeed, since the function φ is bounded, the integral

$$s \longmapsto \int_{0}^{1} \varphi(x)x^{s-1}\, dx = \int_{1}^{\infty} \rho(y)y^{-s-1}\, dy$$

is holomorphic for $s \in \mathbb{C}^+$ (Appendix A). Moreover, for the same s, we have

$$\int_{1}^{\infty} \rho(y)y^{-s-1}\, dy = \int_{1}^{\infty} \left(\rho(y) - \frac{1}{2}\right)y^{-s-1}\, dy + \frac{1}{2}\int_{1}^{\infty} y^{-s-1}\, dy$$

$$= \int_{1}^{\infty} \left(\rho(y) - \frac{1}{2}\right)y^{-s-1}\, dy + \frac{1}{2s}.$$

The last improper integral

$$s \longmapsto \int_1^\infty \left(\rho(y) - \frac{1}{2}\right) y^{-s-1} \, dy.$$

converges uniformly in s on the compact sets in $\mathrm{Re}(s) > -1$, and hence can be extended analytically in the named half-plane: indeed, the function

$$R(x) = \int_1^x \left(\rho(y) - \frac{1}{2}\right) dy$$

is bounded for $x > 1$, since $\int_k^{k+1} (\rho(y) - 1/2) \, dy = 0$ for every k and since for any $a > 0$, an integration by parts leads to

$$\int_1^a \frac{\rho(y) - (1/2)}{y^{s+1}} \, dy = R(a)a^{-s-1} + (s+1) \int_1^a \frac{R(y)}{y^{s+2}} \, dy.$$

Consequently, after such an extension, for $-1 < \mathrm{Re}(s) < 0$ we obtain

$$\frac{\zeta(s)}{s} = \frac{1}{s-1} - \int_1^\infty \left(\rho(y) - \frac{1}{2}\right) y^{-s-1} \, dy - \frac{1}{2s}$$

$$= \frac{1}{s-1} - \int_0^\infty \left(\rho(y) - \frac{1}{2}\right) y^{-s-1} \, dy + \int_0^1 \left(\rho(y) - \frac{1}{2}\right) y^{-s-1} \, dy - \frac{1}{2s}$$

$$= -\int_0^\infty \left(\rho(y) - \frac{1}{2}\right) y^{-s-1} \, dy,$$

where we used the fact that

$$\int_0^1 \left(\rho(y) - \frac{1}{2}\right) y^{-s-1} \, dy = \int_0^1 y^{-s} \, dy - \frac{1}{2} \int_0^1 y^{-s-1} \, dy = \frac{1}{1-s} + \frac{1}{2s}. \qquad \blacksquare$$

Theorem 6.1.5 (Euler, 1761; Riemann, 1859) *The function $s \longmapsto \zeta(s)$ can be extended analytically in the entire plane $\mathbb{C}$, with the exception of a simple pole at $s = 1$, where it satisfies the Euler–Riemann functional equation*

$$\zeta(s) = \frac{1}{\pi}(2\pi)^s \sin \frac{\pi s}{2} \Gamma(1-s)\zeta(1-s),$$

or

$$\xi(s) = \xi(1-s),$$

where Γ is the Euler Γ function and

$$\xi(s) = \frac{1}{2}s(s-1)\pi^{-s/2}\Gamma(s/2)\zeta(s).$$

Proof (Hardy, 1922; Titchmarsh, 1951) By Exercise 2.8.3(b), the Fourier series of

$$\rho\left(\frac{x}{2\pi}\right) - \frac{1}{2} = \frac{x - \pi}{2\pi}$$

is

$$-\sum_{k\geq 1} \frac{\sin(kx)}{k\pi},$$

and thus

$$\rho(y) - \frac{1}{2} = -\sum_{k\geq 1} \frac{\sin(2\pi ky)}{\pi k}, \quad y \in \mathbb{R} \setminus \mathbb{Z}.$$

The series converges everywhere, and its partial sums are uniformly bounded (see Exercise 5.6.2(b)). Let $a > 0$ and $s \in \mathbb{C}$, $-1 < \mathrm{Re}(s) < 0$. By Corollary 6.1.4,

$$\frac{\zeta(s)}{s} = \lim_{a\to\infty} \int_0^a \left(\sum_{k\geq 1} \frac{\sin(2\pi ky)}{\pi k}\right) y^{-s-1}\, dy$$

$$= \lim_{a\to\infty} \sum_{k\geq 1} \int_0^a \frac{\sin(2\pi ky)}{\pi k} y^{-s-1}\, dy,$$

and hence the last limit would be equal to

$$\sum_{k\geq 1} \int_0^\infty \frac{\sin(2\pi ky)}{\pi k} y^{-s-1}\, dy$$

(the integrals taken as improper) if we could show that

$$\lim_{a\to\infty} \sum_{k\geq 1} \int_a^\infty \frac{\sin(2\pi ky)}{\pi k} y^{-s-1}\, dy = 0.$$

However,

$$\int_a^\infty \frac{\sin(2\pi ky)}{y^{s+1}}\, dy = -\frac{\cos(2\pi ka)}{2\pi ka^{1+s}} - \frac{s+1}{2\pi k} \int_a^\infty \frac{\cos(2\pi ky)}{y^{s+2}}\, dy,$$

thus

$$\left| \int_a^\infty \frac{\sin(2\pi ky)}{y^{s+1}}\, dy \right| \leq \frac{1}{2\pi k a^{\mathrm{Re}(s)+1}} + \frac{|s+1|}{2\pi k(\mathrm{Re}(s) + 1)a^{\mathrm{Re}(s)+1}},$$

hence the required convergence. We thus obtain

$$\frac{\zeta(s)}{s} = \frac{1}{\pi} \sum_{k\geq 1} \frac{1}{k} \int_0^\infty \frac{\sin(2\pi ky)}{y^{s+1}}\, dy = \frac{1}{\pi} \sum_{k\geq 1} \frac{(2\pi k)^s}{k} \int_0^\infty \frac{\sin(x)}{x^{s+1}}\, dx.$$

The integrals under the summation sign are known and can be calculated with
the aid of the Euler Γ function whose definition for $\mathrm{Re}(z) > 0$ is

$$\Gamma(z) = \int_0^\infty e^{-t} t^{z-1}\, dt.$$

It is well known (see for example Titchmarsh (1939, § 3.1.2.7, § 4.4.1)) that
Γ can be extended to a meromorphic function in $\mathbb{C}$ satisfying the following
equations (among others):

$$\Gamma(z+1) = z\Gamma(z), \quad \Gamma(z)\Gamma(1-z) = \frac{\pi}{\sin(\pi z)},$$

$$\Gamma(2z) = \frac{1}{\sqrt{\pi}} 2^{2z-1}\Gamma(z)\Gamma\left(z + \frac{1}{2}\right),$$

$$\int_0^\infty t^{z-1} \sin(t)\, dt = \Gamma(z)\sin(z\pi/2) \quad (0 < \mathrm{Re}(z) < 1).$$

By using these equations, we come to the first formula stated,

$$\begin{aligned}
\zeta(s) &= \frac{s(2\pi)^s}{\pi}\Gamma(-s)\sin(-\pi s/2) \sum_{k\geq 1}\frac{1}{k^{1-s}}\\
&= \frac{s(2\pi)^s}{\pi}\frac{\Gamma(1-s)}{-s}\sin(-\pi s/2)\zeta(1-s)\\
&= \frac{(2\pi)^s}{\pi}\Gamma(1-s)\sin(\pi s/2)\zeta(1-s),
\end{aligned}$$

at least for the values of s in $-1 < \mathrm{Re}(s) < 0$. This formula shows that ζ can be
extended analytically in the half-plane $\mathrm{Re}(s) < 0$ while continuing to satisfy
the same equation.

By once again using the above identities for Γ it is easy to verify that $\xi(s) =
\xi(1 - s)$ is an equation equivalent to the preceding one.   ∎

Corollary 6.1.6 *ζ has zeros $\zeta(-2n) = 0$ ($n = 1, 2, \dots$); all the other zeros of ζ
(if they exist) are in $\{s \in \mathbb{C}: 0 \leq \mathrm{Re}(s) \leq 1\}$ and are symmetric with respect to
the line $\mathrm{Re}(s) = 1/2$.*

Indeed, this is clear by the equation of Theorem 6.1.5, Corollary 6.1.2 and
the fact that for any z, $\Gamma(z) \neq 0$.   ∎

The $-2n$, $n = 1, 2, \dots$, are called the *trivial zeros* of ζ.

6.1.3 The Riemann Hypothesis (RH), 1859

All the non-trivial zeros of ζ are situated on the line $\mathrm{Re}(s) = 1/2$.

Bernhard Riemann (1826–1866) was a German mathematician, the most influential creative genius in the mathematical renaissance of the second half of the nineteenth century. The ideas of Riemann definitively transformed complex analysis, geometry, and number theory, and also provided a strong impetus for real harmonic analysis. Three of Riemann's four most influential works were "qualifying texts": his doctoral thesis (Göttingen, 1851, under the supervision of Gauss) containing the theory of *Riemann surfaces* and conformal mappings, his habilitation thesis (1853), devoted to Fourier series (with the *Riemann integral* as a tool), and his famous *Habilitationsvortrag* (1854, an inaugural habilitation conference) entitled *Über die Hypothesen, welche der Geometrie zu Grunde liegen* (chosen by Gauss from the three themes proposed by Riemann). These three masterpieces were published posthumously. The fourth work was "Über die Anzahl der Primzahlen unter einer gegebenen Grösse" (1859) where Riemann introduced his ideas on the role of the ζ function in the complex plane in the distribution of prime numbers (this was his only opus devoted to number theory). These contributions of Riemann became – and remain – absolutely fundamental for the mathematics and physics of the nineteenth to twenty-first centuries.

An astronomical number of publications are devoted to the development of Riemann's ideas and results. For a presentation intended for the general public, see for example *Bernhard Riemann 1826–1866: Turning Points in the Conception of Mathematics* by Detlef Laugwitz (Birkhäuser, 2008), *Riemann, Le géomètre de la nature* by Rossana Tazzioli (vol. 12 of Pour la Science, 2002), or "Riemann" by Hans Freudenthal (in *Dictionary of Scientific Biography*, 2008). As remarked in the last of these, "Riemann's evolution was slow and his life short." He only managed to write around 15 mathematical manuscripts, but these rare works opened a new era in mathematics. Riemann's name is associated with dozens of

Riemann's seminal contributions to geometry likely inspired Lewis Carroll, otherwise known as Oxford mathematics lecturer Charles Dodgson, when he wrote *Alice's Adventures in Wonderland* and *Through the Looking Glass*. Carroll's self-caricature is (presumably) entitled "Me when I am lecturing" – perhaps on Riemann's curved spaces and imaginary numbers.

important concepts: *Riemannian geometry, Cauchy–Riemann equations, Riemann surfaces*, the *Riemann integral, Riemann conformal mapping theorem, Riemann–Hilbert problem and method, Riemann hypothesis, Riemann–Lebesgue lemma, Riemann sphere*, etc. In particular, Riemannian geometry was fundamental to the creation of general relativity – and also in the inspiration of a certain mathematician Charles Dodgson (better known under his literary pseudonym Lewis Carroll) for his ingenious *Alice's Adventures in Wonderland* (1865) and *Through the Looking-Glass* (1871).

Riemann's career was slow and brief: he became a professor at Göttingen only in 1859 (after the death of Dirichlet), and always suffered from a lack of students (his renowned course on Abelian functions was frequented by only three students, including Dedekind). A deterioration in his health (latent tuberculosis?) frequently forced him to seek refuge in Italy (1862–1866). Riemann was married in 1862 to Elise Koch, with whom he had a daughter.

6.2 An Approximation Implying the Riemann Hypothesis

In what follows, the following notation is used:

$$\mathcal{V} = \mathrm{Lin}(\varphi(tx)\colon t > 1), \quad \mathcal{V}_0 = \{f \in \mathcal{V}\colon f(1) = 0\},$$

where, recall, $\varphi(x) = \rho(1/x) = 1/x - [1/x]$ $(x > 0)$. In the following theorem, and for the remainder of this chapter, we consider a space $L^2(E)$, where $E \subset \mathbb{R}$, as the subspace of $L^2(\mathbb{R})$ consisting of the functions of $L^2(\mathbb{R})$ that are zero on the complement $\mathbb{R} \setminus E$.

Theorem 6.2.1 *Let*

$$d = \mathrm{dist}_{L^2(0,1)}(1, \mathcal{V}_0).$$

(1) *The disk*

$$D_d = \{s \in \mathbb{C}: d^2|s|^2 < 2\,\mathrm{Re}(s) - 1\}$$

($\Leftrightarrow (x - 1/d^2)^2 + y^2 < r^2 := 1/d^2(1/d^2 - 1)$ where $s = x + iy$) does not contain any zeros of $s \longmapsto \zeta(s)$. If $d > 0$, then $D_d = D(1/d^2, r)$.

(2) **(Nyman, 1950)** $d = 0 \Leftrightarrow 1 \in \mathrm{clos}_{L^2(0,1)}(\mathcal{V}_0) \Rightarrow$ *(RH) (i.e. all the zeros of ζ in $\mathrm{Re}(s) > 0$ are on the line $\mathrm{Re}(s) = 1/2$).*

Proof (1) Suppose $s \in \mathbb{C}$, $\mathrm{Re}(s) > 0$,

$$\zeta(s) = 0$$

and $f \in \mathcal{V}_0$, $f(x) = \sum_{k=1}^{n} a_k \varphi(t_k x)$, where $t_k \geq 1$ and $0 = f(1) = \sum_{k=1}^{n} a_k/t_k$. Then, by Lemma 6.1.3,

$$\int_0^1 (1 - f(x))x^{s-1}\,dx = \frac{1}{s} - \sum_{k=1}^{n} a_k \int_0^1 \varphi(t_k x)x^{s-1}\,dx$$

$$= \frac{1}{s} - \zeta(s)\sum_{k=1}^{n} a_k/t_k^s - \frac{1}{s-1}\sum_{k=1}^{n} a_k/t_k = \frac{1}{s},$$

and hence

$$\left|\frac{1}{s}\right| = \left|\int_0^1 (1 - f(x))x^{s-1}dx\right| \leq \|1 - f\|_{L^2(0,1)}\left(\int_0^1 |x^{2s-2}|dx\right)^{1/2}$$

$$= \|1 - f\|_{L^2(0,1)}\frac{1}{(2\,\mathrm{Re}(s) - 1)^{1/2}}.$$

Passing to the infimum over f, we obtain $2\,\mathrm{Re}(s) - 1 \leq d^2|s|^2$, i.e. $s \notin D_d$.

(2) If $d = 0$, this is the half-plane $D_d = \{s \in \mathbb{C}: \mathrm{Re}(s) > 1/2\}$, which is free of zeros. The RH follows by the symmetry in Corollary 6.1.6.

$\blacksquare$

Remark 6.2.2   Clearly, with an arbitrary function $f \in \mathcal{V}_0$, we obtain a disk $D_{d'}, d \leq d' = \|1 - f\|_{L^2(0,1)}$ free of zeros of ζ. For d' small enough, the most left-hand point of $D_{d'}$ has for abscissa

$$x_{d'} = \frac{1}{d'^2} - r = \frac{1}{2} + \frac{d'^2}{8} + o(d'^2),$$

and the radius $r = (\sqrt{1 - d'^2})/d'^2$ tends to ∞, hence the disks $D_{d'}$ fill the half-plane $\{z \in \mathbb{C}: \operatorname{Re}(z) > 1/2\}$.

We conclude this section with an interpretation of the approximation of Theorem 6.2.1 in terms of invariant subspaces of the semigroup of dilations.

Lemma 6.2.3 *With the notation of Theorem 6.2.1, the following assertions are equivalent.*

(1) $d = 0$.
(2) $1 \in \operatorname{clos}_{L^2(0,1)}(\mathcal{V}_0)$.
(3) $\chi_{(0,1)} \in \operatorname{clos}_{L^2(0,\infty)}(\mathcal{V}_0)$.
(4) $\chi_{(0,1)} \in \operatorname{clos}_{L^2(0,\infty)}(\mathcal{V})$.
(5) $\operatorname{clos}_{L^2(0,\infty)}(\mathcal{V}_0) = L^2(0,1)$.

Proof The equivalence (1) $\Leftrightarrow$ (2) is evident. Since for $f \in \mathcal{V}$ and $x > 1$ we have $f(x) = f(1)/x$, we thus obtain the equivalences (2) $\Leftrightarrow$ (3) $\Leftrightarrow$ (4) (we consider $L^2(0, 1)$ as a subspace of $L^2(0, \infty)$ consisting of functions $f \in L^2(0, \infty)$ that are zero on $(1, \infty)$).

Evidently, (5) $\Rightarrow$ (3). For the converse, (3) $\Rightarrow$ (5), observe that $\mathcal{V}_0$ is a vector subspace of $L^2(0, 1)$ *invariant by dilatation* $\mathcal{D}_t, t \geq 1$,

$$\mathcal{D}_t f(x) = f(tx) \quad (x > 0, \ t \geq 1),$$

$\mathcal{D}_t \mathcal{V}_0 \subset \mathcal{V}_0, t \geq 1$. Indeed, if $f \in \mathcal{V}$, $f(1) = 0$ then $\mathcal{D}_t f \in \mathcal{V}$ and $\mathcal{D}_t f(1) = f(1)/t$, hence $\mathcal{D}_t f \in \mathcal{V}_0$.

Consequently, a subspace $E = \operatorname{clos}_{L^2(0,\infty)}(\mathcal{V}_0)$ is also $\mathcal{D}_t$-invariant, and by hypothesis $\chi_{(0,1)} \in E$. Then, for every $t \geq 1$, $\mathcal{D}_t \chi_{(0,1)} = \chi_{(0,1/t)} \in E$, and hence every step function on the interval $(0, 1)$ is in E. However, the space of step functions is dense in $L^2(0, 1)$, hence $E = L^2(0, 1)$. $\blacksquare$

6.3 $H^2(\mathbb{C}^+)$ and the "Weak Paley–Wiener Theorem"

This section is a somewhat technical portion of this chapter: here we transfer the space $H^2(\mathbb{D})$ into $H^2(\mathbb{C}_+)$ and $H^2(\mathbb{C}^+)$, and obtain descriptions of the subspaces invariant under dilations and under multiplication by characters.

6.3.1 A Unitary Mapping of $L^2(\mathbb{T})$ onto $L^2(\mathbb{R})$

Let $\mathbb{C}_+ = \{z \in \mathbb{C}: \operatorname{Im}(z) > 0\}$ and $\omega: \mathbb{D} \to \mathbb{C}_+$ the conformal mapping

$$\omega(z) = i\frac{1 + z}{1 - z}.$$

The restriction of the inverse

$$\omega^{-1}(w) = \frac{w - i}{w + i}$$

on the boundary $\partial\mathbb{C}_+ = \mathbb{R}$ is a bijection $\mathbb{R} \to \mathbb{T} \setminus \{1\}$ whose Jacobian is

$$|J(x)| = \frac{2}{1 + x^2}, \qquad x \in \mathbb{R}.$$

Hence the mapping U,

$$Uf(x) = \frac{1}{\sqrt{\pi}(x + i)} \cdot f\left(\frac{x - i}{x + i}\right), \qquad x \in \mathbb{R},$$

is a unitary isomorphism between the spaces $L^2(\mathbb{T})$ and $L^2(\mathbb{R})$,

$$U : L^2(\mathbb{T}) \to L^2(\mathbb{R}).$$

Recall that in this chapter, a space $L^2(E)$, where $E \subset \mathbb{R}$, is regarded as a subspace of $L^2(\mathbb{R})$ containing the functions of $L^2(\mathbb{R})$ that are zero on the complement $\mathbb{R} \setminus E$.

(a) Lemma.

$$UH^2(\mathbb{D}) = \mathrm{span}_{L^2(\mathbb{R})}\left(\frac{1}{x - \bar{\mu}} : \ \mathrm{Im}(\mu) > 0\right).$$

Proof First observe that

$$H^2(\mathbb{D}) = \mathrm{span}_{L^2(\mathbb{T})}\left(\frac{1}{1 - \bar{\lambda}z} : \lambda \in \mathbb{D}\right);$$

the inclusion $\supset$ is evident, and the converse follows from the fact that

$$f \in H^2(\mathbb{D}) \quad \text{and} \quad 0 = \left(f, \frac{1}{1 - \bar{\lambda}z}\right) = f(\lambda) \quad (\forall \lambda \in \mathbb{D})$$

imply $f = 0$. Then, clearly,

$$U\frac{1}{1 - \bar{\lambda}z} = \frac{c_\lambda}{x - \bar{\mu}}$$

where $\mu = \omega(\lambda)$, and μ runs over the half-plane $\mathbb{C}_+$ as λ runs over $\mathbb{D}$. Since $U : L^2(\mathbb{T}) \to L^2(\mathbb{R})$ is unitary, we obtain the stated equality.   ∎

(b) Definition (the Hardy space $H^2(\mathbb{C}_+)$, inner functions). By definition,

$$H^2(\mathbb{C}_+) = UH^2(\mathbb{D}).$$

A function θ in $\mathbb{C}_+$ is said to be *inner (in $\mathbb{C}_+$)* if $\theta \circ \omega = \Theta$ is an inner function in the disk $\mathbb{D}$. Hence, θ is inner if and only if it is holomorphic and bounded, and if its boundary values $\theta(x) = \lim_{y \to 0} \theta(x + iy)$ (which exist by Fatou's theorem and the conformal character of ω) are unimodular a.e. on $\mathbb{R}$.

(c) Lemma (semigroup of characters in $L^2(\mathbb{R})$). *Let*

$$u_s = \exp\left(s \frac{z+1}{z-1}\right), \quad s \in \mathbb{R},$$

and let $F \subset L^2(\mathbb{T})$ *be a closed subspace. The following assertions are equivalent.*

(1) $zF = F$ *(respectively, $zF \subset F$).*
(2) $u_s F \subset F (\forall s \in \mathbb{R})$ *(respectively, $\forall s \geq 0$).*
(3) $e^{isx} UF \subset UF$ *($\forall s \in \mathbb{R}$) (respectively, $\forall s \geq 0$).*

Proof The equivalence (2) $\Leftrightarrow$ (3) is clear, as is the implication (1) $\Rightarrow$ (2) (seen in Exercise 1.8.3(a), and as $u_s \in H^\infty$ for $s \geq 0$).

To show (2) $\Rightarrow$ (1), first observe that (2) implies $\varphi_s F \subset F$ for every $s > 0$, where

$$\varphi_s = \frac{u_s - 1 + s}{u_s - 1 - s}.$$

Furthermore, for every $\zeta \in \mathbb{T}$, we have $\operatorname{Re}(1 - u_s(\zeta)) \geq 0$ and hence $|\varphi_s(\zeta)| \leq 1$. Moreover, since $e^{sw} - 1 = sw + o(s)$ ($\forall w \in \mathbb{C}$) as $s \to 0$, we obtain, for every $\zeta \in \mathbb{T} \setminus \{1\}$, $\varphi_s(\zeta) = \zeta + o(1)$ (as $s \to 0$). By the dominated convergence theorem, for every $f \in F$, $\lim_{s \to 0} \|\varphi_s f - zf\|_2 = 0$, hence $zf \in F$. ∎

(d) Corollary. *Let* $E \subset L^2(\mathbb{R})$ *be a closed subspace.*

(1) $e^{isx} E \subset E, \forall s \in \mathbb{R} \Leftrightarrow E = \chi_A L^2(\mathbb{R})$ *where $A \subset \mathbb{R}$ is a Borel set.*
(2) $e^{isx} E \subset E, \forall s \geq 0$ *(and $\exists s > 0$ such that $e^{isx} E \neq E$) $\Leftrightarrow E = qH^2(\mathbb{C}_+)$ where q is a unimodular function.*
(3) $e^{isx} E \subset E \subset H^2(\mathbb{C}_+), \forall s \geq 0, E \neq \{0\} \Leftrightarrow E = \theta H^2(\mathbb{C}_+)$ *where θ is an inner function in $\mathbb{C}_+$.*

Indeed, to deduce (d) from (c) and the descriptions in Corollaries 1.4.1 and 1.4.3, it suffices to remark that if $f \in L^2(\mathbb{T})$ and $\varphi \in L^\infty(\mathbb{T})$, then $U(\varphi f) = (\varphi \circ \omega)Uf$. ∎

(e) The arithmetic of inner functions and canonical factorization in $H^2(\mathbb{C}_+)$**.**
Clearly the simple change of variables $\Theta \longmapsto \Theta \circ \omega^{-1}$ provides a bijective correspondence between the inner functions in $\mathbb{D}$ and those in $\mathbb{C}_+$; it allows the transfer of all the arithmetic of the inner functions of $\mathbb{D}$ (see § 3.2) to $\mathbb{C}_+$ (divisibility, GCD, LCM, spectrum, rules of calculus of the spectrum,

etc.). Without unnecessary repetition of all these rules, we use them (with the modifications due to the change of variables) as needed. The question of canonical factorization is treated in the same manner: for $f \in H^2(\mathbb{C}_+)$ we write $f = Ug = (g_{in} \circ \omega^{-1})Ug_{out}$, where $g \in H^2(\mathbb{D})$, and define $f_{in} = g_{in} \circ \omega^{-1}$, $f_{out} = Ug_{out}$. In particular, a singular function with singularity at the point $z = 1$,

$$\Theta = \exp\left(-a\frac{1+z}{1-z}\right),$$

is transformed into $\theta(w) = \Theta \circ \omega^{-1} = e^{iaw}$, $w \in \mathbb{C}_+$ (a function singular "at infinity").

6.3.2 Fourier Transforms and the "Weak Paley–Wiener Theorem"

Another unitary mapping, this time of $L^2(\mathbb{R})$ onto $L^2(\mathbb{R})$, is given by the Fourier transform and its inverse (Plancherel's theorem: see Appendix A),

$$\mathcal{F}f(z) = \frac{1}{\sqrt{2\pi}} \int_{\mathbb{R}} f(x)e^{-ixz}\,dx, \quad \mathcal{F}^{-1}f(z) = \frac{1}{\sqrt{2\pi}} \int_{\mathbb{R}} f(x)e^{ixz}\,dx.$$

The next lemma is a "weakened" form (but sufficient for our needs) of an important theorem of Paley and Wiener.

Raymond E. A. C. Paley (1907–1933) was a brilliant English mathematician, who studied at Eton and then at Trinity College, Cambridge, and was elected a Fellow in this prestigious establishment at the age of 23. As he was "inspired by the genius of G. H. Hardy and J. E. Littlewood" (in the words of Norbert Wiener), Paley worked mainly in harmonic analysis but also in probability and graph theory. Over the very short period of his professional career (not even three full years), he collaborated with a group of remarkable mathematicians, including Littlewood, Zygmund, Wiener, and Pólya. With Littlewood he constructed the *Littlewood–Paley decomposition* – a tool of "hard" harmonic analysis that has become classical and indispensable in any serious application of Fourier analysis: to weighted singular integrals, Fourier multipliers, the maximal regularity of semigroups, etc. Another fundamental result is the *Paley–Wiener theorem* (1932, published posthumously) linking the Hardy space in the half-plane with the Fourier transform of $L^2(\mathbb{R}_+)$ (see § 6.3.2 for a "weak form" of the result). According to Wiener, Paley was one of the pioneers in using probabilistic methods in harmonic analysis. He also discovered an interpolation of lacunary Fourier coefficients of a totally new kind, and contributed to graph theory.

Paley's name is linked with several important subjects, such as *Littlewood–Paley theory*, *Paley–Wiener spaces*, the *Paley–Zygmund inequality*, and *Paley graphs*.

Paley was killed by an avalanche in the Canadian Rockies (near Calgary) at an altitude of 3 000 meters, during a skiing weekend. As Wiener wrote in his obituary for the *Bulletin of the American Mathematical Society* (vol. 39 (1933), p. 476), "the impression which Paley had made on American mathematicians is remarkable in the extreme ... his premature death is an irreparable loss to mathematics."

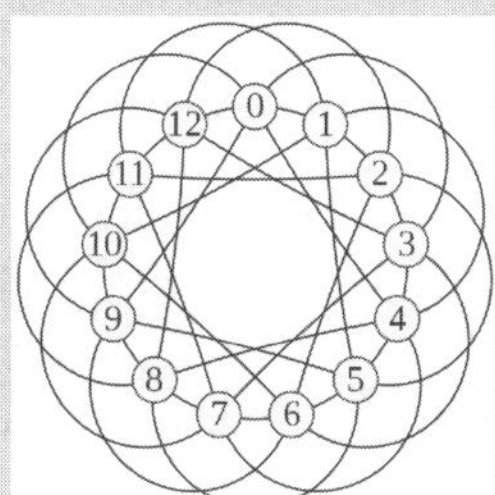

$$\Pr\{Z \geq \theta E(Z)\} \geq \frac{(1-\theta)^2(E(Z))^2}{(1-\theta)^2(E(Z))^2 + \operatorname{Var} Z}$$

Paley's impact on mathematics is very diverse and goes far beyond analysis. Left, the order-13 *Paley graph* allowing graph-theoretic tools to be applied to the number theory of quadratic residues. Right, a useful *Paley–Zygmund inequality* which bounds the probability that a positive random variable is small, in terms of its mean and variance.

(a) Lemma (Paley and Wiener, 1934). $H^2(\mathbb{C}_+) = \mathcal{F}^{-1}L^2(\mathbb{R}_+) = \mathcal{F}L^2(\mathbb{R}_-)$, $\mathbb{R}_+ = (0, \infty)$, $\mathbb{R}_- = (-\infty, 0)$.

Proof We calculate the inverse Fourier transform of an exponential function $\chi_{\mathbb{R}_+} e^{i\lambda x}$, $\lambda \in \mathbb{C}_+$:

$$\mathcal{F}^{-1}(\chi_{\mathbb{R}_+} e^{i\lambda x})(z) = \frac{1}{\sqrt{2\pi}} \int_{\mathbb{R}_+} e^{i\lambda x} e^{ixz}\, dx = \frac{i}{\sqrt{2\pi}} \cdot \frac{1}{z+\lambda}.$$

Since $\mathcal{F}^{-1}$ is an isometric mapping, by the Lemma in § 6.3.1(a) it remains only to verify that the family $\chi_{\mathbb{R}_+} e^{i\lambda x}$, $\lambda \in \mathbb{C}_+$ generates $L^2(\mathbb{R}_+)$,

$$L^2(\mathbb{R}_+) = \operatorname{span}_{L^2(\mathbb{R})}(\chi_{\mathbb{R}_+} e^{i\lambda x} : \lambda \in \mathbb{C}_+).$$

To verify this last equality, suppose there is a function $f \in L^2(\mathbb{R}_+)$ such that for every $\lambda \in \mathbb{C}_+$, $f \perp \chi_{\mathbb{R}_+} e^{i\lambda x}$. Calculating with $\lambda = i + y$, $y \in \mathbb{R}$, we obtain

$$0 = (f, \chi_{\mathbb{R}_+} e^{i\lambda x}) = \mathcal{F}(f\chi_{\mathbb{R}_+} e^{-x})(y) \quad (\forall y \in \mathbb{R}),$$

hence $f\chi_{\mathbb{R}_+} e^{-x} = 0$, and thus $f = 0$. ∎

(b) Lemma. *For every $s \in \mathbb{R}$, we have*

$$\mathcal{F} e^{isx} \mathcal{F}^{-1} = \tau_s$$

on the space $L^2(\mathbb{R})$ where τ_s is the operation of translation, $\tau_s f(x) = f(x - s)$.

Proof Direct computation. ∎

(c) Corollary (Lax, 1959). *Let $E \subset L^2(\mathbb{R})$ be a (closed) subspace.*

(1) *$\tau_s E \subset E, \forall s \in \mathbb{R} \Leftrightarrow E = \mathcal{F}(\chi_A L^2(\mathbb{R}))$ where $A \subset \mathbb{R}$ is a Borel set.*
(2) *$\tau_s E \subset E, \forall s \geq 0$ (and $\exists s > 0$ such that $\tau_s E \neq E$) $\Leftrightarrow E = \mathcal{F}(q H^2(\mathbb{C}_+))$ where q is a unimodular function.*
(3) *$\tau_s E \subset E \subset L^2(\mathbb{R}_+), \forall s \geq 0, E \neq \{0\} \Leftrightarrow E = \mathcal{F}(\theta H^2(\mathbb{C}_+))$ where θ is an inner function in $\mathbb{C}_+$.*

The corollary is immediate by (a), (b) and § 6.3.1(d). ∎

6.3.3 The Mellin Transform and the Group of Dilations

The *Mellin transform* $\mathcal{F}_*$ is the Fourier transform on the multiplicative group $\mathbb{R}_+ = (0, \infty)$, that can be obtained with the aid of the group homomorphism $\varphi : \mathbb{R} \to \mathbb{R}_+$, $\varphi(x) = e^{-x}$:

$$\mathcal{F}_* f = \mathcal{F}(f \circ \varphi), \quad \mathcal{F}_* f(z) = \frac{1}{\sqrt{2\pi}} \int_{\mathbb{R}_+} f(y) y^z \frac{dy}{y}.$$

Robert Hjalmar Mellin (1854–1933) was a Finnish mathematician, a native of Liminka, a remote village of 3 000 inhabitants in the north of Finland. Son of a clergyman, Mellin studied at the University of Helsinki under Mittag-Leffler (submitting a thesis in 1881 on algebraic functions), and then continued in Berlin under the supervision of Weierstrass. After a few years at the University of Stockholm, he obtained a position as professor at the newly founded Technical University of Finland (1908). During this long episode of his career, Mellin made a noble gesture: in 1901, when applying for the post of professor in Helsinki, he withdrew to leave the place to the young prodigy Ernst Lindelöf.

Mellin is principally known for the study of an integral transform that bears his name (see § 6.3.3) and for his philosophical opposition to the theory of relativity, with at least ten articles published on this subject. In a long series of articles, he applied his transform to the study of the gamma and hypergeometric functions, Dirichlet series, the Euler ζ function and other arithmetic functions, and asymptotic developments.

Mellin was one of the founders of the Finnish Academy of Sciences (1908) and was the Finnish representative on the editorial board of *Acta Mathematica*.

Mellin was also known as an activist in the *Fennoman* movement, promoting the use of the Finnish language in place of Swedish, which had been dominant, especially before the attachment of Finland to the Russian Empire in 1809.

Plancherel's theorem states that $\mathcal{F}_*$ is a unitary transform of $L^2(\mathbb{R}_+, dy/y)$ onto $L^2(i\mathbb{R})$, with the same sense given to the integral as in the case of the group $\mathbb{R}$ (see Appendix A). As by tradition the definition of $\mathcal{F}_* f$ is given on the imaginary axis $i\mathbb{R}$, we "turn by $\pi/2$" the definition of the Hardy space, and examine

$$H^2(\mathbb{C}^+) = \{f : f(z) = g(iz), \quad g \in H^2(\mathbb{C}_+)\}.$$

With the notation above, as well as the evident equality

$$L^2(\mathbb{R}_+) = L^2((0, 1), dy/y) \circ \varphi,$$

we obtain the following version of the Paley–Wiener lemma.

(a) Lemma (Paley and Wiener, 1934). $H^2(\mathbb{C}^+) = \mathcal{F}_* L^2((0, 1), dy/y)$.

(b) The *group of dilations*

$$\mathcal{D}_t f(y) = f(ty) \quad (t > 0, \ y > 0)$$

is a unitary group of transforms of the space $L^2(\mathbb{R}_+, dy/y)$ (verify!). We have

$$\mathcal{D}_t\left(L^2\left((0,1), \frac{dy}{y}\right)\right) \subset L^2\left((0,1), \frac{dy}{y}\right), \quad t \geq 1,$$

$$\mathcal{F}_*\mathcal{D}_t f(z) = \frac{1}{\sqrt{2\pi}} \int_{\mathbb{R}_+} f(ty)y^z \frac{dy}{y} = \frac{t^{-z}}{\sqrt{2\pi}} \int_{\mathbb{R}_+} f(y)y^z \frac{dy}{y}$$

$$= t^{-z}\mathcal{F}_* f(z) = e^{-sz}\mathcal{F}_* f(z),$$

where $f \in L^2(\mathbb{R}_+, dy/y)$, $s = \log(t) \in \mathbb{R}$. Hence, $\mathcal{F}_*$ transforms a dilatation $\mathcal{D}_t$ into a multiplication by the character e^{-sz}, $s = \log(t)$. Given the unitary property of $\mathcal{F}_*$, we obtain the following description of the subspaces invariant under dilations.

(c) Corollary (subspaces invariant under dilations). *Let $E \subset L^2(\mathbb{R}_+, dy/y)$ be a closed subspace.*

(1) $\mathcal{D}_t E \subset E, \forall t \in \mathbb{R}_+ \Leftrightarrow \mathcal{F}_* E = \chi_A L^2(\mathbb{R})$, *where* $A \subset \mathbb{R}$ *is a Borel set.*
(2) $\mathcal{D}_t E \subset E, \forall t \geq 1$ *(and* $\exists t > 0$ *such that* $\mathcal{D}_t E \neq E$*)* $\Leftrightarrow \mathcal{F}_* E = qH^2(\mathbb{C}^+)$, *where q is a unimodular function.*
(3) $\mathcal{D}_t E \subset E \subset L^2((0,1), dy/y), \forall t \geq 1, E \neq \{0\} \Leftrightarrow \mathcal{F}_* E = \theta H^2(\mathbb{C}^+)$, *where θ is an inner function in $\mathbb{C}^+$.*

The corollary is immediate by § 6.3.1(d) and/or § 6.3.2(c).   ∎

6.3.4 Completeness of the Characters, the Translations, and/or the Dilations

By a simple comparison of the proposition in § 3.2.2(c) and of points § 6.3.1(d), § 6.3.2(c) and § 6.3.4(c) above, we obtain the following descriptions of the subspaces generated by a given family of functions Φ, $\Phi \neq \{0\}$.

(a) *Let $\Phi \subset H^2(\mathbb{C}_+)$ be a subset of $H^2(\mathbb{C}_+)$ and let*

$$E_\Phi = \operatorname{span}_{H^2(\mathbb{C}_+)} \left(e^{isx}\Phi : s \geq 0\right)$$

be the invariant subspace under $(e^{isx})_{s\geq 0}$ *generated by Φ. Then,*
$E_\Phi = \theta H^2(\mathbb{C}_+)$ *where* $\theta = GCD(\varphi_{in} : \varphi \in \Phi)$.

(b) *Let $\Phi \subset L^2(\mathbb{R}_+)$ and let*

$$E_\Phi = \operatorname{span}_{L^2(\mathbb{R}_+)} \left(\tau_s \Phi : s \geq 0\right)$$

be the subspace invariant under translations $(\tau_s)_{s\geq 0}$ *generated by Φ. Then,*
$E_\Phi = \mathcal{F}(\theta H^2(\mathbb{C}_+))$ *where* $\theta = GCD((\mathcal{F}^{-1}\varphi)_{in} : \varphi \in \Phi)$.

(c) *Let $\Phi \subset L^2((0,1), dy/y)$ and let*

$$E_\Phi = \operatorname{span}_{L^2(\mathbb{R}_+, dy/y)} (\mathcal{D}_t\Phi : t > 1)$$

be the subspace invariant under dilations $(\mathcal{D}_t)_{t \geq 1}$ generated by Φ. Then $\mathcal{F}_* E_\Phi = \theta H^2(\mathbb{C}^+)$ where $\theta = GCD((\mathcal{F}_*\varphi)_{in} : \varphi \in \Phi)$.

6.4 The Nyman Theorem

We are now ready to return to Theorem 6.2.1 and prove that the approximation by dilations mentioned in this theorem is not only sufficient, but is in fact equivalent to the Riemann hypothesis. Recall (see § 6.2): $\mathcal{V} = \operatorname{Lin}(\mathcal{D}_t\varphi : t > 1)$, $\mathcal{V}_0 = \{f \in \mathcal{V} : f(1) = 0\}$, where $\varphi(x) = \rho(1/x) = 1/x - [1/x]$ $(x > 0)$.

Theorem 6.4.1 (Nyman, 1950) *Let*

$$d = \operatorname{dist}_{L^2(0,1)}(1, \mathcal{V}_0).$$

The following assertions are equivalent.

(1) *$d = 0$.*
(2) *$\chi_{(0,1)} \in \operatorname{clos}_{L^2(0,\infty)}(\mathcal{V})$.*
(3) *$\operatorname{clos}_{L^2(0,\infty)}(\mathcal{V}_0) = L^2(0,1)$.*
(4) *(RH) is correct (i.e. all the zeros of ζ in $\operatorname{Re}(s) > 0$ are on the line $\operatorname{Re}(s) = 1/2$).*

Proof The implications (1) $\Leftrightarrow$ (2) $\Leftrightarrow$ (3) $\Rightarrow$ (4) are already known (see Theorem 6.2.1 and Lemma 6.2.3). Let us show that (4) $\Rightarrow$ (3). Denote

$$F = \operatorname{clos}_{L^2(0,\infty)}(\mathcal{V}_0),$$

and let V be the isometry $L^2(0,1) \to L^2((0,1), dy/y)$ defined by $Vf(y) = y^{1/2} f(y)$. Since $\mathcal{D}_t \mathcal{V}_0 \subset \mathcal{V}_0$ for $t \geq 1$, we have $\mathcal{D}_t F \subset F$ and $\mathcal{D}_t(VF) \subset VF$ for every $t \geq 1$. Also note that $V\mathcal{V}_0 = \operatorname{Lin}(\Psi)$, where

$$\Psi = \left\{ \psi_t := y^{1/2} \left(\varphi(ty) - \frac{1}{t}\varphi(y) \right) : t \geq 1 \right\};$$

indeed, if $f = y^{1/2} \sum_k a_k \varphi(t_k y)$ $(t_k > 1)$ and $\sum_k \frac{a_k}{t_k} = 0$, then

$$f = y^{1/2} \sum_k a_k \varphi(t_k y) = y^{1/2} \sum_k (a_k \varphi(t_k y) - \frac{a_k}{t_k}\varphi(y)) = \sum_k a_k \psi_{t_k}.$$

By § 6.3.4(c),

$$\mathcal{F}_*(VF) = \theta H^2(\mathbb{C}^+), \quad \text{where } \theta = GCD((\mathcal{F}_*\psi_t)_{in} : it \geq 1).$$

We calculate $\mathcal{F}_*\psi_t$ using Lemma 6.1.3:

$$
\begin{aligned}
\mathcal{F}_*\psi_t(z) &= \frac{1}{\sqrt{2\pi}} \int_0^1 y^{1/2}\varphi(ty)y^z\frac{dy}{y} - \frac{1}{t\sqrt{2\pi}} \int_0^1 y^{1/2}\varphi(y)y^z\frac{dy}{y} \\
&= \frac{1}{\sqrt{2\pi}} \left\{ \frac{1}{t(z-1/2)} - \frac{\zeta(z+1/2)}{(z+1/2)t^{z+1/2}} - \frac{1}{t(z-1/2)} + \frac{\zeta(z+1/2)}{t(z+1/2)} \right\} \\
&= \frac{1}{\sqrt{2\pi}} \cdot \frac{\zeta(z+1/2)}{z+1/2} \left(\frac{1}{t} - \frac{1}{t^{z+1/2}} \right).
\end{aligned}
$$

To find $(\mathcal{F}_*\psi_t)_{in}$, we use § 3.2 (transferred to the half-plane $\mathbb{C}^+$ with the aid of the conformal mapping $\omega = (1+z)/(1-z)$). The roots of $\mathcal{F}_*\psi_t(z) = 0$ consist of the union

$$
\left\{ z \in \mathbb{C}^+ : \zeta\left(z + \frac{1}{2}\right) = 0 \right\} \cup \left\{ z \in \mathbb{C}^+ : \frac{1}{t^{z-1/2}} = 1 \right\}
$$

$$
= \left\{ z \in \mathbb{C}^+ : \zeta\left(z + \frac{1}{2}\right) = 0 \right\} \cup \left\{ z \in \mathbb{C}^+ : z = \frac{1}{2} + \frac{2\pi i n}{\log(t)}, \ n \in \mathbb{Z} \setminus \{0\} \right\}.
$$

Clearly the common zeros of the $\mathcal{F}_*\psi_t$, $t \geq 1$ (and hence of θ), are exactly

$$
Z = \left\{ z \in \mathbb{C}^+ : \zeta\left(z + \frac{1}{2}\right) = 0 \right\}.
$$

Each function $\mathcal{F}_*\psi_t$ is holomorphic at the boundary $z \in i\mathbb{R}$, thus the singular measure of $(\mathcal{F}_*\psi_t)_{in} \circ \omega$ is situated at the point $\{1\}$. However the logarithmic residue vanishes:

$$
\overline{\lim_{x>0,x\to\infty}} \frac{\log|\mathcal{F}_*\psi_t(x)|}{x} = \overline{\lim_{x>0,x\to\infty}} \frac{\log|1/(x+1/2)|}{x} = 0,
$$

hence the singular part of $(\mathcal{F}_*\psi_t)_{in}$ is trivial (constant).

Conclusion The function θ is a Blaschke product corresponding to the set of the zeros Z (more exactly, $\theta \circ \omega$ is a Blaschke product in the disk $\mathbb{D}$ corresponding to the zeros $\omega^{-1}(Z)$), hence $\mathcal{F}_*(VF) = H^2(\mathbb{C}^+)$ if and only if $Z = \emptyset$. ∎

6.5　The Distance Function and Zero-free Disks of ζ

We introduce here a family of approximation problems which generalize Nyman's approach in § 6.4 and provide a number of disks in $\mathbb{C}^+$ free of zeros of the Euler ζ function. We begin with a few observations on the reproducing kernel of the space $H^2(\mathbb{D})$, and then transfer them to $H^2(\mathbb{C}^+)$ and $L^2((0,1), dy/y)$ to obtain the result.

6.5.1 The Distance Function

Let $\lambda \in \mathbb{D}$; the *reproducing kernel at the point* λ is the function k_λ,

$$k_\lambda(z) = \sum_{k \geq 0} \overline{\lambda}^k z^k = \frac{1}{1 - \overline{\lambda}z} \quad (z \in \mathbb{D}),$$

such that $\varphi_\lambda(f) = (f, k_\lambda)_{H^2}$ for every function $f \in H^2$, where φ_λ is an *evaluation functional at the point* λ,

$$\varphi_\lambda(f) = f(\lambda), \quad f \in H^2$$

(see also Exercise 2.8.2(b) where k_λ is mentioned). Let $E \subset H^2$ be a subspace and

$$\Theta_E(\lambda) = \|\varphi_\lambda|E\|, \quad d_E(\lambda) = \mathrm{dist}_{H^2}(k_\lambda, E), \quad \lambda \in \mathbb{D}.$$

The following properties are evident:

(a) $k_\lambda(z) = 1/(1 - \overline{\lambda}z)$ and $\Theta_{H^2}(\lambda) = \|k_\lambda\| = (1 - |\lambda|^2)^{-1/2}$.
(b) $\Theta_E(\lambda) = d_{E^\perp}(\lambda)$, where $E^\perp$ is the orthogonal complement of E in H^2.
(c) $\Theta_E^2(\lambda) + d_E^2(\lambda) = \|k_\lambda\|^2 = (1 - |\lambda|^2)^{-1}$.
(d) If $E = \Theta H^2$, where Θ is an inner function, then
$\Theta_E(\lambda) = |\Theta(\lambda)|(1 - |\lambda|^2)^{-1/2}$.

Lemma 6.5.1 *Let $E \subset H^2(\mathbb{D})$ be a subspace, $Z(E) = \{z \in \mathbb{D}: f(z) = 0 \,\forall f \in E\}$ (the common zeros of E), $\lambda \in \mathbb{D}$ and*

$$D_\lambda = \{z \in \mathbb{D}: |b_\lambda(z)|^2 < 1 - d_E^2(\lambda)/\|k_\lambda\|^2\},$$

where b_λ is the elementary Blaschke factor. Then, $Z(E) \cap D_\lambda = \emptyset$.

Proof Let $z \in Z(E)$ and $f \in E$. Then, $f = b_z g$ where $g \in H^2$, $\|f\| = \|g\|$. Consequently, $|f(\lambda)| = |b_z(\lambda)| \cdot |g(\lambda)| \leq |b_z(\lambda)| \cdot \|k_\lambda\| \cdot \|f\|$, and then

$$\Theta_E(\lambda) \leq |b_z(\lambda)| \cdot \|k_\lambda\|.$$

Thus $z \notin D_\lambda$, since $1 - d_E^2(\lambda)/\|k_\lambda\|^2 = \Theta_E^2(\lambda)/\|k_\lambda\|^2$. ∎

Theorem 6.5.2 *Let $s \in \mathbb{C}^+$, $\gamma > 0$,*

$$\Psi_\gamma = \mathrm{Lin}\left(\psi_{t,\gamma}(y) := y^\gamma\left(\varphi(ty) - \frac{1}{t}\varphi(y)\right) : t \geq 1\right),$$

and

$$d_\gamma(s) = \mathrm{dist}_{L^2(0,1;dy/y)}(x^s, \Psi_\gamma).$$

Then the disk $D_{s,\gamma}$,

$$D_{s,\gamma} = \gamma + D_s = \gamma + \left\{ z \colon \left| \frac{z-s}{z-s_*} \right|^2 < 1 - 2\operatorname{Re}(s)d_\gamma^2(s) \right\},$$

is free of zeros of the Euler ζ function; here s_* designates the point of $\mathbb{C}$ symmetrical with s with respect to $i\mathbb{R}$.

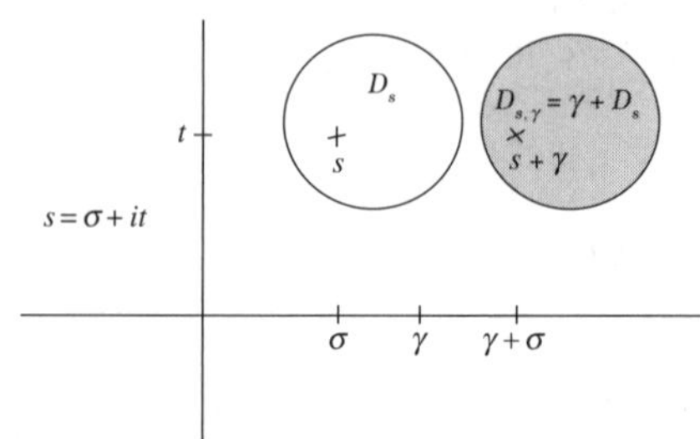

A shifted non-Euclidean disk (shaded) from Theorem 6.5.2 free of zeros of the ζ function (γ and s, the non-Euclidean center, are free parameters).

Proof We begin by noting that the case where $\gamma = s = 1/2$ corresponds to Theorem 6.4.1. We re-use the computation of Theorem 6.4.1 by replacing $1/2$ with $\gamma > 0$. Namely, in the same manner as in Theorem 6.4.1, we have

$$\mathcal{F}_* \psi_{t,\gamma}(z) = \frac{1}{\sqrt{2\pi}} \cdot \frac{\zeta(z+\gamma)}{z+\gamma} \left(\frac{1}{t} - \frac{1}{t^{z+\gamma}} \right).$$

Consequently, setting

$$F_\gamma = \operatorname{clos}_{L^2(0,1;dy/y)}(\Psi_\gamma),$$

we obtain

$$Z(\mathcal{F}_* F_\gamma) = \{ z \in \mathbb{C}^+ \colon \zeta(z+\gamma) = 0 \}.$$

Moreover, the Mellin transform $\mathcal{F}_* \colon L^2((0,1), dy/y) \to H^2(\mathbb{C}^+)$ is isometric, and

$$d_\gamma(s) = \operatorname{dist}_{H^2(\mathbb{C}^+)}(\mathcal{F}_* x^s, \mathcal{F}_* F_\gamma)$$

where $\mathcal{F}_* x^s(z) = \pi^{-1/2}(s+z)^{-1}$ $(z \in \mathbb{C}^+)$ is the reproducing kernel of $H^2(\mathbb{C}^+)$,

$$\|\mathcal{F}_* x^s\|^2_{H^2(\mathbb{C}^+)} = \|x^s\|^2_{L^2(0,1,dy/y)} = 1/2\operatorname{Re}(s).$$

By Lemma 6.5.1 (transferred to $\mathbb{C}^+$), the disk

$$\left\{ z \colon \left| \frac{z-s}{z-s_*} \right|^2 < 1 - 2\operatorname{Re}(s)d_\gamma^2(s) \right\}$$

is free of zeros of $\mathcal{F}_* F_\gamma$, and the result follows.   ∎

Theorem 6.5.3 *Let $\gamma > 0$. With the notation of Theorem 6.5.2, the following assertions are equivalent.*

(1) *The function ζ does not vanish in the half-plane $\{z\colon \mathrm{Re}(z) > \gamma\}$.*
(2) $F_\gamma = L^2((0,1), dy/y)$.
(3) $x^s \in F_\gamma$ ($\exists s \in \mathbb{C}^+$, *or* $\forall s \in \mathbb{C}^+$).

Proof The implications (2) $\Rightarrow$ (3) $\Rightarrow$ (1) follow immediately from Theorem 6.5.2 (with $d_s(\gamma) = 0$).

For (1) $\Rightarrow$ (2) we repeat the portion (4) $\Rightarrow$ (3) of Theorem 6.4.1, but replacing ψ_t by $\psi_{t,\gamma}$, $\zeta(z + 1/2)$ by $\zeta(z + \gamma)$, etc.   ∎

6.6 Completeness of Dilations and the Hilbert Multi-disk

In this section, we study an approximation problem of the integer dilations $\mathcal{D}_n f(x) = f(nx)$, $n = 1, 2, \ldots$, of an arbitrary function f, associated with the completeness (already treated) of all the dilations $\varphi(tx)$, $t \geq 1$. A link with the Riemann hypothesis is established in the following theorem, whose proof lies beyond the elementary framework of this chapter; this theorem shows that in Nyman's Theorem 6.4.1 we can limit ourselves to the *integer dilations $\mathcal{D}_n\varphi$*, $n = 1, 2, \ldots$

Theorem (Báez-Duarte, 2003) *The following assertions are equivalent.*

(1) *(RH) is correct (i.e. all the zeros of ζ in $\mathrm{Re}(s) > 0$ are on the line*
    $\mathrm{Re}(s) = 1/2$).
(2) $\chi_{(0,1)} \in \mathrm{clos}_{L^2(0,\infty)}(\mathcal{N})$, *where* $\mathcal{N} = \mathrm{Lin}(\varphi(nx)\colon n = 1, 2, \ldots)$,
    $\varphi(x) = \rho(1/x) = 1/x - [1/x]$ ($x > 0$).
(3) $1 \in \mathrm{clos}_{L^2(0,1)}(\mathcal{N}_0)$, *where* $\mathcal{N}_0 = \{f \in \mathcal{N}\colon f(1) = 0\}$.

Next, we examine the general question of cyclic vectors of the semigroup of dilations $(\mathcal{D}_n)_{n \geq 1}$ in the space $L^2(0, 1)$.

6.6.1 The Wintner–Beurling Problem

Aurel Wintner (1944) and, independently, Arne Beurling (1945) posed the *problem of the description of functions $f \in L^2(0,1)$ such that $E_f^{\mathcal{D}} = L^2(0,1)$*, where

$$E_f^{\mathcal{D}} = \mathrm{span}_{L^2(0,1)}(\mathcal{D}_n f\colon n = 1, 2, \ldots).$$

A function with such a completeness property is called $(\mathcal{D}_n)$-*cyclic*. Of course, it is necessary to describe precisely how f is defined outside the interval $(0, 1)$. We have selected (following Wintner and Beurling) the extension given by the development of f over the orthonormal trigonometric basis

$$f(x) = \sum_{k \geq 1} a_k \sqrt{2} \sin(\pi k x), \quad x \in (0, 1).$$

The choice of this basis is justified by the fact that this is the only orthonormal basis of $L^2(0, 1)$ of the form $(\mathcal{D}_n e)_{n \geq 1}$, where $e \in L^2(0, 1)$ (see Exercise 6.7.2). Hence, a function $f \in L^2(0, 1)$ extends to $\mathbb{R}$ in an *odd and 2-periodic* manner. Given the result cited above, the Wintner–Beurling problem is clearly linked to the Riemann hypothesis. It remains unsolved to this day (2018).

Aurel Wintner (second on the left in the third and last row) during a seminar at the Niels Bohr Institute in Copenhagen in 1930. In the first row (where four Nobel Prize winners appear) are Oscar Klein, Niels Bohr, Werner Heisenberg, Wolfgang Pauli, George Gamow, Lev Landau, and Hendrik Kramers. Edward Teller is the first on the right in the second row.

Aurel Wintner (1903–1958), a Hungarian–American mathematician, was one of the principal promoters of analysis of the twentieth century, whose heritage is without doubt not yet properly recognized. Wintner published 437 articles (!) in the most renowned journals, as well as nine reference texts, including *Spektraltheorie der unendlichen Matrizen: Einführung in den analytischen Apparat der Quantenmechanik* (1929),

Analytical Foundations of Celestial Mechanics (1941, 1947), *The Fourier Transforms of Probability Distributions* (1947), and *An Arithmetical Approach to Ordinary Fourier Series* (1945). It was in fact the first of these monographs that provided the very first rigorous treatment of spectral theory: in particular, the spectral theorem for the normal operators, the fact that the spectrum of a bounded operator is non-empty (long before Gelfand theory), the essential spectrum, etc. The second in the list is a standard reference for the subject. Wintner was one of the founders (with Paul Erdős and Mark Kac) of probabilistic number theory (1940): he introduced the expressions *law of the iterated logarithm, essential spectrum, summing method of Eratosthenes*, etc. (The latter was reinvented independently – but later – by Albert Ingham and introduced into common usage by G. H. Hardy under the erroneous name "Ingham's method.")

Wintner was born in Budapest, and after a long hesitation between the sciences and music (he showed a rare talent for the violin), he registered at the University of Budapest (1920) for studies in astronomy and physics. He was forced to leave in 1924 because of the galloping inflation that induced chaos in the finances of his family (and the country). During the following three years (hence, between the ages of 21 and 23) he published about 20 articles in prestigious journals (in celestial mechanics and mathematics). He obtained his doctorate under the supervision of Leon Lichtenstein (Leipzig, 1927), then collaborated with Levi-Civita in Rome and Strömberg in Copenhagen. Next, Wintner moved to the USA (Princeton, Harvard, MIT, and then the Johns Hopkins University) to work with Birkhoff, Erdős, Kac, and Wiener. In particular, he was the very first to propose the use of the approximation technique of harmonic analysis for problems in arithmetic (partially presented in § 6.6).

Editor of the *American Journal of Mathematics*, Wintner played a principal role (with André Weil) in giving this journal its dominant stature and creating its irreproachable standards.

During a meeting of the American Mathematical Society (Columbus, 1940), Wiener and Wintner invented a satirical "journal," *Trivia Mathematica*, and amused themselves by proposing titles of articles deemed "acceptable."

In 1930 Wintner married Irmgard Hölder (daughter of Otto Hölder) and had one child. He died suddenly of a heart attack while he was at the peak of his mathematical productivity.

6.6.2 Change of Orthonormal Basis: The Semigroup $\mathcal{T} = (T_n)$ on H_0^2

Clearly, nothing would be changed if we replace the series

$$f = \sum_{k\geq 1} a_k \sqrt{2} \sin(\pi k x)$$

with

$$\|f\|_2^2 = \sum_{k\geq 1} |a_k|^2$$

and the semigroup

$$\mathcal{D}_n f = \sum_{k\geq 1} a_k \sqrt{2} \sin(\pi k n x)$$

with the power series

$$f = \sum_{k\geq 1} a_k z^k$$

with $\|f\|_2^2 = \sum_{k\geq 1} |a_k|^2 < \infty$ and the semigroup $\mathcal{T} = (T_n)_{n\geq 1}$,

$$T_n f(z) = f(z^n).$$

Then the Wintner–Beurling problem becomes a problem of the functions

$$f \in H_0^2 = \{f \in H^2(\mathbb{T}) : f(0) = 0\},$$

cyclic with respect to the semigroup $\mathcal{T}$, i.e. such that $E_f^{\mathcal{T}} = H_0^2$, where

$$E_f^{\mathcal{T}} = \operatorname{span}_{H^2(\mathbb{T})}(T_n f : n = 1, 2, \dots).$$

Clearly (as with the dilations $(\mathcal{D}_n)$) (T_n) is a multiplicative semigroup (or more precisely, a representation $n \longmapsto T_n$ of $\mathbb{N}$ in H_0^2):

$$T_n T_m = T_{mn}.$$

6.6.3 The Reproduction of Variables and the Bohr Transform

The decomposition into prime numbers $n = p_1^{\alpha_1} p_2^{\alpha_2} \dots p_k^{\alpha_k}$, where $\alpha_j \in \mathbb{Z}_+ = \mathbb{N} \bigcup \{0\}$ and $k = k(n)$ (see § 6.1.1), gives

$$T_n = T_{p_1}^{\alpha_1} \cdots T_{p_k}^{\alpha_k},$$

and suggests considering $z^{p_1}, \dots, z^{p_k}, \dots$ as independent variables.

To formalize this idea we will need some new notation. Namely, for a natural number

$$n = p_1^{\alpha_1} p_2^{\alpha_2} \cdots p_s^{\alpha_s}$$

represented in its Euclidean decomposition, we associate a unique multi-index

$$\alpha(n) = (\alpha_1, \alpha_2, \alpha_3, \dots),$$

where $\alpha_j \in \mathbb{Z}_+ = \mathbb{N} \cup \{0\}$ (and after a certain rank $\alpha_j = 0$). In fact, the mapping $\alpha: n \longmapsto \alpha(n)$ is a bijection of $\mathbb{N}$ onto the set

$$\mathbb{Z}_+(\infty) = \bigcup_{k \geq 1} \mathbb{Z}_+^k$$

of the finitely supported sequences of non-negative integers, and moreover, this is a homomorphism of the multiplicative semigroup $\mathbb{N}$ in $\mathbb{Z}_+(\infty)$. To shorten the notation with the multi-indices we systematically replace $\sum_{\alpha \in \mathbb{Z}_+(\infty)}$ with $\sum_{\alpha \geq 0}$.

Now, we can define what is known as the *Bohr transform*: for

$$f = \sum_{n \geq 1} \hat{f}(n) z^n \in H_0^2,$$

set

$$Uf(\zeta) = \sum_{n \geq 1} \hat{f}(n) \zeta^{\alpha(n)},$$

where $\zeta = (\zeta_1, \zeta_2, \dots) \in \mathbb{D}^\infty$ is such that the series converges absolutely, and a multi-power ζ^α, defined by

$$\zeta^\alpha = \zeta_1^{\alpha_1} \zeta_2^{\alpha_2} \cdots \zeta_s^{\alpha_s} \cdots .$$

This is a finite product because the multi-index α has finite support. For every series $Uf(\zeta)$ to be well-defined at the point $\zeta = (\zeta_1, \zeta_2, \dots) \in \mathbb{D}^\infty$, it is necessary and sufficient that

$$\sum_{\alpha \geq 0} |\zeta^\alpha|^2 < \infty$$

(indeed, to have $\sum_{\alpha \geq 0} |a_\alpha b_\alpha| < \infty$ for every $a = (a_\alpha) \in l^2$ it is necessary and sufficient that $\sum_{\alpha \geq 0} |b_\alpha|^2 < \infty$: see Appendix A).

Lemma 6.6.1 *Let* $\zeta = (\zeta_1, \zeta_2, \dots) \in \mathbb{D}^\infty$. *Then*

$$\sum_{\alpha \geq 0} |\zeta^\alpha|^2 = \prod_{k \geq 1} \frac{1}{1 - |\zeta_k|^2},$$

and hence $\sum_{\alpha \geq 0} |\zeta^\alpha|^2 < \infty$ *if and only if* $\sum_{k \geq 1} |\zeta_k|^2 < \infty$.

Proof We repeat (almost) the proof of § 6.1.2: for a series of positive terms we can write

$$\sum_{\alpha \geq 0} |\zeta^\alpha|^2 = \lim_k \sum_{\alpha \in \mathbb{Z}_+^k} |\zeta^\alpha|^2,$$

$$\sum_{\alpha \in \mathbb{Z}_+^k} |\zeta^\alpha|^2 = \sum_{\alpha_2 \geq 0 \ldots \alpha_k \geq 0} \sum_{\alpha_1 \geq 0} |\zeta_1^{\alpha_1}|^2 |\zeta_2^{\alpha_2} \cdots \zeta_k^{\alpha_k}|^2$$

$$= \sum_{\alpha_2 \geq 0 \ldots \alpha_k \geq 0} (1 - |\zeta_1|^2)^{-1} |\zeta_2^{\alpha_2} \cdots \zeta_k^{\alpha_k}|^2 = \prod_{j=1}^k (1 - |\zeta_j|^2)^{-1},$$

and the result follows. ∎

6.6.4 The Hilbert Multi-disk $\mathbb{D}_2^\infty$ and the Space $H^2(\mathbb{D}_2^\infty)$

In 1909, Hilbert published a sketch of the theory of infinitely many complex variables in a *Hilbert multi-disk* $\mathbb{D}_2^\infty$,

$$\mathbb{D}_2^\infty = \{\zeta = (\zeta_1, \zeta_2, \ldots) : \zeta \in l^2, |\zeta_j| < 1 \text{ for all } j \geq 1\}.$$

We will need an analog of the Hardy space in $\mathbb{D}_2^\infty$, which we define as the space of power series with coefficients in l^2:

$$H^2(\mathbb{D}_2^\infty) := \left\{ F = \sum_{\alpha \in \mathbb{Z}_+(\infty)} c_\alpha(F) \zeta^\alpha : \|F\|_2^2 = \sum_{\alpha \in \mathbb{Z}_+(\infty)} |c_\alpha(F)|^2 < \infty \right\}.$$

Lemma 6.6.2

(1) *For every function $f \in H_0^2$ and $\zeta \in \mathbb{D}_2^\infty$, the series $Uf(\zeta) = \sum_{n \geq 1} \hat{f}(n) \zeta^{\alpha(n)}$ converges absolutely, and*

$$|Uf(\zeta)| \leq \|f\|_2 \prod_{k \geq 1} \left(\frac{1}{1 - |\zeta_k|^2} \right)^{1/2}.$$

(2) *U is a unitary mapping of H_0^2 onto the Hardy space of the multi-disk,*

$$U : H_0^2 \rightarrow H^2(\mathbb{D}_2^\infty),$$

transforming the orthonormal basis $(z^n)_{n \geq 1}$ of H_0^2 to the orthonormal basis of multi-powers $(\zeta^\alpha)_{\alpha \geq 0}$ of $H^2(\mathbb{D}_2^\infty)$.

(3) *For every function $f \in H_0^2$, $\zeta \in \mathbb{D}_2^\infty$ and $n \in \mathbb{N}$,*

$$(UT_n f)(\zeta) = \zeta^{\alpha(n)}(Uf)(\zeta),$$

where $\alpha(n) = (\alpha_1(n), \ldots, \alpha_k(n), \ldots)$ is defined in § 6.6.3.

(4)

$$\mathrm{Lat}(T_n) = U^{-1}\,\mathrm{Lat}(M_\zeta),$$

which means that a closed subspace $E \subset H_0^2$ is $\mathcal{T}$-invariant (i.e.
T_n-invariant for every $n \in \mathbb{N}$) if and only if UE is M_ζ-invariant (i.e.
ζ_k-invariant for every $k \in \mathbb{N}$: $f \in UE \Rightarrow \zeta_k f \in UE$ for every $k \in \mathbb{N}$).
(5) *A function $f \in H_0^2$ is $\mathcal{T}$-cyclic if and only if Uf is M_ζ-cyclic in $H^2(\mathbb{D}_2^\infty)$,*
    i.e.

$$E_{Uf} = H^2(\mathbb{D}_2^\infty), \quad \text{where } E_{Uf} := \mathrm{span}(\zeta^\alpha Uf : \alpha \in \mathbb{Z}_+(\infty))$$

and span *denotes the closed linear hull in $H^2(\mathbb{D}_2^\infty)$.*

Proof **(1)** By Cauchy–Schwarz and Lemma 6.6.1,

$$|Uf(\zeta)|^2 \le \|f\|_2^2 \sum_{\alpha \ge 0} |\zeta^\alpha|^2 = \|f\|_2^2 \prod_{k \ge 1}\left(\frac{1}{1 - |\zeta_k|^2}\right).$$

(2) By definition, $Uz^n = \zeta^{\alpha(n)}$, for every $n \in \mathbb{N}$. Moreover, α is a bijection of
$\mathbb{N}$ on $\mathbb{Z}_+(\infty)$, and the result follows.
(3) Since α is a homomorphism, for every $k \in \mathbb{N}$, we have

$$(UT_n z^k)(\zeta) = (Uz^{kn})(\zeta) = \zeta^{\alpha(kn)} = \zeta^{\alpha(n)}\zeta^{\alpha(k)} = \zeta^{\alpha(n)}(Uz^k)(\zeta).$$

Moreover, the functional $f \longmapsto Uf(\zeta)$ is linear and bounded on H_0^2 (see
(1)), which implies the result.
(4) Evident by (3).
(5) Evident by (4).

■

Corollary 6.6.3 (Beurling, 1945) *If $f \in H_0^2$ is $\mathcal{T}$-cyclic, then $Uf(\zeta) \ne 0$ for*
every $\zeta \in \mathbb{D}_2^\infty$.

Indeed, if $Uf(\zeta) = 0$, then for every $\alpha \in \mathbb{Z}_+(\infty)$, $\zeta^\alpha(Uf)(\zeta) = 0$, and by
Lemma 6.6.2(1) $g(\zeta) = 0$ for every $g \in E_{Uf}$. Hence, $1 \notin E_{Uf}$.   ■

6.6.5 A Few Initial Observations

In general, the necessary condition of Corollary 6.6.3 is not sufficient for
the M_ζ-cyclicity of Uf in $H^2(\mathbb{D}_2^\infty)$ (as we have already seen in the case of
the M_z-cyclicity in the space $H^2(\mathbb{D})$, for example, see Exercise 1.8.3(b)). In
what follows, we will see that for certain classes of functions f the converse
of Corollary 6.6.3 is nonetheless correct. But first, we make a few technical
preparations.

(a) The *space $H^2(\mathbb{D}^n)$ in the polydisk $\mathbb{D}^n$* is defined as the space of power series in $z = (z_1, \ldots, z_n) \in \mathbb{C}^n$ with coefficients in l^2:

$$H^2(\mathbb{D}^n) := \left\{ F = \sum_{\alpha \in \mathbb{Z}_+^n} c_\alpha(F) \zeta^\alpha : \|F\|_2^2 = \sum_{\alpha \in \mathbb{Z}_+^n} |c_\alpha(F)|^2 < \infty \right\}.$$

Another description: $H^2(\mathbb{D}^n)$ is the subspace of $H^2(\mathbb{D}_2^\infty)$ consisting of the functions F depending only on the variables $\zeta_1, \ldots, \zeta_n$ (more precisely, consisting of the functions $F \in H^2(\mathbb{D}_2^\infty)$ such that $c_\alpha(F) = 0$ for any $\alpha = (\alpha_1, \ldots, \alpha_j, \ldots)$ such that $\alpha_j > 0$ for an index $j > n$). It is easy to see that for every function $F \in H^2(\mathbb{D}^n)$ (and with $0 < r < 1$),

$$\|F\|_2^2 = \lim_{r \to 1} \sum_{\alpha \in \mathbb{Z}_+^n} r^{2|\alpha|} |c_\alpha(F)|^2 = \lim_{r \to 1} \int_{\mathbb{T}^n} |F(r\zeta)|^2 \, dm_n(\zeta),$$

where $m_n = m \times m \times \cdots \times m$ is the Lebesgue measure on the circle $\mathbb{T}^n$.

(b) An integral formula for the norm. Let $F \in H^2(\mathbb{D}_2^\infty)$. Then, by (a),

$$\|F\|_2^2 = \lim_{n \to \infty} \sum_{\alpha \in \mathbb{Z}_+^n} |c_\alpha(F)|^2 = \lim_{n \to \infty} \lim_{r \to 1} \int_{\mathbb{T}^n} |F(r\zeta)|^2 \, dm_n(\zeta).$$

(c) Reproducing kernel. By Lemma 6.6.2(1), for every $\lambda \in \mathbb{D}_2^\infty$, the mapping

$$F \longmapsto F(\lambda)$$

is a continuous linear functional on $H^2(\mathbb{D}_2^\infty)$, and hence there exists a unique function $K_\lambda \in H^2(\mathbb{D}_2^\infty)$ such that

$$F(\lambda) = (F, K_\lambda)$$

for every function $F \in H^2(\mathbb{D}_2^\infty)$. It is called the *reproducing kernel of $H^2(\mathbb{D}_2^\infty)$* at the point $\lambda \in \mathbb{D}_2^\infty$. It is easy to find: $F(\lambda) = \sum_{\alpha \in \mathbb{Z}_+(\infty)} c_\alpha(F) \lambda^\alpha$, and thus, by uniqueness,

$$K_\lambda(\zeta) = \sum_{\alpha \in \mathbb{Z}_+(\infty)} \overline{\lambda}^\alpha \zeta^\alpha, \quad \zeta \in \mathbb{D}_2^\infty.$$

The series converge absolutely, and hence by taking successive summations over α_1, α_2, etc. we obtain

$$K_\lambda(\zeta) = \prod_{j \geq 1} k_{\lambda_j}(\zeta_j) \quad \text{where } k_{\lambda_j}(z) = \frac{1}{1 - \overline{\lambda}_j z} \quad (z \in \mathbb{C}).$$

(d) Fact. $H^2(\mathbb{D}_2^\infty) = \text{span}\,(K_\lambda : \lambda \in \mathbb{D}_2^\infty)$, where the span is taken in the space $H^2(\mathbb{D}_2^\infty)$.

Proof If $F \in H^2(\mathbb{D}_2^\infty)$ and $F \perp K_\lambda$ for every $\lambda \in \mathbb{D}_2^\infty$, then $F(\lambda) = 0$ $(\forall \lambda)$, hence $F = 0$. $\blacksquare$

(e) Weak convergence in $H^2(\mathbb{D}_2^\infty)$. *A sequence (F_n) converges weakly to $F \in H^2(\mathbb{D}_2^\infty)$ if and only if $\sup_n \|F_n\|_2 < \infty$ and, for every $\lambda \in \mathbb{D}_2^\infty$, $\lim_n F_n(\lambda) = F(\lambda)$.*

Proof Clear by the Banach–Steinhaus theorem. $\blacksquare$

(f) The *space $H^\infty(\mathbb{D}_2^\infty)$* is, by definition,

$$H^\infty(\mathbb{D}_2^\infty) = \left\{ F \in H^2(\mathbb{D}_2^\infty) : \|F\|_\infty = \sup_{\zeta \in \mathbb{D}_2^\infty} |F(\zeta)| < \infty \right\}.$$

It is clear by (b) above that, for all functions $G \in H^\infty(\mathbb{D}_2^\infty)$, $F \in H^2(\mathbb{D}_2^\infty)$, we have $FG \in H^2(\mathbb{D}_2^\infty)$ and

$$\|FG\|_2 \le \|F\|_2 \|G\|_\infty.$$

The property below will be useful in what follows.

Fact *For every function $G \in H^\infty(\mathbb{D}_2^\infty)$, there exists a sequence of polynomials (p_n) such that $\|p_n\|_\infty \le \|G\|_\infty$ and $\lim_n p_n(\lambda) = G(\lambda)$, $\lambda \in \mathbb{D}_2^\infty$.*

Proof Let

$$G_{(n)}(\lambda) = \sum_{\alpha \in \mathbb{Z}_+^n} c_\alpha(G) \lambda^\alpha$$

be the restriction of G to $\mathbb{D}^n$. Since the series $G(\lambda) = \sum_{\alpha \in \mathbb{Z}_+(\infty)} c_\alpha(G) \lambda^\alpha$ converges absolutely, we have $\lim_n G_{(n)}(\lambda) = G(\lambda)$ for every $\lambda \in \mathbb{D}_2^\infty$, and (of course) $\|G_{(n)}\|_\infty \le \|G\|_\infty$. We define the p_n as the Fejér polynomials of $G_{(n)}$ of degree $N = N(n)$ sufficiently large:

$$p_n = G_{(n)} * \Phi_{N,n},$$

where $\Phi_{N,n}$ is the Fejér kernel on $\mathbb{T}^n$ (the product of the Fejér kernels on $\mathbb{T}$: see Appendix A). Clearly, for every n, the restrictions of the Fejér approximation $f \longmapsto (f - \Phi_{N,n} * f)$, $N = 1, 2, \ldots$ converge uniformly on the compact subsets of $\mathbb{D}^n$: precisely, for any compact subset $\Delta \overline{\mathbb{D}}^n$, $0 < \Delta < 1$, the mappings

$$\Phi_{N,n,\Delta} : if \longmapsto (f - \Phi_{N,n} * f) | \Delta \mathbb{D}^n$$

tend to zero *for the operator norm $H^2(\mathbb{D}^n) \to H^\infty(\Delta \mathbb{D}^n)$*, i.e.

$$\lim_{N \to \infty} \|\Phi_{N,n,\Delta}\| = \lim_{N \to \infty} \sup_{\|f\|_2 \le 1} \|f - \Phi_{N,n} * f\|_{H^\infty(\Delta \mathbb{D}^n)} = 0.$$

Using this property, we obtain, for $p_n(\zeta) = \Phi_{N,n} * G_{(n)}(\zeta)$,

$$|G(\zeta) - p_n(\zeta)| \le |G(\zeta) - G_{(n)}(\zeta)| + |G_{(n)}(\zeta) - \Phi_{N,n} * G_{(n)}(\zeta)|$$
$$\le \|G - G_{(n)}\|_2 \|K_\zeta\|_2 + \|\Phi_{N,n,\Delta}\| \cdot \|G_{(n)}\|_2,$$

where $\Delta = \Delta(\zeta) = \max_j |\zeta_j| < 1$. Now, clearly there exists a sequence $N = N(n) \to \infty$ such that $\lim_n |G(\zeta) - p_n(\zeta)| = 0$ for every $\zeta \in \mathbb{D}_2^\infty$, uniformly on the sets $\{\zeta \in \mathbb{D}_2^\infty : \|K_\zeta\| \le A, \Delta(\zeta) \le \Delta < 1\}$. $\blacksquare$

(g) The *invariant subspaces* $E_F = \mathrm{span}(\zeta^\alpha F : \alpha \in \mathbb{Z}_+(\infty))$ generated in the space $H^2(\mathbb{D}_2^\infty)$ by a function $F \in H^2(\mathbb{D}_2^\infty)$ satisfy the following property:

$$H^\infty(\mathbb{D}_2^\infty) \cdot F \subset E_F.$$

Proof Indeed, for a function $G \in H^\infty(\mathbb{D}_2^\infty)$ and with the polynomials p_n of (f), we have $\|p_n F\|_2 \le \|p_n\|_\infty \|F\|_2 \le \|G\|_\infty \|F\|_2$ and $\lim_n p_n(\lambda)F(\lambda) = G(\lambda)F(\lambda)$, $\lambda \in \mathbb{D}_2^\infty$. By (e) above, we have $\lim_n p_n F = GF$ for the weak convergence of $H^2(\mathbb{D}_2^\infty)$, and hence $FG \in E_F$. $\blacksquare$

6.6.6 Cyclic Polynomials

We now show that the necessary condition of Corollary 6.6.3 is also sufficient for the *cyclicity of polynomials*. Remark that a function $f \in H_0^2$ is a polynomial (in $\mathbb{D}$) if and only if its image Uf (after a reproduction of the variables) is a polynomial in $\mathbb{D}_2^\infty$. See § 6.6.5 for a larger class of functions $F \in H^2(\mathbb{D}_2^\infty)$ for which the same description remains correct.

(a) Theorem (Neuwirth, Ginsberg, and Newman, 1970). *Let f be a polynomial in H_0^2. The following assertions are equivalent.*

(1) f *is $\mathcal{T}$-cyclic in H_0^2.*
(2) Uf *is M_ζ-cyclic in $H^2(\mathbb{D}_2^\infty)$.*
(3) $Uf(\zeta) \ne 0$ *for every $\zeta \in \mathbb{D}_2^\infty$.*

For the proof, we will need the following lemma, itself of independent interest.

(b) Lemma (Neuwirth, Ginsberg, and Newman, 1970). *Let F be a polynomial in $\mathbb{D}_2^\infty$ such that $F(\zeta) \ne 0$ for every $\zeta \in \mathbb{D}_2^\infty$. Then*

$$\left| \frac{F(\zeta)}{F(r\zeta)} \right| \le 2^{\deg(F)}$$

for every $0 \le r \le 1$ *and* $\zeta \in \mathbb{D}_2^\infty$. *The example* $F(z) = (1+z)^n$ *shows (for* $z \to 1$ *and* $r \to 0$*) that this upper estimate is sharp.*

Proof Since a polynomial depends on only a finite number of the variables $\zeta_1, \zeta_2, \ldots$ (say m variables), we can restrict ourselves to $\zeta \in \mathbb{D}^m$. First suppose that $m = 1$. Then $F(z) = A(z - z_1) \ldots (z - z_d)$ where $|z_j| \ge 1$. If $z \in \overline{\mathbb{D}}$, then $|z - rz| \le 1 - r \le |z_j| - r|z| \le |rz - z_j|$, and hence

$$\left| \frac{z - z_j}{rz - z_j} \right| = \left| 1 + \frac{z - rz}{rz - z_j} \right| \le 2.$$

In the general case, we take $\zeta = (\zeta_1, \ldots, \zeta_m) \in \mathbb{D}^m$ and consider the polynomial in a single variable $P_\zeta(z) = F(z\zeta)$, $z \in \mathbb{D}$. Applying the inequality already proved for $m = 1$, we obtain

$$\left| \frac{P_\zeta(z)}{P_\zeta(rz)} \right| \le 2^{\deg(F)} \quad \text{for all } z \in \overline{\mathbb{D}}.$$

It only remains to select $z = 1$. $\blacksquare$

(c) *Proof of Theorem (a)* Clearly $(1) \Leftrightarrow (2) \Rightarrow (3)$. To prove $(3) \Rightarrow (2)$, denote $F = Uf$. Since F depends on a finite number of the variables ζ_j, $j \ge 1$, we have, for any r, $0 < r < 1$,

$$\inf_{\zeta \in \mathbb{D}_2^\infty} |F(r\zeta)| > 0,$$

hence $1/F_r \in H^\infty(\mathbb{D}_2^\infty)$, and then $F/F_r \in E_F$, by § 6.6.5(g). By Lemma (b),

$$\left\| \frac{F}{F_r} \right\|_2 \le \left\| \frac{F}{F_r} \right\|_\infty \le 2^{\deg(F)} \quad \text{for all } 0 < r < 1,$$

and thus $\lim_{r \to 1} (F/F_r) = 1$ weakly in $H^2(\mathbb{D}_2^\infty)$ (see § 6.6.5(e)). Consequently, $1 \in E_F$, and hence $E_F = H^2(\mathbb{D}_2^\infty)$. $\blacksquare$

6.6.7 Other Classes of (T_n)-cyclic Functions of H_0^2

We conclude this chapter with the statement of a theorem proved in Nikolski (2012) and examine certain consequences. We will deduce from it that every reproducing kernel K_λ is a cyclic function in $H^2(\mathbb{D}_2^\infty)$ and, in particular, the functions $f_\gamma = \sum_{k \ge 1} k^{-\gamma} z^k$ $(\mathrm{Re}(\gamma) > 1/2)$ are (T_n)-cyclic in H_0^2, or, equivalently, the functions $\varphi_\gamma = \sum_{k \ge 1} k^{-\gamma} \sin(\pi k x)$ $(\mathrm{Re}(\gamma) > 1/2)$ are $(\mathcal{D}_n)$-cyclic in $L^2(0, 1)$ (Wintner, 1944). Recall (see Lemma 6.6.2) that a function $f \in H_0^2$ is (T_n)-cyclic in H_0^2 if and only if Uf is M_ζ-cyclic in $H^2(\mathbb{D}_2^\infty)$.

(a) Theorem. *Let F be a function of $H^2(\mathbb{D}_2^\infty)$ such that for every $\zeta \in \mathbb{D}_2^\infty$, $F(\zeta) \ne 0$. Each of the following conditions implies the M_ζ-cyclicity of F in $H^2(\mathbb{D}_2^\infty)$.*

(1) *$1/F \in H^\infty(\mathbb{D}_2^\infty)$.*
(2) *There exist $\epsilon > 0$ and $N \ge 1$ such that $F^{1+\epsilon} \in H^2(\mathbb{D}_2^\infty)$ and $1/F^{1/N} \in H^2(\mathbb{D}_2^\infty)$.*
(3) *$\operatorname{Re}(F(\zeta)) \ge 0$ for every $\zeta \in \mathbb{D}_2^\infty$.*
(4) *F depends on only a finite number of variables ζ_j, $1 \le j \le m$, and $F \in \operatorname{Hol}(r\mathbb{D}^m)$, where $r > 1$.*
(5) *$F = Uf$ where $f \in H_0^2(\mathbb{D})$ has its Fourier spectrum $\sigma(f) = \{k \in \mathbb{N}: \hat{f}(k) \ne 0\}$ in a finitely generated multiplicative semigroup, and $\hat{f}(k) = o(k^{-\epsilon})$ as $k \to \infty$ ($\epsilon > 0$).*
(6) *$F = Uf$, $f \in H_0^2$ with $\sigma(f) \subset \{n^k : k \in \mathbb{Z}_+\}$, $n > 1$, and the function $\varphi = \sum_{k \ge 0} \hat{f}(n^k) z^k$ is a Beurling outer function. (In this special case, the last condition is also necessary for the cyclicity.)*

(b) Corollary. *Each reproducing kernel K_λ, $\lambda \in \mathbb{D}_2^\infty$, is (M_ζ)-cyclic in $H^2(\mathbb{D}_2^\infty)$, or, equivalently, every function*

$$f_\lambda = \sum_{n \ge 1} \lambda^{\alpha(n)} z^n$$

is (T_n)-cyclic in $H_0^2(\mathbb{D})$.

Indeed,

$$K_\lambda(\zeta) = \sum_{\alpha \in \mathbb{Z}_+(\infty)} \overline{\lambda}^\alpha \zeta^\alpha = \prod_{s \ge 1} k_{\lambda_s}(\zeta_s)$$

where $k_a(z) = (1 - \overline{a}z)^{-1}$ $(a, z \in \mathbb{D})$. Moreover, $\|k_a\|_{H^p(\mathbb{T})}^p = 1 + |pa/2|^2(1 + o(1))$ when $a \to 0$ ($\forall p < \infty$; see Exercise 6.7.3), and hence

$$\|K_\lambda\|_p^p = \prod_{s \ge 1} \|k_{\lambda_s}\|_{H^p(\mathbb{T})}^p < \infty \quad \text{for all } \lambda,\ \lambda = (\lambda_s) \in \mathbb{D}_2^\infty.$$

Similarly,

$$\|1/K_\lambda\|_2^2 = \prod_{s \ge 1} \|1 - \overline{\lambda}_s \zeta_s\|_{H^2(\mathbb{T})}^2 < \infty.$$

Then the cyclicity of K_λ follows from Theorem (a), point (2).   ∎

(c) Corollary (Wintner, 1944). *Every function*

$$f_a = \sum_{k \ge 1} k^{-a} z^k, \quad \operatorname{Re}(a) > 1/2$$

is (T_n)-cyclic in H_0^2.

Indeed, $Uf_a = K_\lambda$ where $\lambda = (\lambda_s)_{s\geq 1}$, $\lambda_s = p_s^{-a}$ (p_s are primes). ∎

6.7 Exercises

6.7.1 Multipliers of the Space $H^2(\mathbb{D}_2^\infty)$

The space of multipliers of $H^2(\mathbb{D}_2^\infty)$ is defined by

$$\mathrm{Mult}(H^2(\mathbb{D}_2^\infty)) := \{\varphi \colon F \in H^2(\mathbb{D}_2^\infty) \Rightarrow \varphi F \in H^2(\mathbb{D}_2^\infty)\}$$

equipped with the operator norm.

Show that $\mathrm{Mult}(H^2(\mathbb{D}_2^\infty)) = H^\infty(\mathbb{D}_2^\infty)$ and that the norms coincide: $\|\varphi\|_{\mathrm{Mult}} = \|\varphi\|_\infty$.

SOLUTION: We have already mentioned at the beginning of § 6.6.5(f) that every function $\varphi \in H^\infty(\mathbb{D}_2^\infty)$ is a multiplier with $\|\varphi\|_{\mathrm{Mult}} \leq \|\varphi\|_\infty$. The converse is true in general: a multiplier on a function space is bounded (whatever the set where the space is defined). In our case, we have

$$|\varphi(\lambda)| = |(\varphi^n \cdot 1)(\lambda)|^{1/n} \leq (\|\varphi\|_{\mathrm{Mult}}^n \|1\|_2 \|K_\lambda\|_2)^{1/n}$$

for every $\lambda \in \mathbb{D}_2^\infty$, which implies $\varphi \in H^\infty(\mathbb{D}_2^\infty)$ and $\|\varphi\|_\infty \leq \|\varphi\|_{\mathrm{Mult}}$. ∎

6.7.2 Orthogonal Dilations

Here, we consider the functions f of the space $L^2(0, 1)$ extended to $\mathbb{R}$ so as to be odd and 2-periodic (as in § 6.6).

Prove the following properties.

(1) *Let $g \in L^2(0, 1)$. The sequence $(\mathcal{D}_n g)_{n\geq 1}$ is an orthogonal basis of $L^2(0, 1)$ if and only if $g(x) = a \cdot \sin(\pi x)$, $a \neq 0$.*
(2) *Let $f \in H_0^2$. The sequence $(T_n f)_{n\geq 1}$ is an orthogonal basis of H_0^2 if and only if $f = a \cdot z$, $a \neq 0$.*
(3) *Let $F \in H^2(\mathbb{D}_2^\infty)$. The family $(\zeta^\alpha F)_{\alpha\in\mathbb{Z}_+(\infty)}$ is an orthogonal basis of $H^2(\mathbb{D}_2^\infty)$ if and only if $f = \mathrm{constant} = a$, $a \neq 0$.*

SOLUTION: As explained in § 6.6, the three sequences are unitarily equivalent (with the correspondences $\sqrt{2}\sin(\pi n x) \to z^n \to \zeta^{\alpha(n)}$), hence it suffices to show (3). As multiplication by ζ^α is an isometry on $H^2(\mathbb{D}_2^\infty)$, we can suppose that $\|F\|_2 = 1$. Then the orthonormal bases (ζ^α) and $(\zeta^\alpha F)$ are unitarily equivalent, which means that the multiplication $h \longmapsto hF$ is a unitary mapping of $H^2(\mathbb{D}_2^\infty)$ onto itself. By Exercise 6.7.1, $F \in H^\infty(\mathbb{D}_2^\infty)$ and $\|F\|_\infty = 1$. However, the inverse of this mapping is also a unitary multiplier, thus $\|1/F\|_\infty = 1$. By the maximum principle applied to $F_{(n)}$ (see the definition in § 6.6.5(f)), we obtain consecutively $F_{(1)} = \text{constant}$, $F_{(2)} = \text{constant}$, etc., and clearly all these constants coincide. Since $\lim_n \|F - F_{(n)}\|_2 = 0$, we have $F = \text{constant}$. ∎

6.7.3 Asymptotics of $\|k_a\|_p^p$ as $a \to 0$

Show that for every $p > 0$,

$$\|k_a\|_p^p = \int_{\mathbb{T}} \frac{dm}{|1 - \bar{a}z|^p} = 1 + \left|\frac{pa}{2}\right|^2 (1 + (o(1))), \quad as\ a \to 0.$$

SOLUTION: Recall that for any $\gamma \in \mathbb{C}$ and $w \to 0$, we have

$$(1 + w)^\gamma = 1 + \gamma w + \frac{\gamma(\gamma - 1)}{2} w^2 (1 + o(1)).$$

Applying this with $\gamma = -p/2$ and $|1 - \bar{a}z|^2 = 1 + (|a|^2 - 2\,\mathrm{Re}(\bar{a}z))$, and taking into account

$$\int_{\mathbb{T}} \mathrm{Re}(\bar{a}z)\, dm = 0, \qquad \int_{\mathbb{T}} (\mathrm{Re}(\bar{a}z))^2\, dm = |a|^2/2,$$

we obtain

$$\|k_a\|_p^p = \int_{\mathbb{T}} (1 + \gamma(|a|^2 - 2\,\mathrm{Re}(\bar{a}z)) + (\gamma(\gamma - 1)/2)(|a|^2 - 2\,\mathrm{Re}(\bar{a}z))^2 (1 + o(1)))\, dm$$

$$= 1 + \int_{\mathbb{T}} (\gamma|a|^2 + (\gamma(\gamma - 1)/2))4|a|^2/2)(1 + o(1))\, dm$$

$$= 1 + \gamma^2|a|^2(1 + o(1)) = 1 + \left|\frac{pa}{2}\right|^2 (1 + (o(1))). \quad ∎$$

6.7.4 Particular Features of the Multi-disk $\mathbb{D}_2^\infty$

(a) *Let $\lambda = (\lambda_1, \lambda_2, \dots) \in \mathbb{D}_2^\infty$. Show that $K_\lambda \in H^\infty(\mathbb{D}_2^\infty)$ if and only if $\lambda \in l^1$.*

SOLUTION: By § 6.6.7(b),

$$K_\lambda(\zeta) = \sum_{\alpha \in \mathbb{Z}_+(\infty)} \bar{\lambda}^\alpha \zeta^\alpha = \prod_{s \geq 1} k_{\lambda_s}(\zeta_s),$$

hence

$$\|K_\lambda\|_{H^\infty(\mathbb{D}_2^\infty)} \leq \prod_{s \geq 1} \|k_{\lambda_s}\|_{H^\infty(\mathbb{D})} = \prod_{s \geq 1} (1 - |\lambda_s|)^{-1},$$

$$\|K_\lambda\|_{H^\infty(\mathbb{D}_2^\infty)} \geq \lim_n \|K_\lambda\|_{H^\infty(\mathbb{D}^n)} = \lim_n \prod_{s=1}^n \|k_{\lambda_s}\|_{H^\infty(\mathbb{D})} = \prod_{s\geq1}(1-|\lambda_s|)^{-1},$$

thus $\|K_\lambda\|_{H^\infty(\mathbb{D}_2^\infty)} = \prod_{s\geq1}(1-|\lambda_s|)^{-1}$, and the result follows. ∎

(b) *Show with an example that there exists a function $F \in H^2(\mathbb{D}_2^\infty)$ such that for any r, $0 < r < 1$, we have $F_r \notin H^\infty(\mathbb{D}_2^\infty)$ where $F_r(\zeta) = F(r\zeta)$.*

SOLUTION: $F = K_\lambda$ where $\lambda \in \mathbb{D}_2^\infty \setminus l^1$, for example $\lambda = (1/2, 1/3, 1/4, \dots)$. Then, by (a), $(K_\lambda)_r = K_{r\lambda} \notin H^\infty(\mathbb{D}_2^\infty)$. ∎

(c) *Show that $U(H^\infty(\mathbb{D})) \not\subset H^\infty(\mathbb{D}_2^\infty)$.*

Hint We admit the following profound theorem of Green and Tao (2008): the sequence $P = (p_j)_{j\geq1}$ of prime integers contains arbitrarily long (finite) arithmetic progressions.

SOLUTION: It is easy to see (verify!) that if the inclusion $U(H^\infty(\mathbb{D})) \subset H^\infty(\mathbb{D}_2^\infty)$ were to hold, the mapping $U \colon H^\infty(\mathbb{D}) \to H^\infty(\mathbb{D}_2^\infty)$ would be closed, hence bounded (closed graph theorem: Appendix E), and thus there would exist $C > 0$ such that $\|Uf\|_{H^\infty(\mathbb{D}_2^\infty)} \leq C\|f\|_{H^\infty(\mathbb{D})}$ for every function $f \in H^\infty(\mathbb{D})$. Show that the last majoration is impossible: let $J \subset P = (p_j)_{j\geq1}$ be a finite subset of primes and $f = \sum_{j\in J} a_j z^j$, then $\|Uf\|_{H^\infty(\mathbb{D}_2^\infty)} = \sup_{|\zeta_j|<1}\left|\sum_{j\in J} a_j\zeta_j\right| = \sum_{j\in J}|a_j|$. Using the result of Green and Tao cited above, we define $J = \{m + qj\colon 0 \leq j \leq 2n\} \subset P$ and $f_n = z^m p_n(z^q)$ where p_n is a polynomial of Exercise 5.6.2(c). Then $\|f_n\|_\infty \leq 5$ and $\|Uf_n\|_{H^\infty(\mathbb{D}_2^\infty)} = \|p_n\|_{W_a(\mathbb{D})} \geq \log(n+1)$ (by Exercise 5.6.2(c)). Conclusion: $U(H^\infty(\mathbb{D})) \not\subset H^\infty(\mathbb{D}_2^\infty)$. ∎

6.7.5 A Few Cyclic Functions in $H^2(\mathbb{D}_2^\infty)$

(a) *Let a function $F \in H^2(\mathbb{D}_2^\infty)$ be such that $\mathrm{Re}(F(\zeta)) \geq 0$ for every $\zeta \in \mathbb{D}_2^\infty$. Show that F is M_ζ-cyclic.*

SOLUTION: For any $\epsilon > 0$, the function $\zeta \mapsto 1/(F(\zeta)+\epsilon)$ is in $H^\infty(\mathbb{D}_2^\infty)$. By § 6.6.5(g), $F/(F + \epsilon) \in E_F$. Since $|F(\zeta)/(F(\zeta) + \epsilon)| \leq 1$ and $\lim_{\epsilon\to0}(F(\zeta)/(F(\zeta) + \epsilon)) = 1$ for every $\zeta \in \mathbb{D}_2^\infty$, we have (for the weak topology) $1 = \lim_{\epsilon\to0}(F/(F + \epsilon)) \in E_F$. ∎

(b) *Let $F \in H^2(\mathbb{D}^n) \subset H^2(\mathbb{D}_2^\infty)$. The following assertions are equivalent.*

(i) *F is M_ζ-cyclic in $H^2(\mathbb{D}_2^\infty)$.*
(ii) *the function F is $(\zeta^\alpha)_{\alpha\in\mathbb{Z}_+^n}$-cyclic in $H^2(\mathbb{D}^n)$.*

SOLUTION: (ii) $\Rightarrow$ (i) is evident. For the converse, observe that for every $\alpha \in \mathbb{Z}_+(\infty) \setminus \mathbb{Z}_+^n$ we have $\zeta^\alpha \perp H^2(\mathbb{D}^n)$, and hence for every polynomial $q = q(\zeta) = \sum_{\alpha \in \mathbb{Z}_+(\infty)} c_\alpha \zeta^\alpha$ we have $P(qF) = P(q)F$ where P is the orthogonal projection on $H^2(\mathbb{D}^n)$, and in particular, $P(q)(\zeta) = \sum_{\alpha \in \mathbb{Z}_+^n} c_\alpha \zeta^\alpha$. Moreover, if $\lim_k \|q_k F - 1\|_2 = 0$ then $\lim_k \|P(q_k F) - 1\|_2 = 0$. $\blacksquare$

(c) *Let $f \in H^2(\mathbb{D})$, $\alpha \in \mathbb{Z}_+(\infty)$ and $F(\zeta) = f(\zeta^\alpha)$. The following assertions are equivalent.*

(i) *F is M_ζ-cyclic in $H^2(\mathbb{D}_2^\infty)$.*
(ii) *the function f is outer (thus, cyclic in $H^2(\mathbb{D})$).*

In particular, a function $f = \sum_{k\geq 0} a_k z^{n^k}$, where $n \in \mathbb{N} \setminus \{1\}$, is (T_n)-cyclic if and only if the function $\varphi = \sum_{k \geq 0} a_k z^k$ is outer.

SOLUTION: The same reasoning as for (b), but replacing $\mathbb{Z}_+^n$ by $\{\alpha k : k \in \mathbb{Z}_+\}$. $\blacksquare$

(d) *By using the Theorem of § 6.6.6(a), show the following.*

(i) *$f = a_1 z + a_2 z^2 + a_3 z^3$ is (T_n)-cyclic if and only if*

$$|a_1| \geq |a_2| + |a_3|.$$

(ii) *$f = a_1 z + a_2 z^2 + a_3 z^3 + a_4 z^4$ is (T_n)-cyclic in H_0^2 if and only if*

$$q(\mathbb{D}) \cap a_3 \mathbb{D} = \emptyset,$$

where $q(z) = a_1 + a_2 z + a_4 z^2$. In particular, the condition
$|a_1| \geq |a_2| + |a_3| + |a_4|$ is sufficient, but not necessary (consider the case
$a_3 = 0$).

(iii) *$f = a_1 z + a_2 z^2 + a_3 z^3 + a_4 z^4 + a_5 z^5$ is (T_n)-cyclic if and only if*

$$q(\mathbb{D}) \cap (|a_3| + |a_5|)\mathbb{D} = \emptyset,$$

where $q(z) = a_1 + a_2 z + a_4 z^2$.

(iv) *$f = z(\lambda - z)^N$, where $|\lambda| > 1$ and $N \in \mathbb{N}$, is (T_n)-cyclic if*

$$N < \frac{\log 2}{\log(1 + 1/|\lambda|)},$$

and is not cyclic if $N > \lambda > 0$.

(v) *The polynomials*

$$p_1 = a_0 z + a_1 z^2 + a_2 z^4 + a_3 z^8 + a_4 z^{16}$$

and

$$p_2 = a_0 z + a_1 z^{12} + a_2 z^{144} + a_3 z^{1728} + a_4 z^{20736}$$

are (T_n)-cyclic or not simultaneously (and if and only if the function $\varphi = \sum_{k=0}^{4} a_k z^k$ is outer, i.e. has all its roots in $\mathbb{C} \setminus \mathbb{D}$). However – in general – this is not the case for the pair p_1, p_3, where

$$p_3 = a_0 z + a_1 z^2 + a_2 z^3 + a_3 z^4 + a_4 z^5.$$

6.7.6 A Function $(\mathcal{D}_n)$-cyclic in $L^2(0, 1)$
(Kozlov, 1950; Akhiezer, 1965)

Let f be an odd 2-periodic function such that $f(x) = 1$, $0 < x < 1$. Show that f is $(\mathcal{D}_n)$-cyclic in $L^2(0, 1)$.

SOLUTION: We have

$$f = \sum_{k \geq 0} \frac{4}{\pi(2k + 1)} \sin(2k + 1)\pi x,$$

hence after the change of basis in § 6.6.2 we obtain

$$f = \sum_{k \geq 0} \frac{2\sqrt{2}}{\pi(2k + 1)} z^{2k+1} \quad \text{and} \quad Uf = \frac{2\sqrt{2}}{\pi} \sum_{\alpha \geq 0} \zeta^\alpha \lambda^\alpha,$$

where $\lambda = (0, 1/p_2, 1/p_3, \dots)$. Thus Uf is M_ζ-cyclic in $H^2(\mathbb{D}_2^\infty)$ by § 6.6.7(b). ∎

6.8 Notes and Remarks

The name "Riemann ζ function" (dominant in mathematics) is questionable. Indeed, the author of the definition of the function $\zeta(s) = \sum_{n \geq 1}(1/n^s)$, as well as of its fundamental properties (its multiplicative representation in § 6.1.2 and the functional equation of Theorem 6.1.5) is known: it was Leonhard Euler. A presentation by Gelfond (1958) given in a colloquium devoted to Euler shows in a few lines that Euler's computation in his 1761 note is equivalent (for s real) to the functional equation rediscovered by Riemann in 1859. The fact that Riemann had extended the definition to the plane $\mathbb{C}$ and thereby found profound links with the prime numbers, adds nothing to the question of the discovery. For example, nobody would dream of attributing the invention of the airplane to Willy Messerschmitt (1944) rather than to the brothers Wilbur and Orville Wright (1903) under the pretext that jet planes – whose first series production was the work of Messerschmitt – now dominate the sky; or, to remain within the subject of this book, the Fourier transform does not cease to be Fourier's simply because in the framework of the space $H^2(\mathbb{C}^+)$ it is essentially complex. The name *Riemann ζ function* is apparently due to Helge von Koch

(1902), "Ueber die Riemann'sche Primzahlfunction" (according to the website "Earliest Known Uses of Some of the Words of Mathematics," http://jeff560 .tripod.com/mathword.html). One can hope that this historical error will soon be corrected. For very convincing mathematical and historical arguments for the same point of view see also Blagouchine (2018) and the references therein.

The properties in § 6.1.1–6.1.2 and Lemma 6.1.3 form the standard basis of the theory of the function ζ: the product in § 6.1.2 is due to Euler (1737, published in 1743), and the integral representations of $\zeta(s)$ originate with Riemann (1859). Today numerous integral formulas are known (the web page http://functions.wolfram.com/Zeta contains 198 representative formulas for zeta). The functional equation is implicitly presented by Euler, especially for integer s, but also for certain rationals (1761: see the explanations of Gelfond (1958, pp. 89–90), mentioned above), and then reappears with Riemann (1859), in the complex domain and with two different proofs. Today several proofs are known; that of § 6.1.5 is a combination of a proof by Titchmarsh (1951, Ch. II, § 1) and one by Hardy (1922). In total, Titchmarsh (1951) gives seven different proofs of the functional equation.

The Riemann hypothesis (§ 6.1.3) is the most celebrated unsolved problem in mathematics. It was part of Hilbert's famous list of 23 problems for the twentieth century (presented in his speech at the 1900 International Congress of Mathematicians in Paris); more precisely, problem no. 8 of the list mentions the Riemann hypothesis, the Goldbach conjecture, and the twin prime problem. According to an anecdote, Hilbert jokingly stated that if 500 years after his death he had the right to return to this world for 30 seconds, he would use it to pose the question "Has RH been resolved?"

The RH provoked an enormous amount of activity in mathematics throughout the twentieth century, and in the year 2000 was included in the list of seven "Millennium Prize Problems" (problem no. 4 of the list; the resolution of each of the problems is rewarded with a prize of a million US dollars offered by the Clay Mathematics Institute (Cambridge, USA)). The literature on the problem is immense; the classical sources remain the books by Titchmarsh (1951), Landau (1927), and Hardy and Wright (1938). For modern surveys see, for example, the official presentation for the Millennium Prize by Enrico Bombieri (2000) and a summary of the latest advances by Peter Sarnak (2005); both are available on the website of the Clay Institute, www.claymath.org/millennium-problems/riemann-hypothesis. See also Conrey (2003), a synthesis article by Pérez-Marco (2011), and for a "light literary" history of the RH, Sabbagh (2002).

The approach to the RH by approximation, presented in § 6.2–6.5, is one of the dozens of equivalent forms of the RH; several of them are collected on the website of the American Institute of Mathematics (Palo Alto, California), www.aimath.org/WWN/rh/; for more, see Pérez-Marco (2011). Among the more curious conjectures equivalent to the RH are

- $LCM(1, 2, 3, \ldots, n) \le e^{n + \frac{1}{8\pi} \sqrt{n} \log^2 n}$ for every $n \ge 74$ (Schoenfeld, 1976);
- $\sum_{1 \le j \le A} (b_j - j/A)^2 = O(N^{-1+\epsilon})$, $\forall \epsilon > 0$, as $N \to \infty$; here $1/N = b_1 < b_2 < \cdots < b_A = 1$ are the fractions of John Farey (1816, an English geologist) defined by $\{b_1 < b_2 < \cdots < b_A\} = \{h/k \colon 0 < h \le k \le N\}$ (Franel, 1924).

These last two forms of the RH, as well as commentaries, references, and 17 further equivalent forms can be found in a survey by Balazard (2010).

The approach of § 6.2–6.5 was proposed by Nyman (1950). (Note that Bertil Nyman remains a very enigmatic figure: after a brilliant thesis, he completely disappeared from the mathematical world.) Several other conjectures equivalent to RH are expressed in the language of approximations. For example, Norbert Wiener (him again!) mentioned in his famous work on the Tauberian theorems (Wiener, 1932) that for every σ, $0 < \sigma < 1$, the completeness of the translations $(\tau_s f_\sigma)_{s \in \mathbb{R}}$ in the space $L^1(\mathbb{R})$, where

$$f_\sigma(x) = e^{(\sigma-1)x} \left(\frac{e^x}{e^{e^x} - 1} \right)',$$

is equivalent to the fact that, for every $t \in \mathbb{R}$, $\zeta(\sigma + it) \ne 0$. Then, Salem (1953) showed that this last property is also equivalent to the completeness of the dilations $(\mathcal{D}_t g_\sigma)_{t>0}$ in $L^1(\mathbb{R}_+)$, where

$$g_\sigma(x) = x^{\sigma-1}(e^x + 1)^{-1}.$$

Levinson (1956) provided a similar criterion for the absence of zeros of $\zeta(s)$ for $1/2 \le \sigma_1 < \mathrm{Re}(s) < \sigma_2 \le 1$. The form of Nyman's criterion is particularly advantageous because it introduces the ability to use the classical techniques of the space L^2 on a compact interval. Moreover, the criterion of Theorem 6.4.1 was considerably reinforced by Báez-Duarte (2003), as is mentioned in § 6.6, bringing it closer to the Wintner–Beurling problem treated in § 6.6.1–6.6.7.

Part (2) of Theorem 6.2.1 is taken from Nyman (1950); the quantitative part (1) is a simple clarification of Nyman's reasoning. It is nonetheless interesting to compare it numerically (as well as with the estimations in § 6.5) with the best bounds previously known, obtained by a "heavy artillery" approach due to Korobov and Vinogradov (1958; see Ford (2002) for a modern presentation), specifically with the fact that $\zeta(s) \ne 0$ in the domain

$$\Omega = \left\{ s = x + iy \colon x = \mathrm{Re}(s) \ge 1 - \frac{1}{(57.54)(\log |y|)^{2/3}(\log \log |y|)^{1/3}} \right\}, \quad |y| \ge 3.$$

It seems that at the level $y = 3$ the disk

$$D\left(\frac{1}{d^2}, r\right), \quad r = \frac{1}{d}\sqrt{\frac{1}{d^2} - 1}$$

of Theorem 6.2.1(1) is better than Ω if $0 < d < 1/\sqrt{10} \approx 1/3$.

The classical Paley–Wiener theorem § 6.3.2–6.3.3(a) is presented following Nikolski (1980, 1986), but under a "weak form," hence without a description of the spaces $H^2(\mathbb{C}_+)$, $H^2(\mathbb{C}^+)$ by quadratic means:

$$H^2(\mathbb{C}_+) = \left\{ f \in \mathrm{Hol}(\mathbb{C}_+) : \sup_{y>0} \int_{\mathbb{R}} |f(x + iy)|^2\, dx < \infty \right\}.$$

As is well known, to reach such a description we need a kind of embedding theorem, for example that of Gabriel and Zygmund mentioned in § 2.9. The assertions of § 6.3.2(c), due to Peter Lax, are heavily used in diffusion theory: see Lax and Phillips (1967).

The corollaries of § 6.3.4 touch on a very important theme in analysis, with several applications to harmonic analysis, signal processing, and stochastic processes; see, for example, Nikolski (2002) and Rozanov (1963). We must point out that, as in the case of the circle $\mathbb{T}$ (see Theorems 2.7.4 and 2.7.5), there exists in addition the question of the completeness of the exponentials, the translations, and the dilations in a corresponding "bilateral" space ($L^2(\mathbb{R})$ or $L^2(\mathbb{R}_+, dx/x)$). For example, in the same manner as in Chapter 2, using the results of this chapter we obtain: (i) the translations $(\tau_s f)_{s\in\mathbb{R}}$ generate the space $L^2(\mathbb{R})$ if and only if $\mathcal{F}f \neq 0$ a.e. on $\mathbb{R}$; (ii) the translations $(\tau_s f)_{s\in\mathbb{R}_+}$ generate the space $L^2(\mathbb{R})$ if and only if $\mathcal{F}f \neq 0$ a.e. on $\mathbb{R}$ and

$$\int_{\mathbb{R}} \frac{\log|\mathcal{F}f|}{1 + x^2}\, dx = -\infty$$

(Paley and Wiener, 1934).

Section 6.5 follows Nikolski (1995).

Section 6.6 is an excerpt from Nikolski (2012). The problem of the completeness of the integer dilations $(\mathcal{D}_n f)_{n\geq 1}$ appears naturally in view of the theorem of Báez-Duarte (2003) cited in § 6.6. We mention that Bagchi (2006) further simplified the form of this last criterion for RH by reducing it to the following proposition.

Theorem (Bagchi, 2006) *Let $l^2(1/n^2)$ be the weighted space*

$$l^2(1/n^2) = \left\{ x = (x_n)_{n\geq 1} : \sum_{n\geq 1} |x_n|^2 n^{-2} < \infty \right\},$$

and $x_l = (\rho(n/l))_{n\geq 1}$ ($\rho(x) = x - [x]$), $x_l \in l^2(1/n^2)$, $l = 2, 3, \ldots$ The following assertions are equivalent.

(1) *The RH is correct.*

(2) $\operatorname{span}_{l^2(1/n^2)}(x_l : l = 2, 3, \ldots) = l^2(1/n^2)$.

(3) $1 = (1, 1, \ldots) \in \operatorname{span}_{l^2(1/n^2)}(x_l : l = 2, 3, \ldots)$.

(4) *The same as (2) and/or (3) but for x_l with l not containing any squares.*

For its proof, this even more elementary form of Nyman's theorem requires nonetheless additional techniques of the theory of the function ζ: a decomposition in a Dirichlet series of $1/\zeta(s)$, and then the theorems of Lindelöf, Littlewood, the functional equation of ζ, etc.

In reality, the problem of completeness of the integer dilations $(\mathcal{D}_n f)_{n\geq 1}$ appeared long before these results, namely in Wintner (1944), motivated by certain problems in Diophantine analysis and where the first profound results were obtained. Independently, Beurling (1945) presented this problem to a seminar in Uppsala. The importance of the Wintner–Beurling problem was widely recognized in the 1950s (see, for example, the important publications of Bourgin (1946) and Kozlov (1948, 1950) who, of course, had no knowledge of Beurling's seminar) but virtually forgotten for another 40 years. Concerning the renewal of interest in the completeness of the dilations (which can, *in fine*, help clarify the RH), see Hedenmalm et al. (1997, 1999) and references therein.

The Bohr transform, § 6.6.3, was introduced in Bohr (1913), but for a study of the Dirichlet series $\sum_{n\geq 1}(a_n/n^s)$ (always linked to the cluster of ideas around the RH). The article of Hilbert (1909) mentioned in § 6.6.4 was only a research plan that subsequently was (partially) successful. Hilbert strongly insisted on the absolute convergence of the power series of an infinite number of variables, without which the subject becomes quite fuzzy. Corollary 6.6.3 is the principal result of Beurling's presentation in Uppsala (Beurling, 1945), which was proved without passing to the space $H^2(\mathbb{D}_2^\infty)$. The last remark also applies to the result § 6.6.6(a) (Neuwirth et al., 1970). The lemma § 6.6.6(b) is a somewhat improved form of a result from the last article (which was later rediscovered by several authors). Corollary § 6.6.7(c) is from the founding article by Wintner (1944) (with a different proof), where Wintner was motivated by problems linked to the Sieve of Eratosthenes (hence, by the distribution of prime numbers).

Most of the propositions of § 6.7 are borrowed from Nikolski (2012). The example §6.7.6 is a special case of the results of Kozlov (1950), where the following problem was posed. *Let θ, $0 < \theta \leq 1$, and let f_θ be an odd 2-periodic function such that for $0 < x < 1$, $f_\theta(x) = \chi_{(0,\theta)}(x)$; give a criterion of $(\mathcal{D}_n)$-*

cyclicity of f_θ (as a function of θ). Kozlov stated that f_θ is cyclic for $\theta = 1$ (this is §6.7.6; in fact, this theorem was proved in Akhiezer (1965, Section "Additions and Problems," §I.23), but with a (fairly long) proof completely different to ours), $\theta = 1/2$ and $\theta = 2/3$, and is not cyclic for $\theta = 1/3$ (and for all the θ in a neighborhood of $1/3$), as well as for θ admitting a representation $\theta = q/p$ where $p > 2$ is a prime and q odd such that $\tan^2(q\pi/2p) < 1/p$ (but this condition is not satisfied for $q = 1$, $p = 3$, which corresponds to $\theta = 1/3$).

Appendix A
Key Notions of Integration

A.1 Measures

Let Ω be a set, and $\mathcal{A}$ a σ-algebra on Ω, so that $(\Omega, \mathcal{A})$ is a measurable space. A *positive measure* on $\mathcal{A}$ (or on Ω) is a countably additive mapping $\mu \colon \mathcal{A} \to \overline{\mathbb{R}}_+ = \mathbb{R}_+ \cup \{\infty\}$, such that $\mu(\bigcup A_j) = \sum_j \mu(A_j)$ for every disjoint sequence (A_j) $(A_j \cap A_k = \emptyset, j \neq k)$. A triplet $(\Omega, \mathcal{A}, \mu)$ is called a *measure space*.

A *complex measure* is a countably additive mapping $\mu \colon \mathcal{A} \to \mathbb{C}$. The set of complex measures is denoted $\mathcal{M}(\Omega)$ (when the σ-algebra is evident).

- A function $\mu \colon \mathcal{A} \to \overline{\mathbb{R}}_+$ which is additive and countably sub-additive (i.e. $\mu(\bigcup A_j) \leq \sum_j \mu(A_j)$ for every sequence (A_j)) is a measure.

- A complex measure μ is always bounded, $\sup_{A \in \mathcal{A}} |\mu(A)| < \infty$, and is of *finite total variation* $\|\mu\| = \mathrm{Var}(\mu) := \sup \sum_j |\mu(A_j)|$ where the sup is taken over all disjoint finite families (A_j). The *variation of μ* is a measure $|\mu|$ defined by $|\mu|(A) = \|\mu|A\|$, $A \in \mathcal{A}$. If $\mu \geq 0$ then $|\mu| = \mu$.

- A real measure μ admits a unique representation of the form $\mu = \mu_1 - \mu_2$, where $\mu_j \geq 0$ and there exists $A \in \mathcal{A}$ such that $\mu_1(A) = 0$, $\mu_2(\Omega \backslash A) = 0$. Moreover, $|\mu| = \mu_1 + \mu_2$.

- A complex measure μ admits a representation of the form $\mu = \mu_1 - \mu_2 + i\mu_3 - i\mu_4$, where $\mu_j \geq 0$.

- The set of positive measures on Ω is a *lattice*: for every sequence (μ_n), $\mu_n \geq 0$, there exists a unique $\mu = \sup_n \mu_n$ such that for every n, $\mu_n \leq \mu$, and if $\mu_n \leq \nu$ for every n, then $\mu \leq \nu$. The measure μ is given by

$$\mu A = \sup \sum_{k \geq 1} \mu_k A_k,$$

where the sup is taken over all disjoint partitions $A = \bigcup_k A_k$ ($A_k \cap A_j = \emptyset$, $k \neq j$). Similarly, there exists $\nu = \inf_n \mu_n$, with

$$\nu A = \inf \sum_{k \geq 1} \mu_k A_k.$$

- Let $P(x)$ be a property defined for $x \in \Omega$; P is said to hold μ-*almost everywhere* (abbreviated as μ-*a.e.*) if $\mu(x \in \Omega : P(x)$ does not hold$) = 0$.

- *Measures on a topological space.* A topological space Ω, by default, is equipped with the Borel σ-algebra, $\mathcal{A} = \mathcal{B}$: this is the σ-algebra generated by the open subsets (or, likewise, by the closed subsets) of Ω. A measure on $\mathcal{B}$ is said to be a Borel measure. It is called *regular* if $\forall B \in \mathcal{B}$, $\forall \epsilon > 0$, $\exists A$ closed, $\exists C$ open, such that $A \subset B \subset C$ and $\mu(C \setminus A) < \epsilon$.

- If the space Ω is locally compact and σ-compact and if μ is a Borel measure which is finite on every compact subset, then μ is regular; in particular, any locally finite measure in $\mathbb{R}^n$ is regular.

- The *(closed) support* $\mathrm{supp}(\mu)$ of a Borel measure on Ω is defined by $\Omega \setminus \mathrm{supp}(\mu) = \bigcup O$ (O runs over all the open subsets such that $\mu(O) = 0$).

A.2 The Lebesgue Integral

Let $(\Omega, \mathcal{A}, \mu)$ be a measure space. A function $f : \Omega \to \mathbb{C}$ is said to be *measurable* if $f^{-1}(B) \in \mathcal{A}$ for every Borel set $B \subset \mathbb{C}$ (or, equivalently, for every rectangle (a product of intervals) $B \subset \mathbb{R}^2 = \mathbb{C}$). The *integral* of a positive measurable function is

$$\int_\Omega f \, d\mu := \sup \left\{ \sum_{k=1}^n c_k \mu A_k : 0 \leq c_k \leq f(x), x \in A_k; A_k \cap A_j = \emptyset (k \neq j) \right\},$$

f is *integrable* if $\int_\Omega f \, d\mu < \infty$. A function $f : \Omega \to \mathbb{R}$ is *integrable* if $f^\pm := \max(0, \pm f)$ are integrable, and we set

$$\int_\Omega f d\mu = \int_\Omega f^+ \, d\mu - \int_\Omega f^- \, d\mu.$$

A function $f : \Omega \to \mathbb{C}$ is *integrable* if $\mathrm{Re}(f)$ and $\mathrm{Im}(f)$ are integrable, and we set $\int_\Omega f \, d\mu = \int_\Omega \mathrm{Re}(f) \, d\mu + i \int_\Omega \mathrm{Im}(f) \, d\mu$.

Henri Lebesgue (1875–1941) was a French mathematician and creator of the modern theory of integration, which changed the face of mathematics. With origins in a modest provincial background, he followed (thanks to the efforts of his mother) the complete cycle of French education, including the preparatory classes at the Lycée Louis-le-Grand in Paris, and then the École Normale Supérieure (1897). Upon graduation, he obtained only a modest position as a high school teacher (in the Lycée Central in Nancy), in 1899–1902, when he wrote his famous article "Sur une généralisation de l'intégrale définie" (*Comptes rendus de l'Académie des sciences* (1902)) and prepared his thesis "Intégrale, longueur, aire" (130 pages) submitted in Paris in 1902. Measure and integration theory were thus created, stimulating an explosion of developments in harmonic analysis and in mathematics in general. Even though his work met a fairly hostile reception in France (he had a prolonged rivalry with Baire, and only obtained his first university position in Paris, *maître de conférences* at the Sorbonne, in 1910), his new theory rapidly gained ground internationally. After 10–15 years, those areas of mathematics that required the Lebesgue integral (with the enthusiastic participation of Hardy, Littlewood, Frigyes and Marcel Riesz, Hausdorff, Steinhaus, Borel, Denjoy, Fatou, Nikodym, Banach, Plancherel, Luzin, Kolmogorov, Radon, Saks, Haar, etc.) had changed beyond all recognition. Lebesgue published two important monographs on the subject: *Leçons sur l'intégration et la recherche des fonctions primitives* (1904) and *Leçons sur les séries trigonométriques* (1906). But he was not content to limit himself to the pure theory of integration – after all, he was the author of the famous saying: *Réduites aux théories*

générales, les mathématiques deviendraient une belle forme sans contenu ("Reduced to general theories, mathematics would become a beautiful framework without content"). Lebesgue made a variety of important contributions: in topology, in potential theory, on the Dirichlet problem, in the calculus of variations, in set theory, and in the theory of dimensions. In 1922 he published a summary of his 90 articles and books, *Notice sur les travaux scientifiques de M. Henri Lebesgue* – a work of synthesis with certain evaluations of his major results.

Paul Montel described his final days: "At the beginning of 1941, Henri Lebesgue gave his last annual course at the Collège de France. Already, the sickness that took him a few months later added to the low morale caused by the defeat and the enemy occupation. He could barely walk, and the city was severely lacking in surface transport. In order to give his lectures, he had to rely on the wheelchairs and bicycles that were used to transport the sick."

Lebesgue was elected member of several Academies: l'Académie des Sciences (Paris), the Royal Society, the Belgian Académie Royale, the Academy of Bologna, the Accademia dei Lincei (Rome), the Romanian Academy of Science, and the Krakow Academy of Sciences.

- The *set of integrable functions* $\mathcal{L}^1(\mu)$ is a vector space and the integral

$$f \longmapsto \int_\Omega f \, d\mu$$

  is a linear functional on $\mathcal{L}^1(\mu)$ satisfying

$$\left| \int_\Omega f \, d\mu \right| \le \int_\Omega |f| \, d\mu.$$

  If $f, g \in \mathcal{L}^1(\mu)$ are real and $f(x) \le g(x)$ μ-a.e., then

$$\int_\Omega f \, d\mu \le \int_\Omega g \, d\mu.$$

- If $\mu \in \mathcal{M}(\Omega)$ with the decomposition $\mu = \mu_1 - \mu_2 + i\mu_3 - i\mu_4$, where $\mu_j \ge 0$ (see above), and if $f \in \mathcal{L}^1(|\mu|)$, then we set

$$\int_\Omega f \, d\mu = \int_\Omega f \, d\mu_1 - \int_\Omega f \, d\mu_2 + i \int_\Omega f \, d\mu_3 - i \int_\Omega f \, d\mu_4.$$

- *Passage to the limit.*

(1) The *Beppo Levi Theorem.* If $f_n(x) \nearrow f(x)$ and $f_n(x) \geq 0$ μ-a.e., then

$$\lim_n \int_\Omega f_n d\mu = \int_\Omega f \, d\mu.$$

The same holds for $f_n \searrow f$ if we suppose that $\int_\Omega f_1 \, d\mu < \infty$.

(2) The *Lebesgue dominated convergence theorem.* If $\lim_n f_n(x)$ exists for almost all $x \in \Omega$, and for every n, $|f_n| \leq f$ μ-a.e. with $f \in \mathcal{L}^1(\mu)$, then

$$\lim_n \int_\Omega f_n \, d\mu = \int_\Omega (\lim_n f_n) \, d\mu.$$

(3) *Fatou's lemma.* If $f_n \geq 0$, then

$$\int_\Omega (\underline{\lim}_n f_n) \, d\mu \leq \underline{\lim}_n \int_\Omega f_n \, d\mu.$$

- *Integrals depending on a parameter.* Let K be a metric space, and let $f: \Omega \times K \to \mathbb{C}$ be a mapping such that, for all $t \in K$, $f(\cdot, t) \in \mathcal{L}^1(\mu)$, and

$$F(t) = \int_\Omega f(x, t) \, d\mu(x), \quad t \in K.$$

(1) *Continuity.* If there exists a function $h \in \mathcal{L}^1(\mu)$ such that $\forall t \in K$, $|f(x, t)| \leq h(x)$ μ-a.e. and $\lim_{t \to t_0} f(x, t) = f(x, t_0)$ μ-a.e., then F is continuous at the point t_0.

(2) *Differentiability.* Let $K \subset \mathbb{R}$ be an open set, and suppose that for every $(x, t) \in \Omega \times K$ there exist functions

$$g(x, t) := \frac{\partial f(x, t)}{\partial t}$$

and $h \in \mathcal{L}^1(\mu)$ such that $\forall t \in K$, $|g(x, t)| \leq h(x)$ μ-a.e. Then F is differentiable on K, and $F'(t) = \int_\Omega g(x, t) \, d\mu(x)$.

(3) *Holomorphy.* Let K be an open subset of $\mathbb{C}$, $t \longmapsto f(x, t)$ holomorphic in K and $|f(x, t)| \leq h(x)$ for every $(x, t) \in \Omega \times K$ where $h \in \mathcal{L}^1(\mu)$. Then F is holomorphic on K.

- *Primitive of a integrable function and Lebesgue points.* Let $I \subset \mathbb{R}$ be an interval and $f \in \mathcal{L}^1(I, dx)$.

(i) For almost every point $x \in I$, $\lim_{h \to 0} h^{-1} \int_0^h |f(x) - f(x + t)| \, dt = 0$ (such an x is called a *Lebesgue point* of f).

(ii) At every Lebesgue point x, $f(x) = \lim_{h \to 0} h^{-1} \int_0^h f(x + t) \, dt$.

A.3 Lebesgue Decomposition and the Radon–Nikodym Theorem

Let $(\Omega, \mathcal{A}, v)$ be a measure space.

- *Lebesgue decomposition.* Let $\mu \in M(\Omega)$. There exists a unique decomposition $\mu = \mu_a + \mu_s$ where μ_a, μ_s are two measures such that:

  (i) $\forall A \in \mathcal{A}$, $v(A) = 0 \Rightarrow \mu_a(A) = 0$; μ_a is said to be *absolutely continuous* with respect to v; this is denoted by

  $$\mu_a \ll v.$$

  (ii) $\mu_s \perp v$ in the sense that there exists $A \in \mathcal{A}$ such that $|\mu_s|(A) = 0$, $v(\Omega \setminus A) = 0$ (μ_s is said to be *singular* with respect to v). In fact, $\mu_a = \chi_A \mu$, $\mu_s = (1 - \chi_A)\mu$.

- The *Radon–Nikodym theorem.* Let $\mu \in M(\Omega)$. Then $\mu \ll v \Leftrightarrow$ there exists a function $h \in \mathcal{L}^1(\mu)$ such that $\mu(A) = \int_A h\,dv$ ($\forall A \in \mathcal{A}$); μ is called a *measure with density* h and can be written $\mu = hv$ (or $h = d\mu/dv$). We have $\int_\Omega f\,d\mu = \int_\Omega fh\,dv$ for every $f \in \mathcal{L}^1(|\mu|)$.

  We always have $\mu \ll |\mu|$ and $\epsilon = d\mu/d|\mu|$ is unimodular $|\mu|$-a.e.

A.4 The Riesz Representation Theorem

Let Ω be a compact space and $C(\Omega)$ the space of continuous functions on Ω equipped with the *uniform norm* $\|f\|_\infty = \sup_\Omega |f|$, and let φ be a linear functional on $C(\Omega)$. The following assertions are equivalent.

(1) φ is continuous (bounded).

(2) There exists a complex measure $\mu \in M(\Omega)$ such that, for every function $f \in C(\Omega)$, $\varphi(f) = \int_\Omega f\,d\mu$.

Such a measure μ is unique, and $\|\varphi\| = \|\mu\|$.

Note that Frigyes Riesz proved the theorem for $\Omega = [0, 1]$ (1909), Banach for metric spaces Ω (1933), and Kakutani for the general case (1941); see Rudin (1998) for comments.

- For a locally compact space Ω, the same statement holds for

  $$C_0(\Omega) = \{f \in C(\Omega) : \forall \epsilon > 0, \exists K \text{ a compact set such that} |f| < \epsilon \text{ on } \Omega \setminus K\}.$$

A.5 The Lebesgue $L^p(\mu)$ Spaces

Let $(\Omega, \mathcal{A}, \mu)$ be a measure space and $0 < p < \infty$. We define

$$\mathcal{L}^p(\mu) = \mathcal{L}^p(\Omega, \mu) := \left\{ f : \Omega \to \mathbb{C} \text{ measurable} : \int_\Omega |f|^p \, d\mu := N_p(f)^p < \infty \right\}.$$

$\mathcal{L}^p(\mu)$ is a vector space and, if $p \geq 1$, N_p is a seminorm on $\mathcal{L}^p(\mu)$.

- *The Lebesgue space* is the normed space $L^p(\mu) = \mathcal{L}^p(\mu)/R$ of $\mathcal{L}^p(\mu)$ modulo the equivalence relation $R(f, g) \Leftrightarrow f = g$ μ-a.e. It is a complete normed space, and hence a Banach space, equipped with the norm $\|F\|_p = N_p(f)$, $\forall f \in F \in L^p(\mu)$. For $p = \infty$,

$$\mathcal{L}^\infty(\mu) := \{ f : \Omega \to \mathbb{C} \text{ measurable} : N_\infty(f) = \inf\{\lambda > 0 : \mu(|f| > \lambda) = 0\} < \infty \}.$$

- *The distribution function and weak $\mathcal{L}^p$ spaces.* Let $f : \Omega \to \mathbb{C}$ be a measurable function and

$$\lambda_f(t) = \mu(x \in \Omega : |f(x)| \geq t), t > 0.$$

Then, $N_p(f)^p = p \int_0^\infty t^{p-1} \lambda_f(t) \, dt$, and if $f \in \mathcal{L}^p(\mu)$, then $\lambda_f(t) = o(t^{-p})$ when $t \to \infty$. The space $\mathcal{L}^{p,\infty}$ ("$\mathcal{L}^p$ weak") is defined as the set of functions f such that $\lambda_f(t) = o(t^{-p})$ when $t \to \infty$.

- *The L^p spaces and the Lebesgue decomposition (Radon–Nikodym).* Let $\mu = \mu_a + \mu_s$, where $\mu_a = \chi_A \mu$ and $\mu_s = (1 - \chi_A)\mu$, be the Lebesgue decomposition (see above). For a function $f \in L^p(\mu)$, by setting $f_a = \chi_A f$, $f_s = (1 - \chi_A)f$, we obtain

$$f_a \in L^p(\mu_a), \quad f_s \in L^p(\mu_s) \quad \text{and} \quad f = f_a + f_s, \ \|f\|^p_{L^p(\mu)} = \|f_a\|^p_{L^p(\mu_a)} + \|f_s\|^p_{L^p(\mu_s)}.$$

This is clearly a direct decomposition, denoted

$$L^p(\mu) = L^p(\mu_a) \oplus L^p(\mu_s) \quad \text{(direct sum of type } l^p\text{)}.$$

- *Hölder's inequality.* If $f_j \in \mathcal{L}^{p_j}(\mu)$, $p_j > 0$ $(j = 1, \ldots, n)$, and

$$\frac{1}{s} = \sum_1^n \frac{1}{p_j},$$

then $\prod_1^n f_j \in \mathcal{L}^s(\mu)$ and

$$\left\| \prod_1^n f_j \right\|_s \leq \prod_1^n \|f_j\|_{p_j}.$$

The classical special case: for $n = 2$, $1 = 1/p + 1/p'$, then

$$f \in \mathcal{L}^p(\mu), g \in \mathcal{L}^{p'}(\mu) \Rightarrow \|fg\|_1 \leq \|f\|_p \|g\|_{p'}.$$

For $p = 2$, this becomes the Cauchy–Schwarz inequality: $\|fg\|_1 \leq \|f\|_2 \|g\|_2$.

- *The converse of Hölder's inequality.* Let f be a measurable function and $1 \leq p \leq \infty$, then

$$f \in \mathcal{L}^p(\mu) \Leftrightarrow (fg \in \mathcal{L}^1(\mu), \forall g \in \mathcal{L}^{p'}(\mu));$$

  moreover, $\sup\{|\int_\Omega fg \, d\mu| : \|g\|_{p'} \leq 1\} = \|f\|_p$.

- *Jensen's convexity inequality.* If φ is a convex function defined on an interval $I \subset \mathbb{R}$ where a real function f takes its values ($f(\Omega) \subset I$), then for every positive finite measure μ,

$$\varphi\left(\frac{1}{\mu(\Omega)} \int_\Omega f \, d\mu\right) \leq \frac{1}{\mu(\Omega)} \int_\Omega \varphi \circ f \, d\mu.$$

  (This follows from the fact that $\varphi(x) = \sup\{L(x) : L \text{ linear and } L \leq \varphi\}$ and for linear L, the inequality is a trivial equality.)

- *Density of the polynomials.* Let μ be a finite Borel measure with compact support in $\mathbb{R}^n$. Then the polynomials $f(x) = \sum_{\alpha \geq 0} a_\alpha x^\alpha$, $x^\alpha = x_1^{\alpha_1} x_2^{\alpha_2} \ldots x_n^{\alpha_n}$, $\alpha = (\alpha_1, \ldots, \alpha_n) \in \mathbb{Z}_+^n$, are dense in $\mathcal{L}^p(\mu)$, $p < \infty$.

Outline of a direct proof: (a) the polynomials are dense in the space $C = C(\mathrm{supp}(\mu))$ (theorem of (Stone–)Weierstrass), hence it only remains to show that $\mathrm{clos}_{\mathcal{L}^p(\mu)}(C) = \mathcal{L}^p(\mu)$; (b) we show that for any compact set F, $\chi_F \in \mathrm{clos}_{\mathcal{L}^p(\mu)}(C)$ ($\chi_F = \lim_n f_n$ where $f_n(x) = (1 - \min(\mathrm{dist}(x, F), 1))^n$); then (c) for every $A \in \mathcal{B}$, $\chi_A \in \mathrm{clos}_{\mathcal{L}^p(\mu)}(C)$ (by the regularity of μ); finally (d) $\mathcal{L}^p(\mu) = \mathrm{clos}_{\mathcal{L}^p(\mu)}(C)$. ∎

A.6 Convolution and the Fourier Transform

If G is a locally compact commutative group (such as $\mathbb{T}^n$, $\mathbb{Z}^n$, $\mathbb{R}^n$), the *convolution* of two measures $\mu, \nu \in M(G)$ is defined using the Riesz representation theorem as the measure $\mu * \nu$ such that, for every $\varphi \in C_0(G)$,

$$\int_G \varphi \, d(\mu * \nu) = \int_G \int_G \varphi(x + y) \, d\mu(x) \, d\nu(y).$$

- We have $\mu * \nu = \nu * \mu$, $\|\mu * \nu\| \leq \|\mu\| \cdot \|\nu\|$, hence $M(G)$ is a commutative Banach algebra (see Appendix D) with unit δ_0, the *Dirac delta at the origin*.

- *The Fourier transform.* Let $\hat{G}$ be the dual group of the unimodular continuous multiplicative characters of G and $\mu \in M(G)$; the *Fourier transform of μ* is defined by

$$\mathcal{F}\mu(\gamma) = \int_G \gamma(-x)\, d\mu(x), \quad \gamma \in \hat{G}.$$

Remark For a reason of normalization (linked especially to Plancherel's theorem, see below), when we apply the definition to a measure $\mu = fm$ absolutely continuous with respect to the invariant measure m (Haar measure), we use an embedding $\mathcal{L}^1(m) \subset M(G)$, $f \longmapsto cfm$, selecting a constant c in order to have $\mathcal{F}(\mathcal{F}f) = f$ for certain test functions. In particular, $c = (2\pi)^{-n/2}$ in the case of $G = \mathbb{R}^n$, $c = (2\pi)^{-n}$ in the case of $\mathbb{T}^n = \mathbb{R}^n/\mathbb{Z}^n$, $c = 1$ in the case of $\mathbb{Z}^n$, hence

$$\mathcal{F}f(t) = \frac{1}{\sqrt{2\pi}} \int_{\mathbb{R}} f(x)e^{-ixt}\, dx, \quad f \in \mathcal{L}^1(\mathbb{R}), \ t \in \mathbb{R},$$

$$\mathcal{F}f(n) = \hat{f}(n) = \frac{1}{2\pi} \int_{\mathbb{R}/\mathbb{Z}} f(x)e^{-ixn}\, dx, \quad f \in \mathcal{L}^1(\mathbb{T}), \ n \in \mathbb{Z}.$$

- For every $\mu \in M(G)$, $\mathcal{F}\mu$ is bounded and uniformly continuous; for $f \in \mathcal{L}^1(m)$, $\lim_{\gamma \to \infty} \mathcal{F}f(\gamma) = 0$ (the *Riemann–Lebesgue lemma*, correct for (at least) the classical groups $\mathbb{T}^n$, $\mathbb{R}^n$).

- *Transfer formula.* For every $\mu \in M(G)$ and $v \in M(\hat{G})$, $\int_{\hat{G}} \mathcal{F}\mu\, dv = \int_G \mathcal{F}v\, d\mu$. This follows from Fubini's theorem.

- *Uniqueness theorem.* $\mathcal{F}\mu = 0 \Rightarrow \mu = 0$. (In the case $G = \mathbb{T}^n$, this follows from the preceding formula and Weierstrass's theorem.)

- For every $\mu, v \in M(G)$ and every $\gamma \in \hat{G}$, $\mathcal{F}(\mu * v)(\gamma) = \mathcal{F}\mu(\gamma)\mathcal{F}v(\gamma)$.

- The *Fourier–Plancherel transform.* With a proper normalization (mentioned above), $\mathcal{F}: (L^1(G) \cap L^2(G)) \to L^2(G)$ is an isometric mapping with a dense image, hence it can be extended in a unique manner to a unitary operator

$$\overline{\mathcal{F}}: L^2(G) \to L^2(G)$$

such that $\overline{\mathcal{F}}(\overline{\mathcal{F}}f)(x) = f(-x)$, $\forall f$, and hence $\overline{\mathcal{F}}^4 = \mathrm{id}$. For every $f \in L^2(G)$, $\lim_K \|\overline{\mathcal{F}}f - \mathcal{F}(f\chi_K)\|_2 = 0$, where K runs over the compact subsets "filling G" (for example, in the case of $\mathbb{R}$, K running over the intervals $[-t, t], t > 0$).

- *Convolution in $L^p(G)$.* Let $f \in L^p(G)$ and $\tau_s f(x) = f(x - s)$, $s \in G$; then the convolution $f * \mu = \int_G \tau_s f \, d\mu(s)$ is well-defined in $L^p(G)$, and $\|f * \mu\|_p \leq \|f\|_p \|\mu\|$.
- *Approximate identities, Fejér polynomials.*

(i) If $(\mu_k)_{k \geq 1}$ are measures on $\mathbb{T}$ (or $\mathbb{T}^n$) such that $\sup_k \|\mu_k\| < \infty$, and if for every $n \in \mathbb{Z}$, $\lim_k \hat{\mu}_k(n) = 1$, then for every $f \in L^p(\mathbb{T})$, $1 \leq p < \infty$, $\lim_k \|f - f * \mu_k\|_p = 0$. (This follows from the density of the trigonometric polynomials in $L^p(\mathbb{T})$.)

(ii) In particular, for $\mu_k = \Phi_k m$, where

$$\Phi_k(e^{ix}) = \sum_{|j| \leq k}(1 - |j|/k)e^{ijx} = k^{-1}(\sin(kx/2)/\sin(x/2))^2$$

(Fejér kernel), $\lim_k \|f - f * \Phi_k\|_p = 0$ $(\forall f \in L^p(\mathbb{T}))$.

Appendix B
Key Notions of Complex Analysis

B.1 Analytic Functions and Holomorphic Functions

Let Ω be an open subset of the complex plane $\mathbb{C}$ and $f : \Omega \to \mathbb{C}$ a function in Ω. The following assertions are equivalent.

(1) *f is analytic in Ω*: $\forall z \in \Omega$, $\exists r > 0$ such that $D(z, r) \subset \Omega$ and for every $\zeta \in D(z, r)$ $f(\zeta) = \sum_{k \geq 0} a_k (\zeta - z)^k$ (absolute convergence).
(2) *f is holomorphic in Ω*: $f \in C^1(\Omega)$ and

$$\frac{\partial f}{\partial \bar{z}} = 0 \quad \text{in } \Omega,$$

where

$$\frac{\partial}{\partial \bar{z}} = \frac{1}{2}\left(\frac{\partial}{\partial x} + i \frac{\partial}{\partial y} \right), \quad z = x + iy \in \Omega.$$

This equation is called the *Cauchy–Riemann (C-R) equation*. In particular, a holomorphic function f is in $C^\infty(\Omega)$; its derivative $\partial f / \partial z = \frac{1}{2}(\partial f / \partial x - i(\partial f / \partial y))$ is denoted $f'(z)$ (*complex derivative* of f). By separating the real part $u = \mathrm{Re}(f)$ and the imaginary part $v = \mathrm{Im}(f)$, we obtain another form of the C-R equation:

$$\frac{\partial u}{\partial x} = \frac{\partial v}{\partial y}, \quad \frac{\partial v}{\partial x} = -\frac{\partial u}{\partial y}.$$

- The set $\mathrm{Hol}(\Omega)$ of holomorphic functions in Ω is a vector space.

B.2 Harmonic Functions, Forms, and Primitives

A function $u \in C^2(\Omega)$ is said to be *harmonic* if $\Delta u = 0$ in Ω, where $\Delta = \partial^2 / \partial x^2 + \partial^2 / \partial y^2$ is the Laplacian operator. The set of harmonic functions on Ω

is a vector space. Given C-R, the real and imaginary parts $\mathrm{Re}(f)$ and $\mathrm{Im}(f)$ of a holomorphic function f are harmonic, and hence so is f.

- Two real harmonic functions u, v are called *harmonic conjugates* if there exists a function $f \in \mathrm{Hol}(\Omega)$ such that $u = \mathrm{Re}(f)$, $v = \mathrm{Im}(f)$ (or, equivalently, u, v satisfy the C-R system).

- Recall that a differential form $\alpha = P\,dx + Q\,dy$ (where $P, Q \in C^1(\Omega)$) is said to be closed if $d\alpha = 0$, where $d\alpha = (\partial P/\partial y - \partial Q/\partial x)\,dx \wedge dy$, and exact if there exists a primitive $v \in C^2(\Omega)$ of α, i.e. v such that $dv = \alpha$, where $dv = (\partial v/\partial x)\,dx + (\partial v/\partial y)\,dy$. An exact form is always closed.

- For an open subset Ω of the complex plane $\mathbb{C}$ the following assertions are equivalent.

(1) Every closed form in Ω is exact.
(2) Every real harmonic function in Ω admits a harmonic conjugate.
(3) Every holomorphic function in Ω admits a *holomorphic primitive*:
     $f \in \mathrm{Hol}(\Omega) \Rightarrow \exists F \in \mathrm{Hol}(\Omega)$ such that $F'(z) = f(z)$, $z \in \Omega$.
(4) Ω is *simply connected* (i.e. every continuous closed curve is *homotopic* to a point: "there are no holes in Ω").

Remark (2) follows from (1) by applying it to $\alpha = (\partial u/\partial y)\,dx - (\partial u/\partial x)\,dy$ where u is harmonic ($\Delta u = 0$). The standard example of a harmonic function without a conjugate is $u(z) = \log|z|$, $z \in \Omega = \mathbb{C} \setminus \{0\}$, and that of a holomorphic function without a primitive is $f(z) = 1/z$, $z \in \Omega = \mathbb{C} \setminus \{0\}$.

B.3 Integral Formulas

If $f \in \mathrm{Hol}(\Omega)$ and if γ is a closed curve homotopic to a point in Ω, then $\int_\gamma f(z)\,dz = 0$ (a form $\alpha = f(z)\,dz$ is closed).

- If $f \in \mathrm{Hol}(\Omega)$, Ω is simply connected, and γ is a simple closed curve in Ω, then for every $\zeta \in \mathrm{int}(\gamma)$,

$$f(\zeta) = \frac{1}{2\pi i} \int_\gamma \frac{f(z)\,dz}{z - \zeta}$$

(*Cauchy's formula*).

- If u is harmonic in Ω and $D(\zeta, r) \subset \Omega$ ($r > 0$), then

$$u(\zeta) = \frac{1}{2\pi r} \int_{\partial D(\zeta,r)} u(z)|dz| = \frac{1}{\pi r^2} \int \int_{D(\zeta,r)} u(x + iy)\,dx\,dy$$

(*mean-value formulas*).

B.4 Major Principles of Complex Analysis

Let Ω be an open subset of $\mathbb{C}$.

- *Principle of isolated zeros.* If Ω is connected, then for every function $f \in \text{Hol}(\Omega)$, $f \neq 0$ and any $\zeta \in \Omega$ there exists $r > 0$ such that $D(\zeta, r) \subset \Omega$ and $f(z) \neq 0$ for every $0 < |z - \zeta| < r$. Consequently, the set of zeros of f is either finite, or else form a sequence tending to the boundary $\partial\Omega$.

Remark Let $f, g \in \text{Hol}(\Omega)$ and let γ be a simple continuous closed curve in a simply connected Ω; let $N(f, \gamma)$ denote the number of zeros of f in $\text{int}(\gamma)$. We cite *Rouché's theorem* for the zeros of a "perturbed" function: $N(f + g, \gamma) = N(f, \gamma)$ if $|g(z)| < |f(z)|$ for $z \in \gamma$.

- *Maximum principle.* If Ω is bounded, then for every harmonic function u (in particular, for every holomorphic function) and any $\zeta \in \Omega$,

$$|u(\zeta)| \leq \sup_{\lambda \in \partial\Omega} \left(\varlimsup_{z \to \lambda, z \in \Omega} |u(z)| \right),$$

and equality holds only if $u = $ constant on the connected component of Ω containing ζ.

- *The compactness principle (Montel).* If there is a sequence $(f_n) \subset \text{Hol}(\Omega)$, uniformly bounded on every compact subset K of Ω, that is,

$$\|f_n\|_{C(K)} = \sup_{z \in K} |f_n(z)| \leq c_K < \infty \quad \text{for every } n = 1, 2, \ldots,$$

then there exists a subsequence (f_{n_j}) converging uniformly on any compact subset $K \subset \Omega$ to a function $f \in \text{Hol}(\Omega)$: $\lim_j \|f - f_n\|_{C(K)} = 0$.
Principle of conformal mappings (Riemann). Every connected and simply connected open set $\Omega \subset \mathbb{C}$, $\Omega \neq \mathbb{C}$ is *conformally equivalent* to the unit disk $\mathbb{D} = D(0, 1)$ (and hence they are all conformally equivalent to each other): there exists a bijective and biholomorphic mapping (said to be *conformal*) $\varphi \colon \Omega \to \mathbb{D}$, $\varphi \in \text{Hol}(\Omega)$, $\varphi^{-1} \in \text{Hol}(\mathbb{D})$.

Remarks (1) A Jordan domain Ω is a bounded open set whose boundary $\partial\Omega$ is homeomorphic to the unit circle $\mathbb{T}$ ($\Leftrightarrow$ it is a simple, continuous, and closed curve); every conformal mapping $\varphi \colon \mathbb{D} \to \Omega$ on a Jordan domain can be extended to a homeomorphism of $\overline{\mathbb{D}}$ onto $\overline{\Omega}$ (and in particular, $\varphi \in C(\overline{\mathbb{D}})$) (Carathéodory, 1913).
(2) Every conformal mapping φ of $\mathbb{D}$ onto itself is of the form

$$\varphi(z) = \epsilon \frac{z - \lambda}{1 - \bar{\lambda}z}$$

where $|\epsilon| = 1$, $|\lambda| < 1$. Every conformal mapping of $\mathbb{C}_+ = \{z: \mathrm{Im}(z) > 0\}$ on $\mathbb{D}$ is of the form

$$\varphi(z) = \epsilon \frac{z - \lambda}{z - \bar{\lambda}}$$

where $|\epsilon| = 1$, $\lambda \in \mathbb{C}_+$.

B.5 Holomorphic Extensions

Let $f \in \mathrm{Hol}(\Omega)$ and $\lambda \in \partial\Omega$. The function f is said to be *(holomorphically) extendable* at a point λ if there exists $r > 0$ and $g \in \mathrm{Hol}(D(\lambda, r))$ such that $f = g$ on $\Omega \cap D(\lambda, r)$.

- A function $f \in \mathrm{Hol}(\Omega)$ is extendable at the point $\lambda \in \partial\Omega$ if and only if there exists $\zeta \in \Omega$ such that the radius of convergence R of the local development $f(z) = \sum_{k \geq 0} a_k (z - \zeta)^k$ satisfies $R > |\zeta - \lambda|$.'

- Let $f \in \mathrm{Hol}(\Omega)$, and let $\lambda \in \partial\Omega$ be an isolated point of the boundary $\partial\Omega$. Then f is extendable at the point λ if and only if $\sup_{0 < |z - \lambda| < r} |f(z)| < \infty$ for a certain $r > 0$ (removable singularity).

B.6 Infinite Products

Recall that, by definition, a numerical product $\prod_{k=1}^{\infty} c_k$, $c_k \in \mathbb{C}$, converges if the limit $\lim_n \prod_{k=1}^{n} c_k \in \mathbb{C} \setminus \{0\}$ exists ($\Leftrightarrow$ the series $\sum_k \log(c_k)$ converges, $|\arg(c_k)| < \pi$).

- Let $\prod_{k=1}^{\infty} f_k$ where $f_k \in \mathrm{Hol}(\Omega)$, $k = 1, 2, \ldots$ The product is said to *converge uniformly on compact subsets* if, for every compact $K \subset \Omega$, there exists N such that $f_j(z) \neq 0$ for every $j > N$ and every $z \in K$ and such that the (numerical) product $\prod_{k > N} f_k(z)$ converges uniformly with respect to $z \in K$ (hence, the sequence $(\prod_{N < k \leq n} f_k)_{n > N}$ converges in the space $C(K)$).

  If such a convergence takes place, the result

$$f(z) = \prod_{k=1}^{\infty} f_k(z) = \prod_{k=1}^{N} f_k(z) \cdot \prod_{k > N} f_k(z)$$

is a holomorphic function on Ω. The set of the zeros of f is the union of the zeros of the f_k, $k = 1, 2, \ldots$.

Appendix C
Key Notions of Hilbert Spaces

In this chapter, every vector space is over the field $\mathbb{C}$ of complex numbers. For the properties shared by all Banach spaces, see Appendix D.

C.1 Scalar Products and Hilbert Spaces

Let H be a vector space. A complex function $(\cdot, \cdot) = (\cdot, \cdot)_H$ on $H \times H$ is called a *scalar product* if it satisfies the following properties:

  (i) $x \longmapsto (x, y)$ is a linear functional on H for any $y \in H$,
 (ii) $(x, y) = \overline{(y, x)}$ for every $x, y \in H$,
(iii) $(x, x) \geq 0$ for every $x \in H$,
(iv) $(x, x) = 0 \Leftrightarrow x = 0.$

- *Cauchy–Schwarz inequality.* $|(x, y)|^2 < (x, x) \cdot (y, y)$, except in the case where x and y are collinear (where equality holds in place of the inequality).

- Given a scalar product $(\cdot, \cdot)$, the function $x \longmapsto \|x\| = (x, x)^{1/2}$ is a norm on H. A vector space H equipped with a scalar product $(\cdot, \cdot) = (\cdot, \cdot)_H$ and with the associated norm is called a *pre-Hilbert* (or *Hermitian*) space; if it is complete (as a normed space, see Appendix D), it is said to be a *Hilbert space*.

- *Example.* $H = L^2(\Omega, \mu)$ with $(f, g) = \int_\Omega f\bar{g}\, d\mu$ $(f, g \in L^2(\Omega, \mu))$. In particular,

$$l^2(J) = \left\{ (x_j)_{j \in J} : x_j \in \mathbb{C}, \sum_j |x_j|^2 < \infty \right\}$$

with $(x, y) = \sum_{j \in J} x_j \bar{y}_j.$

In what follows, H always denotes a Hilbert space.

C.2 Orthogonal Decompositions

Let $x, y \in H$. An element x is said to be *orthogonal* to y (written $x \perp y$) if $(x, y) = 0$. Subspaces $E, F \subset H$ are orthogonal ($E \perp F$) if $x \perp y$ for every $x \in E$, $y \in F$.

- The *Pythagorean theorem (580–495 BCE)*. If $x_j \in H$ and $x_j \perp x_k$ $(j \neq k)$, then $\| \sum_1^n x_j \|^2 = \sum_1^n \|x_j\|^2$.

- *Corollary.* A vector sum of closed and orthogonal subspaces is closed: if $E, F \subset H$ are closed and $E \perp F$ then $E + F$ is closed (this is not necessarily the case for arbitrary E, F).

- *The orthogonal complement* of a vector subspace $E \subset H$ is $E^\perp = \{y \in H : x \perp y \ \forall x \in E\}$. If E is closed, then $E = (E^\perp)^\perp$ and $H - E + E'$ (often written as $H - E \oplus E'$ to highlight the orthogonality), hence every $x \in H$ can be uniquely written in the form $x = x' + x''$ where $x' \in E$, $x'' \in E^\perp$.

  The mapping $P_E : x \longmapsto x'$ is called the *orthogonal projection onto* E. Clearly P_E is linear, with $P_E^2 = P_E$, and for every $x \in H$, $\|P_E x\| \leq \|x\|$.

- *Corollary.* Let $A \subset H$. Then, $\mathrm{span}_H(A) = H \Leftrightarrow (x \perp A \Rightarrow x = 0)$.

- *Convergence of an orthogonal series.* Let $x_j \in H$ $(j = 1, 2, \dots)$ and $x_j \perp x_k$ $(j \neq k)$. The series $\sum_{j \geq 1} x_j$ converges in H if and only if $\sum_j \|x_j\|^2 < \infty$; in this case, $\|x\|^2 = \sum_j \|x_j\|^2$ where $x = \sum_j x_j$. An orthogonal series $\sum_j x_j$ *converges* unconditionally (if it converges): i.e. for any $\epsilon > 0$ there exists a finite set $\sigma_\epsilon \subset \mathbb{N}$ such that for every finite $\sigma \supset \sigma_\epsilon$, $\|x - \sum_{j \in \sigma} x_j\| < \epsilon$

- *Orthogonal decomposition.* Let $H_j \subset H$ $(j = 1, 2, \dots)$ and $H_j \perp H_k$ $(j \neq k)$. Then the closed linear hull of the family (H_j) is

$$\mathrm{span}_H (H_j : j = 1, 2, \dots) = \left\{ x = \sum_{j \geq 1} x_j : x_j \in H_j (\forall j) \text{ and } \sum_j \|x_j\|^2 < \infty \right\}.$$

This is denoted $\sum_{j \geq 1} \oplus H_j$.

- *Orthogonal decomposition (continued).* We have $\sum_{j \geq 1} \oplus H_j = H$ if and only if $(x \perp H_j, \forall j \Rightarrow x = 0)$, and if this is the case, then for every $x \in H$,

$$x = \sum_j P_{H_j} x, \quad P_{H_j} x \in H_j, \quad \|x\|^2 = \sum_j \|P_{H_j} x\|^2$$

(*Parseval's identity*).

C.3 Orthogonal Bases

A special case of the preceding decompositions is when H_j is generated by a single vector $e_j \neq 0$; hence (e_j) is an orthogonal sequence, $(e_j, e_k) = 0$, $j \neq k$. The sequence (e_j) is said to be *complete in H* if $(x \perp e_j, \forall j \Rightarrow x = 0)$. In this case, for every $x \in H$ there exists a unique convergent series of the form $\sum_j a_j e_j$ whose sum is x; indeed,

$$a_j e_j = P_{H_j} x = \frac{(x, e_j)}{\|e_j\|^2} e_j,$$

thus

$$\forall x \in H: \quad x = \sum_j \frac{(x, e_j)}{\|e_j\|^2} e_j, \quad \|x\|^2 = \sum_j \frac{|(x, e_j)|^2}{\|e_j\|^2}.$$

Such an orthogonal and complete sequence is called *an orthogonal basis of H*; if $\forall j$ we have $\|e_j\| = 1$, it is said to be *an orthonormal basis*.

- *The existence of an orthonormal basis (Gram–Schmidt orthogonalization theorem).* Let $(x_j)_{j\geq 1} \subset H$ be a "free" sequence ($\forall k$, $x_k \notin \mathrm{Lin}(x_j: j \neq k)$). Then there exists a unique *orthonormal sequence* $(e_j)_{j\geq 1}$ satisfying the following properties.

  (i) For every $n = 1, 2, \ldots$, $\mathrm{Lin}(e_j: 1 \leq j \leq n) = \mathrm{Lin}(x_j: 1 \leq j \leq n) := L_n$.
  (ii) For every j, $(x_j, e_j) > 0$. The explicit formula is

$$e_n = \frac{x_n - P_{L_{n-1}} x_n}{\|x_n - P_{L_{n-1}} x_n\|}, \quad P_{L_{n-1}} x = \sum_{j=1}^{n-1} (x, e_j) e_j \quad (\forall x \in H).$$

- *Corollary.* In every separable Hilbert space there exists an orthonormal basis.

- *Example.* Let μ be a finite Borel measure on $\mathbb{T}$, such that $\mathrm{supp}(\mu)$ is an infinite set. Then there exists a unique orthonormal basis $(\varphi_k)_{k\geq 1}$ of trigonometric polynomials φ_k such that $\deg(\varphi_k) = [k/2]$, $k = 1, 2, \ldots$. (We apply the theorem to $(x_k)_{k\geq 1} = (1, e^{ix}, e^{-ix}, e^{2ix}, e^{-2ix}, \ldots)$; $\mathrm{span}_{L^2(\mu)}(x_k)_{k\geq 1} = L^2(\mu)$ by § A.5 above.) The φ_k are called the *orthogonal polynomials* with respect to μ.

- *Corollary.* All separable Hilbert spaces of the same dimension are unitarily isomorphic: if $\dim H_1 = \dim H_2$ (and the H_j are separable) there exists a unitary (linear bijective isometric) $U: H_1 \to H_2$.

C.4 The Riesz Representation Theorem

Every linear continuous (bounded) functional φ on a Hilbert space H is of the form $\varphi(x) = (x, y)$ ($\forall x \in H$); such a $y \in H$ is unique and we have $\|\varphi\| = \|y\|$.

Appendix D
Key Notions of Banach spaces

In this chapter, every vector space is over the field $\mathbb{C}$ of complex numbers.

D.1 Normed Spaces and Banach Spaces

Let X be a vector space. A function $x \longmapsto \|x\|$ on X is called a *norm* if it satisfies:

(i) $\|x + y\| \leq \|x\| + \|y\|$ for every $x, y \in X$ ($\| \cdot \|$ is subadditive),
(ii) $\|\lambda x\| = |\lambda| \cdot \|x\|$ for every $x \in X$ and any $\lambda \in \mathbb{C}$,
(iii) $\|x\| = 0 \Leftrightarrow x = 0$.

If $\| \cdot \|$ is a norm, $\rho(x, y) = \|x - y\|$ is a distance on X (associated with $\| \cdot \|$).

X equipped with a norm (and with the associated distance) is said to be a *normed space*. If X is complete as a metric space, X is called a *Banach space*.

In what follows, X denotes a normed space, equipped with a norm $\| \cdot \| = \| \cdot \|_X$.

- A normed space is complete if and only if every *absolutely convergent series* $\sum_{k \geq 0} x_k$ $(x_k \in X)$ (i.e. $\sum_{k \geq 0} \|x_k\| < \infty$) *converges* in X (i.e. there exists $x \in X$ such that $\lim_n \|x - \sum_{k=0}^{n} x_k\| = 0$).

D.2 The Baire Category Theorem

Every Banach space X (and moreover every complete metric space) is of *Baire second category* (i.e. for every sequence $(X_n)_{n \geq 1}$, $X_n \subset X$, of closed subsets with empty interior, we have $X \neq \bigcup_{n \geq 1} X_n$; the subsets that are unions of this last type are said to be of *Baire first category*).

D.3 Duality

For a normed space X, we denote X^* its dual space, i.e. the space of bounded linear functionals φ equipped with the norm

$$\|\varphi\| = \sup\{|\varphi(x)| : x \in X, \|x\| \le 1\}.$$

X^* is always a Banach space. For reasons of symmetry, we also use the notation

$$\langle x, \varphi \rangle = \varphi(x) \quad (x \in X, \ \varphi \in X^*).$$

- *The Hahn–Banach Theorem* (1932)

  (i) Let $E \subset X$ be a vector subspace and $\varphi_0 \in E^*$. Then there exists $\varphi \in X^*$ such that $\varphi|E = \varphi_0$ and $\|\varphi\| = \|\varphi_0\|$.
  (ii) Let $E \subset X$ be a vector subspace and $x \in X$. Then, for every functional $\varphi \in X^*$ such that $\varphi|E = 0$ we have $|\langle x, \varphi \rangle| \le \mathrm{dist}_X(x, E) \cdot \|\varphi\|$. If $x \notin \overline{E}$ then there exists a functional $\varphi \in X^*$ such that $\varphi|E = 0$ and $1 = |\langle x, \varphi \rangle| = \mathrm{dist}_X(x, E) \cdot \|\varphi\|$.

- *Corollary.* For every $x \in X$, $\|x\| = \sup\{|\langle x, \varphi \rangle| : \varphi \in X^*, \|\varphi\| \le 1\}$.

- *Corollary.* Let $A \subset X$, $x \in X$. The following assertions are equivalent:

  (i) $x \in \mathrm{span}_X(A)$.
  (ii) $\forall \varphi \in X^*, \varphi|A = 0 \Rightarrow \varphi(x) = 0$.

- *Corollary.* Let $\varphi \in X^*$, $E \subset X$ be a vector subspace and $E^\perp = \{\psi \in X^* : \psi|E = 0\}$ (the polar subspace of E). Then

$$\|\varphi|E\| = \mathrm{dist}_{X^*}(\varphi, E^\perp).$$

- *Weak topologies.* A *base of the weak topology* $\sigma(X, X^*)$ is defined by

  $\{x \in X : |\langle x - x_0, \varphi_j \rangle| < \epsilon, j = 1, \ldots, n\}$ where $n \in \mathbb{N}, \epsilon > 0, \varphi_j \in X^*, x_0 \in X$.

  A *base of the weak-star topology* $\sigma(X^*, X)$ is defined by

  $\{\varphi \in X^* : |\langle x_j, \varphi - \varphi_0 \rangle| < \epsilon, j = 1, \ldots, n\}$ where $n \in \mathbb{N}, \epsilon > 0, x_j \in X, \varphi_0 \in X^*$.

- *Weak-star convergence.* Let X be a Banach space and $A \subset X$ such that $\mathrm{span}_X(A) = X$. Then a countable sequence (φ_k) converges $\sigma(X^*, X)$ to 0 if and only if $\sup_k \|\varphi_k\| < \infty$ and $\lim_k \varphi_k(x) = 0 \ \forall x \in A$. If X is separable, then the unit ball $\{\varphi \in X^* : \|\varphi\| \le 1\}$ is $\sigma(X^*, X)$-compact.

- *Reflexivity.* For every $x \in X$, the formula $j(x)\varphi = \langle x, \varphi \rangle, \varphi \in X^*$, defines a functional $j(x) \in (X^*)^*$ such that $\|j(x)\| = \|x\|$. X is said to be *reflexive* if $j(X) = (X^*)^*$.

- A Banach space X is reflexive if and only if the ball $\{x \in X : \|x\| \leq 1\}$ is $\sigma(X, X^*)$-compact.

D.4 Examples of Duality

For $1 \leq p < \infty$, $(L^p(\Omega, \mu))^* = L^{p'}(\Omega, \mu)$, $1/p + 1/p' = 1$, with respect to the (bilinear) form realizing the duality

$$\langle f, g \rangle = \int_\Omega fg \, d\mu, \quad f \in L^p(\Omega, \mu), g \in L^{p'}(\Omega, \mu).$$

Hence L^p, with $1 < p < \infty$, is reflexive. If $\mathrm{supp}(\mu)$ is infinite, neither L^1 nor L^∞ are reflexive.

- If K is compact, then $(C(K))^* = \mathcal{M}(K)$, while if Ω is locally compact then $(C_0(\Omega))^* = \mathcal{M}(\Omega)$, with respect to the dualities

$$\langle f, \mu \rangle = \int_K f \, d\mu, \quad f \in C(K), \mu \in \mathcal{M}(K),$$

and the analog for $C_0(\Omega)$ (see § A.4).

D.5 Schauder Bases (1927)

A sequence $(e_k)_{k \geq 1}$ is called a *Schauder basis* of a space X if $\forall x \in X$, $\exists$ a unique sequence (a_k), $a_k \in \mathbb{C}$ such that

$$\lim_n \left\| x - \sum_{k=1}^n a_k e_k \right\| = 0.$$

The sums $P_n x = \sum_{k=1}^n a_k e_k$ are called the *partial sums*.

- In a Banach space X, a sequence $(e_k)_{k \geq 1}$ is a Schauder basis if and only if

(i) $\mathrm{span}_X(e_k : k \geq 1) = X$,
(ii) the projections $P_n(\sum_k a_k e_k) := \sum_{k=1}^n a_k e_k$ are well-defined and continuous on $\mathrm{Lin}(e_k : k \geq 1)$, and
(iii) $\sup_n \|P_n\| < \infty$.

- *Remark.* There exist separable Banach spaces (and even subspaces of $l^p = l^p(\mathbb{N})$, $p \neq 2$) without a Schauder basis (Enflo, 1972).

Appendix E
Key Notions of Linear Operators

E.1 Bounded Operators

Let X, Y be normed spaces, and let $T \colon X \to Y$ be a linear mapping. The following assertions are equivalent.

(i) T is continuous.

(ii) T is continuous at the point 0.

(iii) T is *bounded*: $\exists$ a constant $C > 0$ such that $\forall x \in X$,
$$\|Tx\| = \|Tx\|_Y \le C\|x\| = C\|x\|_X.$$

(iv) $\|T\| = \|T\|_{Op} := \sup\{\|Tx\| : x \in X, \|x\| \le 1\} < \infty$ (the best constant C of (iii)).

The set of bounded linear operators $X \to Y$, denoted $L(X, Y)$ (and with $L(X) = L(X, X)$), is a normed space (with the norm $\|\cdot\|_{Op}$), complete if Y is complete.

- *Adjoint operator.* If $T \in L(X, Y)$ and $y' \in Y^*$, the functional $T^* y'$ is defined by the requirement $\langle x, T^* y' \rangle = \langle Tx, y' \rangle$, $\forall x \in X$. The mapping $y' \longmapsto T^* y'$ is linear and bounded, hence $T^* \in L(Y^*, X^*)$; T^* is said to be the adjoint operator of T. Clearly $\|T^*\| = \|T\|$.

E.2 Three Fundamental Principles

(1) *Closed graph theorem.* Let X, Y be Banach spaces, and let $T \colon X \to Y$ be a linear mapping. The following assertions are equivalent.

(i) T is continuous.

(ii) The graph $G(T) = \{(x, y) \in X \times Y : y = Tx\}$ is closed.

(iii) $\lim_n \|x_n\|_X = 0$ and $\lim_n \|Tx_n - y\|_Y = 0$ implies $y = 0$.

(2) The *Banach–Steinhaus theorem (1927, principle of equicontinuity, or of uniform boundedness).* Let X, Y be Banach spaces, $A \subset X$ such that $\mathrm{span}_X(A) = X$ and let $T_n \in L(X, Y)$ be a sequence of continuous linear mappings. The following assertions are equivalent.

  (i) For every $x \in X$, there exists $\lim_n T_n x := T x$.

  (ii) $\sup_n \|T_n\| < \infty$, and for every $x \in A$ the limit $\lim_n T_n x$ exists.

The limit T in (i) is always bounded.

(3) *Open Mapping Theorem (Banach and Schauder, 1932).* Let X, Y be Banach spaces and $T \in L(X, Y)$. The following assertions are equivalent.

  (i) The image $T(G)$ of any open set $G \subset X$ is open (we say "T is open").

  (ii) $TX = Y$.

  (iii) There exists a constant $c > 0$ such that for every $y' \in Y^*$ we have
$$\|T^* y'\|_{X^*} \ge c \|y'\|_{Y^*}.$$

If (i)–(iii) hold, then $\forall y \in Y$, $\exists x \in X$ such that $T x = y$ and $\|x\| \le \frac{1}{c}\|y\|$.

- *Remark.* For a bijective operator T the equivalence of properties (i) and (iii) is obvious (with $c = 1/\|T^{-1}\|$); the proof of (3) uses the quotient operator $T \colon X/\mathrm{Ker}(T) \to Y$, already bijective.

- *Corollary.* Let X, Y be Banach spaces.

    (i) If $T \in L(X, Y)$ is bijective, it is a homeomorphism.

    (ii) If $E, F \subset X$ are closed subspaces such that $E \cap F = \{0\}$, $E + F = X$, then the projection $P_{E\|F}(x + y) = x$ ($x \in E, y \in F$) is bounded.

    (iii) If $T \colon X \to Y$ is linear and continuous for a separable topology τ on Y, then $T \in L(X, Y)$ (for example, it could be that $\tau = \sigma(Y, Y^*)$).

    (iv) The *Riemann–Lebesgue lemma.* If $T_n f = \hat{f}(n)$, $f \in L^1(\mathbb{T})$, then $\|T_n\| = 1$ and $\lim_n T_n(f) = 0$ for every trigonometric polynomial f. By (2), $\lim_n T_n(f) = 0$ for every $f \in L^1(\mathbb{T})$.

Remark Other corollaries similar to (iv) can easily be produced.

E.3 The Spectrum

Let A be a Banach space equipped with a multiplication operation $(x, y) \longmapsto x \cdot y = xy$ which transforms A into an algebra with unit $e \in A$ satisfying $\|xy\| \le \|x\| \cdot \|y\|$ (for every $x, y \in A$) and $\|e\| = 1$. Such an algebra A is called a *Banach algebra*.

- *Examples.* $A = L(X)$ where X is a Banach space; $A = C(K)$ or $A = L^\infty(\Omega, \mu)$ (equipped with the norm $\| \cdot \|_\infty$).

Let A^{-1} denote the set of invertible elements of A. The spectrum $\sigma(a) = \sigma_A(a)$ of an element $a \in A$ is defined by

$$\sigma(a) = \{\lambda \in \mathbb{C} : \lambda e - a \notin A^{-1}\}.$$

- *Immediate properties of the spectrum.* Let A be a Banach algebra with unit e.

  (i) For every $a \in A$, $\sigma(a)$ is a non-empty compact set.
  (ii) The spectral radius $r(a) := \max\{|\lambda| : \lambda \in \sigma(a)\}$ coincides with $\lim_n \|a^n\|^{1/n}$ (Gelfand's formula).
  (iii) For $A = L(X)$, the *point spectrum of an operator T* (the *eigenvalues of T*) $\sigma_p(T) = \{\lambda \in \mathbb{C} : \mathrm{Ker}(\lambda I - T) \neq \{0\}\}$ is contained in $\sigma(T)$.
  (iv) For $A = L(X)$ and for a bilinear duality between X and X^*, we have $\sigma(T) = \sigma(T^*)$.
  (v) *Spectral mapping theorem.* For every polynomial f, $\sigma(f(T)) = f(\sigma(T))$.

E.4 Invariant Subspaces

Let X be a Banach space, $E \subset X$ a closed subspace, and $T \in L(X)$. E is said to be an *invariant subspace* for T if $x \in E \Rightarrow Tx \in E$ (in brief, $TE \subset E$). The set of all invariant subspaces is denoted $\mathrm{Lat}(T)$.

- $E \in \mathrm{Lat}(T) \Leftrightarrow E^\perp \in \mathrm{Lat}(T^*)$.
- $\mathrm{Lat}(T)$ is a lattice with respect to the set operations $\cup, \cap$.

Remark There exists a $T \in L(l^1)$ with the trivial lattice $\mathrm{Lat}(T) = \{\{0\}, l^1\}$ (Read, 1984, inspired by Enflo, 1976) and there exist Banach spaces X where $\mathrm{Lat}(T) \neq \{\{0\}, X\}$, $\forall T \in L(X)$ (Argyros–Haydon, 2009). For a Hilbert space, the question of the existence of a T with trivial $\mathrm{Lat}(T)$ remains open.

E.5 In a Hilbert Space: Self-adjoint, Unitary, Normal Operators

Let H, K be *Hilbert spaces* and $T \in L(H, K)$. We define $T^* : K \to H$ by $(Tx, y)_K = (x, T^*y)_H$ (for every $x \in H$, $y \in K$), which gives a complex conjugate for certain properties of T^*. For example, if $T \in L(H)$, then $\sigma(T^*) = \sigma(T)^* = \{\bar{\lambda}: \lambda \in \sigma(T)\}$.

- An operator $T \in L(H)$ is said to be *self-adjoint* if $T = T^*$, *unitary* if $TT^* = T^*T = \mathrm{id}$ (a modification for $T \in L(H, K)$: $T^*T = \mathrm{id}_H$, $TT^* = \mathrm{id}_K$), and *normal* if $TT^* = T^*T$.

- Operators $A \in L(H)$ and $B \in L(K)$ are said to be *unitarily equivalent* if there exists a unitary operator $U: H \to K$ such that $UA = BU$.

- *Spectral theorem for a normal operator with a simple spectrum (von Neumann, 1929).* Let $T \in L(H)$ be a normal cyclic operator (said to be "with simple spectrum"). Then there exists a Borel measure μ on $\mathbb{C}$, with compact support, such that T is unitarily equivalent to the multiplication operator

$$M_z: L^2(\mu) \to L^2(\mu), \quad M_z f = zf.$$

We have $\sigma(T) = \mathrm{supp}(\mu)$, and the equivalence class of μ (i.e. $\{\nu \geq 0: \nu \ll \mu, \mu \ll \nu\}$) is uniquely defined by T (μ is called the scalar spectral measure of T).

- *Outline of the proof.*

(1) We first show that for any normal operator N, we have $\|N\| = r(N)$ (spectral radius).
(2) By using the spectral mapping theorem (§ E.3), we deduce that for every polynomial in z and $\bar{z}$, $\|f(T)\| = r(f(T)) = \max\{|f(\lambda)|: \lambda \in \sigma(T)\}$, and hence $f \longmapsto (f(T)x, y)$ is a continuous linear functional on $C(\sigma(T))$ (for self-adjoint and/or unitary operators, this step is much simpler than in the general case).
(3) We select a cyclic vector $x \in H$, $H = \mathrm{span}_H(T^n x: n = 0, 1, \dots)$, and observe (by the Riesz representation theorem, § A.4) that there exists a measure $\mu \geq 0$ such that

$$(f(T)x, g(T)x) = \int_{\sigma(T)} f\bar{g}\, d\mu$$

for any polynomials $f = f(z, \bar{z})$ and $g = g(z, \bar{z})$.
(4) Setting $U(f(T)x) = f$, $U: H \to L^2(\mu)$, we obtain the result. ∎

- *Polar decomposition.* For every $T \in L(H, K)$ (where H and K are Hilbert spaces) such that $\dim \mathrm{Ker}(T) = \dim \mathrm{Ker}(T^*)$, there exists a unitary operator $U: H \to K$ such that $T = U|T|$, where $|T| := (T^*T)^{1/2} \geq 0$ is the *modulus* of T, $|T| \in L(H)$.

- *Corollary.* If $\text{Ker}(T) = \{0\}$, $\text{Ker}\, T^* = \{0\}$, then the operators T^*T and TT^* are unitarily equivalent. (Indeed, $TT^* = U(T^*T)U^*$.)

- *Reducing subspaces* of an operator $T \in L(H)$: these are the elements of $\text{Lat}(T) \cap \text{Lat}(T^*)$. We have $E \in \text{Lat}(T) \cap \text{Lat}(T^*) \Leftrightarrow E, E^\perp \in \text{Lat}(T)$ (hence $H = E \oplus E^\perp$ where the two subspaces $E, E^\perp$ are T-invariants). As the closed linear span of a family of reducing subspaces is again in $\text{Lat}(T) \cap \text{Lat}(T^*)$, we deduce that $\forall E \in \text{Lat}(T)$ we have $E = E' \oplus E''$, where $E' \in \text{Lat}(T) \cap \text{Lat}(T^*)$ and $E'' \in \text{Lat}(T)$ but does not contain any T-reducing subspace (E'' is said to be completely non-reducing).

References

A

N. I. Akhiezer (1956), On the weighted approximation of continuous functions by polynomials on the real axis. *Uspekhi Mat. Nauk* **11:4 (70)**, 3–43. English translation: *Amer. Math. Soc. Transl.* (2) **22** (1962), 95–137.

N. I. Akhiezer (1965), *Lectures on Approximation Theory* (in Russian), second edition. Nauka, Moscow. English translation: *Approximation Theory*, Dover, New York (1992).

B

L. Báez-Duarte (2003), A strengthening of the Nyman–Beurling criterion for the Riemann hypothesis. *Rend. Lincei (9) Mat. Appl.* **14**, 5–11.

B. Bagchi (2006), On Nyman, Beurling, and Baez-Duarte's Hilbert space reformulation of the Riemann hypothesis. *Proc. Indian Acad. Sci. (Math. Sci.)* **116:2**, 137–146.

M. Balazard (2010), Un siècle et demi de recherches sur l'hypothèse de Riemann. *Gazette des mathématiciens (Soc. Math. France)* **126**, 7–24.

S. Banach (1932), Théorie des opérations linéaires. Monografie Matematyczne, Warsaw.

A. Baranov, Yu. Belov and A. Borichev (2013), Hereditary completeness for systems of exponentials and reproducing kernels, *Adv. in Math.* **235**, 525–554.

A. Baranov and D. Yakubovich (2016), Completeness and spectral synthesis of nonselfadjoint one-dimensional perturbations of selfadjoint operators, *Adv. in Math.* **302**, 740–798.

K. Barbey and H. König (1977), *Abstract Analytic Function Theory and Hardy Algebras*. Vol. 593 of Lecture Notes in Mathematics, Springer, Berlin.

S. N. Bernstein (1924), Le problème de l'approximation des fonctions continues sur tout l'axe réel et l'une de ses applications. *Bull. Math. Soc. France* **52**, 399–410.

A. Beurling (1945), On the completeness of $\{\psi(nt)\}$ on $L^2(0, 1)$. In *The Collected Works of Arne Beurling*, vol. 2: *Harmonic Analysis*. Contemporary Mathematicians, Birkhäuser, Boston (1989), pp. 378–380.

A. Beurling (1949), On two problems concerning linear transformations in Hilbert space. *Acta Math.* **81**, 79–93.

Ia. Blagouchine (2018), The history of the ζ functional equation, and the role of different mathematicians in its proof, *A seminar talk at POMI seminar on the history of mathematics*, March 1, 2018,
www.mathnet.ru/php/conference.phtml?option_lang=rus&eventID=10&confid=504.

W. Blaschke (1915), Eine Erweiterung des Satzes von Vitali über Folgen analytischer Funktionen. *S.-B. Sächs Akad. Wiss. Leipzig Math-Natur. Kl.* **67**, 194–200.

R. P. Boas (1954), *Entire Functions*. Academic Press, New York.

H. Bohr (1913), Über die Bedeutung der Potenzreihen unendlich vieler Variablen in der Theorie der Dirichletschen Reihen $\sum \frac{a_n}{n^s}$. *Nachr. Ges. Wiss. Göttingen. Math.-Phys. Kl.* **A9**, 441–488.

A. Borichev (2001), On the closure of polynomials in weighted spaces of functions on the real line. *Indiana Univ. Math. J.* **50**, 829–846.

A. Borichev and M. Sodin (2001), Krein's entire functions and Bernstein approximation problem. *Illinois J. Math.* **45:1**, 167–185.

D. G. Bourgin (1946), A class of sequences of functions. *Trans. Amer. Math. Soc.* **60**, 478–518.

P. L. Butzer (1983), A survey of the Whittaker–Shannon sampling theorem and some of its extensions. *J. Math. Res. Exposition* **3**, 185–212.

P. L. Butzer, P. J. S. G. Ferreira, J. R. Higgins, S. Saitoh, G. Schmeisser, R. L. Stens (2011), Interpolation and Sampling: E. T. Whittaker, K. Ogura and Their Followers. *J. Fourier Analysis Appl.* **17:2**, 320–354.

P. L. Butzer, J. R. Higgins, and R. L. Stens (2000), Sampling theory of signal analysis 1950–1995. In *Development of Mathematics 1950–2000* (ed. J.-P. Pier), Birkhäuser, Basel, pp. 193–234.

C

A. P. Calderón (1950), On theorems of M. Riesz and A. Zygmund. *Proc. Amer. Math. Soc.* **1**, 533–535.

L. Carleson (1956), Representations of continuous functions. *Math. Zeit.* **66**, 447–451.

J. B. Conrey (2003), The Riemann hypothesis. *Notices Amer. Math. Soc.* **March 2003**, 341–353.

M. Cotlar and C. Sadosky (1979), On the Helson–Szegő theorem and a related class of modified Toeplitz kernels. In *Harmonic Analysis in Euclidean Spaces*, part 1 (ed. G. Weiss and S. Wainger), vol. 35 of Proceedings of Symposia in Pure Mathematics, American Mathematical Society, Providence, RI, pp. 387–407.

D

L. de Branges (1959), The Bernstein problem. *Proc. Amer. Math. Soc.* **10**, 825–832.

A. Devinatz and M. Shinbrot (1969), General Wiener–Hopf operators. *Trans. Amer. Math. Soc.* **145**, 467–494.

R. A. DeVore and G. G. Lorentz (1993), *Constructive Approximation*. Springer.

J. Duoandikoetxea (2001), *Fourier Analysis.* American Mathematical Society, Providence, RI.

P. L. Duren (1970), *Theory of H^p Spaces.* Academic Press, New York.

F

P. Fatou (1906), Série trigonométriques et séries de Taylor. *Acta Math.* **30**, 335–400.

L. Fejér and F. Riesz (1921), Über einige funktionentheoretische Ungleichungen. *Math. Zeit.* **11**, 305–314.

K. Ford (2002), Vinogradov's integral and bounds for the Riemann zeta function. *Proc. London Math. Soc.* (3) **85**, 565–633.

K. O. Friedrichs (1937), On certain inequalities and characteristic value problems for analytic functions and for functions of two variables. *Trans. Amer. Math. Soc.* **41**, 321–364.

G

T. W. Gamelin (1969), *Uniform Algebras.* Prentice Hall, Englewood Cliffs, New Jersey.

F. R. Gantmacher (1966), *The Theory of Matrices* (in Russian), second edition. Nauka, Moscow. English translation: Chelsea, New York (1960).

J. B. Garnett (1981), *Bounded Analytic Functions.* Academic Press, New York.

A. O. Gelfond (1958), Die Rolle der Arbeiten L. Eulers für die Entwicklung der Zahlentheorie (in Russian, with a summary in German). In *Leonhard Euler (zu 250. Geburtstages)* (ed. M. Lavrentiev, A. Yushkevich, and A. Grigoriyan), Academy of Sciences of the USSR, Moscow, pp. 96–129.

I. M. Glazman and Y. I. Lyubich (1969), *Finite-dimensional Linear Analysis* (in Russian). Nauka, Moscow. English translation: *Finite-dimensional Linear Analysis: A Systematic Presentation in Problem Form*, MIT Press, Cambridge, MA (1974).

G. Golub and C. Van Loan (1996), *Matrix Computations*, third edition. Johns Hopkins University Press, Baltimore and London.

G. M. Goluzin (1966), *Geometric Theory of Functions of a Complex Variable* (in Russian). Nauka, Moscow. English translation: American Mathematical Society, Providence, RI (1969).

B. Green and T. Tao (2008), The primes contain arbitrarily long arithmetic progressions. *Ann. of Math.* **167:2**, 481–547.

H

G. H. Hardy (1913), A theorem concerning Taylor's series. *Quart. J. Pure Math.* **44**, 147–160.

G. H. Hardy (1915), On the mean value of the modulus of an analytic function. *Proc. London Math. Soc.* (2) **14**, 269–277.

G. H. Hardy (1922), On the integration of Fourier series. *Messenger of Math.* **51**, 186–192.

G. H. Hardy (1941), Notes on special system of orthogonal functions (IV): The orthogonal functions of Whittaker's cardinal series. *Proc. Cambridge Phil. Soc.* **37**, 331–348.

G. H. Hardy and J. E. Littlewood (1916), Some problems of Diophantine approximation: a remarkable trigonometrical series. *Proc. Nat. Acad. USA* **2**, 583–586.

G. H. Hardy and J. E. Littlewood (1926), Some new properties of Fourier constants. *Math. Ann.* **97**, 159–209.

G. H. Hardy and E. M. Wright (1938), *An Introduction to the Theory of Numbers*. Sixth edition, Oxford University Press (2008).

V. Havin and B. Jöricke (1994), *The Uncertainty Principle in Harmonic Analysis*. Springer.

H. Hedenmalm, P. Lindquist, and K. Seip (1997), A Hilbert space of Dirichlet series and systems of dilated functions in $L^2(0, 1)$. *Duke Math. J.* **86**, 1–37.

H. Hedenmalm, P. Lindquist, and K. Seip (1999), Addendum to "A Hilbert space of Dirichlet series and systems of dilated functions in $L^2(0, 1)$". *Duke Math. J.* **99**, 175–178.

H. Helson (1964), *Lectures on Invariant Subspaces*. Academic Press, New York.

H. Helson and D. Lowdenslager (1961), Invariant subspaces. In *Proc. Intern. Symp. Linear Spaces, Jerusalem*, Pergamon Press, Oxford, pp. 251–262.

H. Helson and D. Sarason (1967), Past and future. *Math. Scand.* **21**, 5–16.

H. Helson and G. Szegő (1960), A problem of prediction theory. *Ann. Mat. Pura Appl.* **51**, 107–138.

G. Herglotz (1911), Über Potenzreihen mit positiven reellen Teil im Einheitskreise. *Berichte Verh. Kgl.-sächs. Gesellsch. Wiss. Leipzig, Math.-Phys. Kl.* **63**, 501–511.

J. R. Higgins (1985), Five short stories about the cardinal series. *Bull. Amer. Math. Soc.* **12:1**, 45–89.

J. R. Higgins (1996), *Sampling Theory in Fourier and Signal Analysis: Foundations*. Clarendon Press, Oxford, and Oxford University Press, New York.

J. R. Higgins and R. L. Stens, editors (1999), *Sampling Theory in Fourier and Signal Analysis: Advanced Topics*. Clarendon Press, Oxford.

D. Hilbert (1909), Wesen und Ziele einer Analysis der unendlich vielen unabhängigen Variablen. *Rend. Cir. Mat. Palermo* **27**, 59–74.

D. Hilbert (1912), *Gründzüge einer allgemeinen Theorie der linearen Integralgleichungen*. Teubner, Leipzig.

K. Hoffman (1962), *Banach Spaces of Analytic Functions*. Prentice Hall, Englewood Cliffs, New Jersey.

B. Hollenbeck and I. Verbitsky (2000), Best constants for the Riesz projection. *J. Funct. Analysis* **175**, 370–392.

R. Hunt, B. Muckenhoupt, and R. L. Wheeden (1973), Weighted norm inequalities for the conjugate function and Hilbert transform. *Trans. Amer. Math. Soc.* **176**, 227–251.

I

I. A. Ibragimov and Y. A. Rozanov (1970), *Gaussian Stochastic Processes* (in Russian). Nauka, Moscow. English translation: Springer (1978).

A. E. Ingham (1936), A note on Hilbert's inequality. *J. London Math. Soc.* **11**, 237–240.

J

J. L. Jensen. 1899), Sur un nouvel et important théorème de la théorie des fonctions. *Acta Math.* **22**, 219–251.

K

M. Kac (1966), Can one hear the shape of a drum? *Amer. Math. Monthly* **73:4(2)**, 1–23.

J. P. Kahane and Y. Katznelson (1971), Sur le comportement radial des fonctions analytiques. *C. R. Acad. Sci. Paris* Ser. A–B **227**, A718–A719.

J.-P. Kahane and P. G. Lemarié-Rieusset (1998), *Séries de Fourier et ondelettes.* Cassini, Paris.

J.-P. Kahane and R. Salem (1963) *Ensembles parfaits et séries trigonométriques.* Hermann, Paris.

Y. Katznelson (1976), *An Introduction to Harmonic Analysis.* Dover, New York.

C. E. Kenig (1994), *Harmonic Analysis Techniques for Second Order Elliptic Boundary Value Problems.* CBMS Conference series no. 83, American Mathematical Society, Providence, RI.

A. N. Kolmogorov (1925), Sur les fonctions harmoniques conjuguées et les séries de Fourier. *Fund. Math.* **7**, 24–29.

A. N. Kolmogorov (1941), Stationary sequences in Hilbert space (in Russian). *Bull. Moscow Univ. Math.* **2:6**, 1–40.

P. Koosis (1966), Weighted polynomial approximation on arithmetic progressions of intervals or points. *Acta Math.* **116**, 223–277.

P. Koosis (1980), *Introduction to H^p Spaces.* Cambridge University Press.

V. A. Kotelnikov (1933), On the transmission capacity of "aether" and wire in electro-communications (in Russian). *Izdat. Red. Upr. Svyazi RKKA.* English translation: http://ict.open.ac.uk/classics/1.pdf.

V. A. Kotelnikov (1956), *The Theory of Optimum Noise Immunity.* McGraw-Hill (1959). Russian original: Izdat. Radio i Svyaz', Moscow.

V. Y. Kozlov (1948), On the completeness of systems of functions $\{\varphi(nx)\}$ in the space $L^2(0, 2\pi)$ (in Russian). *Doklady Akad. Nauk SSSR* **61**, 977–980.

V. Y. Kozlov (1950), On the completeness of a system of functions of type $\{\varphi(nx)\}$ in the space L^2 (in Russian). *Doklady Akad. Nauk SSSR* **73**, 441–444.

L

E. Landau (1927), *Vorlesungen über Zahlentheorie*, vols 1–3. Hirzel, Leipzig.

P. D. Lax and R. S. Phillips (1967), *Scattering Theory.* Academic Press, New York and London.

B. Y. Levin (1956), *Distribution of Zeros of Entire Functions* (in Russian). GITTL, Moscow. English translation: American Mathematical Society, Providence, RI (1980).

N. Levinson (1956), On the closure problems and the zeros of the Riemann zeta-function. *Proc. Amer. Math. Soc.* **7**, 838–845.

J. Lindenstrauss and L. Tzafriri (1977), *Classical Banach Spaces*, vols I (1977) and II (1979). Springer.

J. E. Littlewood (1925), On inequalities in the theory of functions. *Proc. London Math. Soc.* **23**, 481–519.

J. E. Littlewood (1953), *A Mathematician's Miscellany*. Methuen, London. Revised edition, *Littlewood's Miscellany* (ed. B. Bollobás), Cambridge University Press (1986).

J. E. Littlewood (1970), The "pits effect" for functions in the unit circle. *J. Analyse Math.* **23**, 237–268.

M

T. Makino, (2003), The Mathematician K. Ogura and the "Greater East Asia War". In *Mathematics and War* (ed. B. Booß-Bavnbek and J. Høyrup), Springer, pp. 326–335.

P. Masani (1966), Wiener's contribution to generalized harmonic analysis, prediction theory and filter theory. *Bull. Amer. Math. Soc.* **72:1(2)**, 73–125.

C. A. McCarthy and J. Schwartz (1965), On the norm of a finite Boolean algebra of projections and applications to theorems of Kreiss and Morton. *Comm. Pure Appl. Math.* **18**, 191–201.

O. C. McGehee, L. Pigno, and B. Smith (1981), Hardy's inequality and the L^1-norm of exponential sums. *Ann. of Math.* **113**, 613–618.

S. N. Mergelyan (1956), Weighted approximation by polynomials (in Russian). *Uspekhi Mtem. Nauk* **11:5**, 107–152. English translation: *AMS Transl.* Ser. 2 **10** (1958), 59–106.

Y. Meyer (1992), *Wavelets and Operators*. Cambridge University Press.

N

Z. Nehari (1957), On bounded bilinear forms. *Ann. of Math.* **65**, 153–162.

J. H. Neuwirth, J. Ginsberg, and D. J. Newman (1970), Approximation by $f(kx)$. *J. Funct. Anal.* **5**, 194–203.

J. H. Neuwirth and D. J. Newman (1967), Positive $H^{1/2}$ functions are constant. *Proc. Amer. Math. Soc.* **18**, 958.

F. Nevanlinna and R. Nevanlinna (1922), Über die Eigenschaften analytischer Functionen in der Umgebung einer singulären Stelle oder Linie. *Acta Soc. Sci. Fenn.* **50:5**, 1–46.

N. Nikolski (1980), *Lekzii ob Operatore Sdviga* (in Russian). Nauka, Moscow.

N. Nikolski (1986), *Treatise on the Shift Operator*. Springer.

N. Nikolski (1995), Distance formulae and invariant subspaces, with an application to localization of zeros of the Riemann ζ-function. *Ann. Inst. Fourier* **45:1**, 143–159.

N. Nikolski (2002), *Operators, Functions, and Systems*, vols 1 and 2. American Mathematical Society, Providence, RI.

N. Nikolski (2012), In a shadow of the RH: cyclic vectors of Hardy spaces on the Hilbert multidisc. *Ann. Inst. Fourier* **62:5**, 1601–1626.

N. Nikolski and A. Volberg (1990), Tangential and approximate free interpolation. In *Analysis and Partial Differential Equations* (ed. C. Sadosky), Marcel Dekker, New York, pp. 277–299.

B. Nyman (1950), On the one-dimensional translation group and semi-group in certain function spaces. Thesis, Uppsala University.

O

K. Ogura (1920), On a certain transcendental integral function in the theory of interpolation. *Tôhoku Math. J.* **17**, 64–72.

B. K. Øksendal (1971), A short proof of the F. and M. Riesz theorem. *Proc. Amer. Math. Soc.* **30**, 204.

P

R. E. A. C. Paley and N. Wiener (1934), *Fourier Transforms in the Complex Domain.* Vol. 19 of American Mathematical Society Colloquium Publications, Providence, RI.

A. Papoulis (1984), *Signal Analysis.* McGraw-Hill.

M. Pavlović (2004), *Introduction to Function Spaces on the Disk.* Matematički Institut SANU, Belgrade.

V. V. Peller (2003), *Hankel Operators and their Applications.* Springer.

V. V. Peller and S. V. Khruschev (S. V. Hruschev) (1982), Hankel operators, best approximations and stationary Gaussian processes (in Russian). *Uspekhi Mat. Nauk* **37:1**, 53–124. English translation: *Russian Math. Surveys* **37:1** (1982), 61–144.

R. Pérez-Marco (2011), Notes on the Riemann hypothesis. In *Jornadas sobre los problemas del milenio*, Barcelona 1–3 junio, 2011.

E. Phragmén and E. Lindelöf (1908), Sur une extension d'un principe classique de l'analyse. *Acta Math.* **31**, 381–406.

A. I. Plessner (1927), Über das Verhalten analytischer Funktionen am Rande ihres Definitionsbereichs. *J. Reine Angew. Math.* **158**, 219–227.

G. Pólya and G. Szegő (1925), *Aufgaben und Lehrsätze aus der Analysis*, vols 1, 2. Springer, Berlin. English translation: Springer (1972).

S. C. Power (1982), *Hankel Operators on Hilbert Space.* Vol. 64 of Pitman Research Notes in Mathematics, Pitman.

I. I. Privalov (1941), *Boundary Properties of Analytic Functions* (in Russian). Moscow (second edition 1950). German translation: Deutscher Verlag, Berlin (1956).

R

C. Reid (1970), *Hilbert.* Springer, New York.

F. Riesz (1923), Über die Randwerte einer analytische Funktion. *Math. Z.* **18**, 87–95.

F. Riesz and M. Riesz (1916), Über die Randwerte einer analytische Funktion. In *Quatrième Congrès des Math. Scand.*, Stockholm, pp. 27–44.

F. Riesz and B. Szőkefalvi-Nagy (1955), *Leçons d'analyse fonctionnelle.* Akadémiai Kiado, Szeged.

M. Riesz (1927), Sur les fonctions conjuguées. *Math. Zeit.* **27**, 218–244.

M. Rosenblum (1962), Summability of Fourier series in $L^p(\mu)$. *Trans. Amer. Math. Soc.* **105:1**, 32–42.

Y. A. Rozanov (1963), *Stationary Stochastic Processes* (in Russian). Fizmatgiz, Moscow. English translation: Holden-Day, San Francisco (1967).

W. Rudin (1956), Boundary values of continuous analytic functions. *Proc. Amer. Math. Soc.* **7**, 808–811.

W. Rudin (1962), *Fourier Analysis on Groups.* Wiley, New York.

W. Rudin (1998), *Analyse réelle et complexe*, third edition. Dunod, Paris.

S

K. Sabbagh (2002), *The Riemann Hypothesis: The Greatest Unsolved Problem in Mathematics.* Farrar, Straus and Giroux, New York.

R. Salem (1953), Sur une proposition équivalente à l'hypothèse de Riemann. *C. R. Acad. Sci. Paris* **236**, 1127–1128.

D. Sarason (1994), *Sub-Hardy Hilbert Spaces in the Unit Disk.* University of Arkansas Lecture Notes, no. 10, Wiley, New York.

C. E. Shannon (1948), A mathematical theory of communication. *Bell System Technical Journal* **27**, July and October, 379–423 and 623–656.

C. E. Shannon (1949), Communication theory of secrecy systems. *Bell System Technical Journal* **28**, October, 656–715.

C. E. Shannon (1950), Programming a computer for playing chess. *Philosophical Magazine* (7) **41:314**, 256–275.

J. H. Shapiro (1993), *Composition Operators and Classical Function Theory.* Springer, New York.

B. Simon (2005), *Orthogonal Polynomials on the Unit Circle*, Part 1: *Classical Theory.* American Mathematical Society, Providence, RI.

V. I. Smirnov (1928a), Sur la théorie des polynômes orthogonaux à une variable complexe. *J. Leningrad Fiz.-Mat. Obsch.* **2:1**, 155–179.

V. I. Smirnov (1928b), Sur les valeurs limites des fonctions régulières à l'intérieur d'un cercle. *J. Leningrad Fiz.-Mat. Obsch.* **2:2**, 22–37.

V. I. Smirnov (1932), Sur les formules de Cauchy et Green et quelques problèmes qui s'y rattachent. *Izvestia AN SSSR, ser. fiz.-mat.* **3**, 338–372.

V. I. Smirnov (1988), *Œuvres choisies: Analyse complexe et théorie de diffusion* (in Russian). University of Leningrad.

M. N. Spijker, S. Tracogna, and B. Welfert (2003), About the sharpness of the stability estimates in the Kreiss matrix theorem. *Math. Comp.* **72**, 697–713.

T. P. Srinivasan (1963), Simply invariant subspaces. *Bull. Amer. Math. Soc.* **69**, 706–709.

J. M. Steele (2004), *The Cauchy–Schwarz Master Class: An Introduction to the Art of Mathematical Inequalities.* Cambridge University Press.

E. Stein (1993), *Harmonic Analysis: Real-variable Methods, Orthogonality, and Oscillatory Integrals.* Princeton University Press, Princeton, New Jersey.

G. Szegő (1920), Beiträge zur Theorie der Toeplitzsche Formen, I. *Math. Zeit.* **6:3/4**, 167–202.

G. Szegő (1921), Über die Randwerte einer analytischen Funktion. *Math. Ann.* **84:3/4**, 232–244.

T

J. E. Thomson (1991), Approximation in the mean by polynomials. *Ann. of Math.* (2), **133:3**, 477–507.

E. C. Titchmarsh (1939), *The Theory of Functions.* Oxford Science Publications.

E. C. Titchmarsh (1951), *The Theory of the Riemann Zeta-function.* Oxford Science Publications.

O. D. Tsereteli (1975), Metric properties of conjugate functions (in Russian). *Itogi Nauki i Techniki Sovrem. Probl. Mat.* **7**, 18–57. English translation: *J. Soviet Math.* **7** (1977), 309–414.

V

S. Verblunsky (1936), On positive harmonic functions (second paper). *Proc. London Math. Soc.* (2) **40**, 290–320.

H. von Koch (1902), Ueber die Riemann'sche Primzahlfunction. *Math. Annalen* **55**, 441–464.

W

H. Weyl (1908), Singuläre Integralgleichungen. *Math. Ann.* **66**, 273–324.

E. T. Whittaker (1915), On the functions which are represented by the expansions of the interpolation theory. *Proc. Royal Soc. Edinburgh* Ser. A **35**, 181–194.

E. T. Whittaker (1924), *The Calculus of Observations: A Treatise on Numerical Mathematics.* Blackie, London.

N. Wiener (1930), Generalized harmonic analysis. *Acta Math.* **55**, 117–258.

N. Wiener (1932), Tauberian theorems. *Ann. of Math.* (2) **33**, 1–100.

N. Wiener (1933), *The Fourier Integral and Certain of its Applications.* Cambridge University Press, New York.

N. Wiener (1949), *Extrapolation, Interpolation, and Smoothing of Stationary Time Series: With Engineering Applications.* MIT Press, Cambridge, MA, and Wiley, New York.

N. Wiener and P. R. Masani (1957), The prediction theory of multivariate stochastic processes, I: The regularity condition. *Acta Math.* **98**, 111–150.

N. Wiener and P. R. Masani (1958), The prediction theory of multivariate stochastic processes, II: The linear predictor. *Acta Math.* **99**, 93–137.

A. Wintner (1944), Diophantine approximation and Hilbert's space. *Amer. J. Math.* **66**, 564–578.

H. Wold (1938), *A Study in the Analysis of Stationary Time Series.* Almquist och Wiksell, Uppsala.

Z

A. Zygmund (1959), *Trigonometric Series*, vols I and II. Cambridge University Press.

Notation

Sets and Measures

$\mathbb{C}$ - the complex plane
$\mathbb{T} = \{z \in \mathbb{C}: |z| = 1\}$
$\mathbb{C}_+ = \{z \in \mathbb{C}:\ \mathrm{Im}(z) > 0\}$
$\mathbb{C}^+ = \{z \in \mathbb{C}:\ \mathrm{Re}(z) > 0\}$
$D(z, r) = \{\zeta \in \mathbb{C}: |z - \zeta| < r\}$
$\mathbb{D} = D(0, 1)$
$\mathbb{D}_2^\infty$ - Hilbert multi-disk, § 6.6.5
$\mathcal{P}$ - Theorem 1.2.1
$\mathcal{P}_a$ - Corollary 1.4.4
$\mathcal{P}_n$ - Exercise 5.6.2(f)
$\mathrm{Lat}(T)$, $\mathrm{Lat}(\mathcal{T})$ - § 1.1
$\sigma(\theta)$ - spectrum of a function, Definition 3.2.3
m - normalized Lebesgue measure, § 1.3
μ_a, μ_s - § 1.3

Spaces and Operations

$H^2 = H^2(\mathbb{T})$ - Definition 1.3.4
$H^2(\mathbb{T}, \mu) = H^2(\mu)$ - Definition 1.5.1
$H^p = H^p(\mathbb{T})$ - § 1.3.1
$H^p(\mathbb{T}, \mu) = H^p(\mu)$ - § 2.9
$H^\infty = H^\infty(\mathbb{T})$ - Exercise 1.8.3
$W = W(\mathbb{T})$ - § 5.3.1
H_0^2 - Theorem 1.7.6
$H_0^2(\mu)$ - Lemma 1.6.4
$H^p(\mathbb{D})$ - Definition 2.2.1

$H^\infty(\Omega)$ - § 3.4.1
$C_a(\mathbb{D})$ - § 5.4.1
$W_a(\mathbb{D})$ - Exercise 5.6.2
Mult(X) - multipliers of X, Exercises 1.8.3(a), 4.9.2, 6.7.1
$\mathcal{D}$ - Smirnov class - Definition 3.3.1
$\mathcal{D}(\Omega)$ - § 3.4.1
$\mathcal{N}$ - Nevanlinna class - Definition 3.3.1
$\mathcal{N}(\Omega)$ - § 3.4.1
E_f - invariant subspace generated by f, Corollary 1.4.4
Lin(A) - linear hull of A
$\mathrm{span}_X(A) = \mathrm{span}(A)$ - closed linear hull of A, § 1.1
$\mathrm{clos}_X(A) = \mathrm{clos}(A)$ - the closure (the adherence) of A, § 1.3.1, Corollary 1.4.4

Functions, Constants, and Transforms

χ_A - characteristic function of A (if $x \in A$, $\chi_A(x) = 1$, otherwise $\chi_A(x) = 0$)
f_{in}, f_{out} - Theorem 1.7.2
V_μ - singular function with measure μ, Corollary 2.6.4
Γu - Herglotz transform of u, Exercise 2.8.4(c)
Γ - Euler gamma function, Theorem 6.1.5
$H(u)$ - Hilbert transform of u, Exercise 2.8.4(d)
P_+ - Riesz projection, Exercise 2.8.3(g)
P_E - orthogonal projection on E, Appendix C.2
$P_{L\|M}$ - skew projection, Definition 4.2.1
$A(L, M) = A_H(L, M)$ - the angle between L and M, Definition 4.3.1
$\mathcal{F}$ - Fourier transform, Appendix A.6
$\mathcal{F}_*$ - Mellin transform, § 6.3.3
H_φ - Hankel operator, § 4.7.2
τ_s - translation, Definition 5.1.1, § 6.3.2(b)
$\mathcal{D}_t$ - dilatation (dilation), Lemma 6.2.3
M_z - shift operator, § 1.8.2
$\zeta(s)$ - Euler zeta function, Definition 6.1.1
$\mathrm{sinc}(t)$ - *sinus cardinalis* $(= \frac{\sin(\pi t)}{\pi t})$, § 5.7
$b(X)$ - basis constant, § 4.1.1(e)
$ub(X)$ - unconditional basis constant, § 4.8
$w \in (HS)$ - Helson–Szegő weight, Definition 4.6.2